DOWN TO EARTH

NEW DIRECTIONS IN INTERNATIONAL STUDIES

PATRICE PETRO, SERIES EDITOR

The New Directions in International Studies series focuses on transculturalism, technology, media, and representation, and features the innovative work of scholars who explore various components and consequences of globalization, such as the increasing flow of peoples, ideas, images, information, and capital across borders. Under the direction of Patrice Petro, the series is sponsored by the Center for International Education at the University of Wisconsin–Milwaukee. The Center seeks to foster interdisciplinary and collaborative research that probes the political, economic, artistic, and social processes and practices of our time.

A. ANEESH, LANE HALL, AND PATRICE PETRO, eds.
Beyond Globalization: Making New Worlds in Media, Art, and Social Practices

MARK PHILIP BRADLEY AND PATRICE PETRO, eds.
Truth Claims: Representation and Human Rights

MELISSA A. FITCH
Side Dishes: Latin/o American Women, Sex, and Cultural Production

ELIZABETH SWANSON GOLDBERG
Beyond Terror: Gender, Narrative, Human Rights

LINDA KRAUSE AND PATRICE PETRO, eds.
Global Cities: Cinema, Architecture, and Urbanism in a Digital Age

ANDREW MARTIN AND PATRICE PETRO, eds.
Rethinking Global Security: Media, Popular Culture, and the "War on Terror"

TASHA G. OREN AND PATRICE PETRO, eds.
Global Currents: Media and Technology Now

PETER PAIK AND MARCUS BULLOCK, eds.
Aftermaths: Exile, Migration, and Diaspora Reconsidered

LISA PARKS AND JAMES SCHWOCH, eds.
Down to Earth: Satellite Technologies, Industries, and Cultures

FREYA SCHIWY
Indianizing Film: Decolonization, the Andes, and the Question of Technology

CRISTINA VENEGAS
Digital Dilemmas: The State, the Individual, and Digital Media in Cuba

DOWN TO EARTH

SATELLITE TECHNOLOGIES, INDUSTRIES, AND CULTURES

EDITED BY
LISA PARKS AND **JAMES SCHWOCH**

RUTGERS UNIVERSITY PRESS
New Brunswick, New Jersey, and London

LIBRARY OF CONGRESS CATALOGING-IN-PUBLICATION DATA

Down to Earth : satellite technologies, industries, and cultures / edited by Lisa Parks and
James Schwoch.
   p.   cm.
   ISBN 978-0-8135-5273-6   (hardcover : alk. paper)
   ISBN 978-0-8135-5274-3   (pbk. : alk. paper)
   ISBN 978-0-8135-5333-7   (e-book)
   1. Artificial satellites in telecommunication—Popular works.   2. Artificial satellites—
Popular works.   3. Telecommunication—Social aspects—Popular works.   4. Mass media—
Popular works.   I. Parks, Lisa.   II. Schwoch, James, 1955–

   TK5104.D686   2012
   354.5'1   54.5'1—dc23

                                                                        2011028838

A British Cataloging-in-Publication record for this book is available from the British Library.

Visit our website: http://rutgerspress.rutgers.edu

Manufactured in the United States of America

To our Sputniks: John and Mimi

CONTENTS

II

SATELLITE MEDIASCAPES

III

ORBITAL MATTERS

ACKNOWLEDGMENTS

The editors of this book gratefully acknowledge the support of the Qatar Foundation and of Northwestern University in Qatar, who generously coprovided a subvention to Rutgers University Press in support of the publication of this volume. We are also thankful for the advice and leadership of Leslie Mitchner, Anne Hegeman, Lisa Boyajian, Katie Keeran, Suzanne Kellam, and Rutgers University Press. The guidance and wisdom of Patrice Petro was inspirational. Special thanks also to John Durham Peters for a sympathetic, critical, and effective reading of the manuscript in draft form, to Robert Burchfield for his careful copyediting, and to Betsy Lane of Lane Editorial, whose brilliant and timely assistance rescued this project from a series of late-mission wobbly orbits and took on the needed function of mission control. We also thank the contributors to this volume and are extraordinarily pleased to publish their work.

Lisa Parks would like to thank her wonderful colleagues in the Department of Film and Media Studies at the University of California at Santa Barbara for supporting her research interests in satellite technologies and media over the years. Her thinking about satellites also has been touched and informed by provocative exchanges with a number of researchers and artists, especially Jody Berland, Ursula Biemann, Charlotte Brunsdon, Amelie Hastie, James Hay, Francis Hunger, John Fiske, Joanna Griffin, Lisa Jevbratt, Geert Lovink, Patrick McCray, Angela Melitopoulos, David Morley, Marko Peljhan, Trevor Paglen, Nicole Starosielski, Jonathan Sterne, Ginette Verstraete, and Miha Vipotnik. Parks also has been inspired by the curiosities and questions of students in her Satellite Media courses at UC Santa Barbara and by artists who participated in the Zemos98 Conference in Seville, Spain and the Satellites/Footprints/Borders Workshop at HMKV in Dortmund, Germany. Parks is grateful for the generous support of the Wissenschaftskolleg of Berlin, where she worked as a research fellow in 2006/2007 while this project was getting off the ground. She would also like to thank those who attended her talks at Central European University, Humboldt University, McGill University, University of Southern California, and University of Stockholm, where she received vital feedback on her chapters in this book. Finally, Parks expresses deepest gratitude to her coeditor and collaborator, James Schwoch, for his sweeping satellite intelligence, worldly vision, and stalwart friendship. Working on this book has opened new fields of inquiry for me and altered my course.

James Schwoch is thrilled to have had this opportunity to work with Lisa Parks. Lisa continues to expand my intellectual horizons about satellites and global security (as well as everything else germane to global media studies) and to help me in my never-ending pursuit of originality, rigor, and eloquence in my research, all the while a close friend. I look forward to our future collaborations. In addition to Lisa, I have benefited greatly in my career from the support of archivists, librarians, catalogers, and staffers at various institutions around the world. Archives of particular significance for this book include the National Archives and Records Administration (USA); the National Air and Space Museum; Smithsonian Institution (USA); Wisconsin Historical Society (USA); the Eisenhower, Kennedy, Johnson, and Reagan presidential libraries (all USA); and the British Telecomm (BT) Archives (UK). During the production of this book, I was intellectually boosted on various and general issues of satellites by lively audiences and interlocutors at the 2009 Cités De Telecom Doctoral Seminar held at Pleumeur-Bodou and sponsored by the Sorbonne, Maastricht University, and Orange (formerly France Telecom); a 2009 lecture at the Department of Film and Media Studies, University of California at Santa Barbara; a 2009 session at the Institute for Defense Analyses, Washington D.C.; a 2009 conference paper at the Aleksanteri Conference, University of Helsinki; a conference paper at the 2008 Society for the History of Technology Conference, Lisbon, and a 2008 Seminar Paper at ZiF (Institute of Advanced Study), Bielefeld. During this period of archival research and at these various venues, many individuals gave constructive advice, including Pascal Griset, Andreas Fickers, Richard John, Paul Ceruzzi, Michael Neufeld, W. Patrick McCray, Jason Gallo, and Alexander Geppert. Last but not least—a multitude of thanks to many students in the past twenty years who have generously allowed me to explore satellites and outer space as part of my teaching. Their enthusiasm, ranging from classroom comments to research papers, constitutes a wonderful affirmation of the value and relevance of satellites, global security, and outer space as lively topics for both the classroom and for research.

DOWN TO EARTH

INTRODUCTION

LISA PARKS AND JAMES SCHWOCH

Thousands of satellites have been launched into orbit during the past fifty years. During these launches eager spectators gazed upward in amazement as a fiery plume turned into a delicate white contrail tracing a rocket as it bolted into the sky only to vanish a few minutes later. The scene of a satellite launch is familiar to most. Not only have thousands of people witnessed launches with their own eyes, such scenes have appeared in television news accounts and have been popularized by Hollywood films over the years. While the purpose of a launch is to thrust a satellite into orbit, this book sets out to perform a reverse maneuver and bring the satellite down to Earth. In doing so, it focuses on the material effects and functions of satellites, the countries and companies that develop them, the cultures they generate, the orbital paths they occupy, and the industries of which they are a part.

While many are familiar with the spectacle of the launch, fewer people know about satellites themselves. Thus the act of witnessing a satellite launch can be understood as one small step toward a deeper investigation of satellite technologies, industries, and cultures. Satellite design is a highly specialized field that involves engineers and scientists who tinker behind closed doors in federal science labs and corporate clean rooms. Satellite regulation enlists a scattered web of national and international agencies ranging from the Federal Communications Commission to the European Space Agency and from the United Nations (U.N.) to the International Telecommunications Union. Satellite funding structures are labyrinthine and support everything from the fabrication of the satellite itself to the policies that insure it, from the building of Earth stations to the salaries of employees who manage them. And satellite uses are manifold, engaging all kinds of players from different parts of the world, whether a wildlife biologist tracking a grizzly bear in Alaska, an immigrant worker in Germany downlinking a television show from Turkey, or a U.S. military intelligence analyst monitoring nuclear weapons facilities in Iran.

This book is titled *Down to Earth* to emphasize the material and territorial relations of satellite technologies, industries, and cultures. It is also meant to highlight the vertical stretch between Earth's surface and the outer limits of orbital space, drawing attention to the imperceptible and multiple "spheres" (atmosphere, stratosphere, ionosphere) through which satellite-to-Earth transactions move and world histories unfold. In their efforts to bring the satellite down to Earth, the authors in this book examine various satellite projects (whether Iridium or Sirius/XM) and applications (whether global positioning, broadcasting, remote sensing, or telephony) and approach them with the frameworks and tools of historical investigation and critical analysis. Satellites are enigmatic objects of study that demand methodological experimentation and creativity. Their remoteness and imperceptibility constantly beg the question: how is it possible to study and understand things and processes that cannot readily be seen or sensed (and which, in some cases, are purposefully hidden and suppressed)? Though this may be a common question for a historian looking back in time or a scientist working at the nanoscale, it is perhaps less common for a media or communication scholar. How do we communicate about satellites and work to make them intelligible? How do satellites matter within our global culture? What are the appropriate critical concepts and approaches for studying satellite technologies, industries, and cultures? These are some of the broader questions motivating the research trajectories of this book.

The satellite's position on and beyond Earth has meant that from the get-go concerns about the technology's development, launching, funding, and uses have been quintessentially global. Throughout the 1960s, representatives from countries around the world participated in a series of international dialogues about the appropriate development and use of satellites. At the same time, the technology was being commandeered for Cold War geopolitics as the United States and the Soviet Union deployed their secret satellite espionage programs into space. By the 1980s and 1990s satellite industries—including direct satellite broadcasting, remote sensing, and global positioning—had emerged in different nations and have since expanded to operate in most parts of the world. It is almost impossible to imagine our twenty-first-century world without satellites since financial institutions, television broadcasters, military officials, city planners, oil corporations, airlines, telephone carriers, environmental activists, and cartographers, to name a few, all use them daily for a variety of purposes.

Despite the centrality of satellite technologies and services in so many aspects of contemporary life, the histories and practices key to these systems and services remain, by comparison to other means of global electronic information and entertainment, relatively unknown to most observers. In the field of global media, for example, scholars and students are much more likely to be able to name the major Hollywood film studios, various television and cable networks, the world's leading newspapers, or the most trafficked websites than they are able to name individual satellites, satellite constellations, or orbital paths. Details about the satellite industry, from financing to launch to service applications, remain hazy at best. Public perception of various orbits—even the awareness that there are many possible orbital configurations—is often dim. And how and why satellites help configure contemporary politics and culture at both global and local levels is only glimpsed, rather than understood in detail. Knowledge gaps regarding satellite technologies, industries, and cultures are found in many disciplines and fields addressing global issues. In an effort to address this matter, *Down to Earth* projects, if you will, a variety of footprints to its readers, footprints that represent an overlapping range of social, historical, cultural, geographical, empirical, political, and critical explorations aimed to inform and empower readers. Additionally, the collection encourages researchers to further consider satellites as an important component of many disciplines and fields of scholarship, including global media.

Since their emergence in the late 1950s, satellites have been embroiled in the formation of new global imaginaries, security paradigms, economies, and cultures. Satellites have been fundamental to contemporary conceptualizations of the global and to processes of globalization. Satellites circulate signals across and beyond the sovereign boundaries of nations on Earth and in doing so facilitate the flows of a global economy. As machines that orbit our planet, satellites are uniquely positioned to visually represent Earth, and their images have been composited to construct Earth as "whole." Satellites have also been used to remotely sense the space beyond Earth, and have enabled us to grasp the planet's position in a broader spatial context, whether defined as galactic, universal, or cosmic. While satellite technologies are implicated in the production of contemporary global imaginaries, they have historically been developed and controlled by a relatively small number of nation-states and corporate entities, making the technology's associations with the "global" tenuous, if not specious. Despite the language of the U.N. Outer Space Treaty of 1967, which provided that laws of sovereignty on Earth did not apply in orbit and that all nations of the world should have equal access to uses of outer space no matter their level of wealth, those countries with the financial and scientific capital were the first to assert control over orbital space, and they remain the major players in the world's satellite industry today. While fifty-one countries have sent their own satellites into orbit and eleven countries have launching capabilities, the satellite remains a high-capital technology, and, with a handful of exceptions, has been developed only by and for the richest nations of the world.[1] Given the fluctuation in satellite participation over the years, it is important that the satellite's relationship to the "global" and "globalization" not be thought of as fixed, but rather as a historically shifting field of power relations that must be specified in order to make sense and be critically understood.

In addition to participating in the production of global imaginaries, satellites are embedded within global security agendas. They have a dark side. Or as Trevor Paglen aptly puts it, they occupy a place in the "other night sky."[2] Since the early days of satellite experimentation, global security has served as one of the primary drivers and rationales for the development of these costly, complex machines. Today the satellite is part of an assemblage of technologies of remote control that can at once be organized to observe, communicate about, and target sites on Earth. The satellite represents the ultimate rationalization and instrumentalization of the quest for global security and domination. Because of this, world peace and stability are rarely understood as being the result of diplomacy alone; increasingly, they are thought of as being leveraged on the technological success or failure of satellites. We might call this the "satellitization" of global security. Given this scenario, satellites are useful objects that can draw attention to and expose the reach and intensity of global security strategies. Understanding the satellite in this way involves describing and analyzing security policies and practices that have both shaped and taken shape in relation to the satellite, and that inform its spin-offs, outgrowths, and extensions.

Beyond its role in shaping contemporary global imaginaries and security paradigms, the satellite is key to transnational media economies and cultures. The world is covered with more satellite footprints than sovereign nation-state boundaries. There are more than 400 communication satellites in geostationary orbit with hundreds of footprints on Earth's surface, and there are 195 countries with sovereign boundaries.[3] In other words, there are more legally sanctioned satellite territories on Earth than there are national territories. Satellite operators have carved up the planet and staked out new territories in ways reminiscent of European political leaders during the Treaty of Westphalia centuries ago. Currently, there are no maps sufficient for visualizing the reality that Earth is a satellitized domain, although a number of individual satellite footprint maps do exist. Rapid expansion and activity within the field of satellite broadcasting during the past thirty years seems to have exceeded the capacities of cartographers—and yet there are exceptions. For instance, Viennese graphic artist Michael Paukner's conceptual map, "Big Brothers: Satellites Orbiting Earth" (see Figure I.1), conveys important structural information about orbit by visualizing concentrations of satellite ownership and indicating how many satellites are functional or dysfunctional. The mapping points to the United States and Russia as the Big Brothers—the two nations that have installed the most satellites in orbit and left behind the most debris.

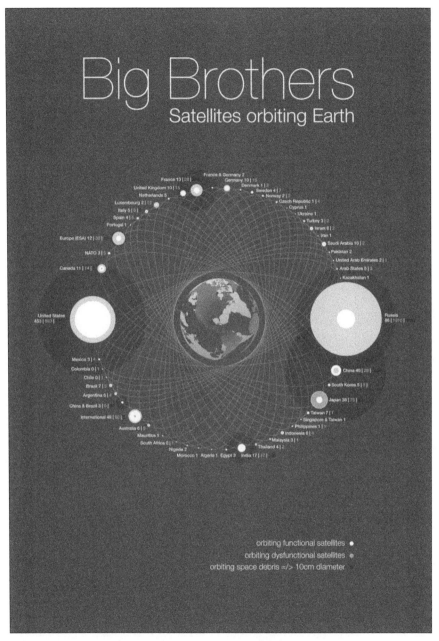

Figure I.1 Michael Paukner's conceptual map, "Big Brothers: Satellites Orbiting Earth," uses different-size circles to represent the number of satellites a nation or institution has in orbit.
Source: Michael Paukner

Despite such visualizations, it is challenging to paint a comprehensive picture of the world's satellite infrastructure, financing, and regulation. Because of this, interdisciplinary scholarship that strives to develop a critical language and agenda for understanding satellite-based broadcasting and telecommunication services is vital. As these services are integrated further within understandings of world history and politics, the satellite's relation to territoriality, marketing, and media flows will take on greater specificity and urgency in critical and historical imaginaries. Satellites demand different ways of thinking about media production, consumption, and distribution. They necessitate revisiting communication theories of time/space and the very definition of "media" itself. Satellites also require a critical approach that sets broadcasting and telecommunication into play with such issues as migration/displacement, geopolitics, natural resource development, and cultural translation. While current maps may not be able to demonstrate the complexity of the world's satellite-based signal traffic, *Down to Earth* brings together a group of scholars who identify the major players, sites, and systems that should be on such a map.

Sketching Satellite History

Many ideas and factors have contributed to the emergence of satellites, including manned and unmanned artificial flight; questions concerning measurement, data-gathering, aerial imagery, and scientific research about Earth and outer space; military and commercial interests in rocketry; the ongoing exploration of new communication technologies as part of the overall growth of global telecommunication systems; and twentieth-century military and geopolitical activities of several industrialized nations. These and other factors leading to the development of satellites were not separate and static, but rather intertwined and dynamic. The confluence of revolutionary movements, world war, and global electronic information networks in the 1910s launched trajectories of interest, tactics, and statecraft fully and permanently into the realm of global electronic surveillance—a realm central to understanding the eventual rise of satellite technologies during the Cold War. Visual observations of Earth from the hot-air balloons of the brothers Montgolfier in the 1780s fascinated scientists, cartographers, governments, militaries, and the general public, and the fascination with aerial images of our planet continues unabated into the present day, as is evident in the wide use of web applications such as Google Earth. Rocket research and production throughout the twentieth century, most notoriously the German rocket and missile laboratory at Peenemünde during World War II, foreshadowed the problematic relationships between military and civilian uses of outer space that still challenge us today. The war ended with the victors in Europe (largely the United States and the USSR) acquiring both technology and personnel from the Peenemünde site, along with a growing recognition that routine access to outer space would soon become a global reality.

The conjunctures driving the eventual emergence of satellites during the Cold War era meant that satellites in the 1950s came to be thought of as a prime example of a "dual-use" technology (although that specific term was not in common use at that time). As satellites entered public consciousness, political jargon, scientific experiments, and military applications, the widespread recognition that satellites—and the rockets that launched them—carried double signification as applied technologies with both military and peaceful purposes became commonplace. Thus from the very first satellite launches (Sputnik, Vanguard), and even in the ten or so years leading up to those launches, various attempts to explain, manage, and put into practice the conceptual cleavage of military and peaceful satellite applications became a sort of grand discursive project of the Cold War period. The continuing legacy of this discursive practice persists in our study, analysis, and understanding of satellites. There are many cases in which the distinct separation of satellites for military uses and for peaceful uses is apparent, but there are also many cases where this discursive and practical cleavage remains uncertain and ambiguous. Classified development of the first spy satellites came with simultaneous public pronouncements—usually highlighting scientific research projects, or the potential for global communications—extolling the imminent future of satellites as the latest, most modern tools for global science and global culture. During the Cold War, especially in the 1950s and 1960s, the ultimate means of global destruction became discursively intertwined with the ultimate means of global communication.[4] Intercontinental ballistic missiles (ICBMs) shared headlines with the International Geophysical Year (IGY). Atomic weapons tests, the Berlin Wall, and the Cuban missile crisis shared the global spotlight with astronauts, cosmonauts, and communication satellites. The superpowers competed for global public opinion and for strategic power through militarization, propaganda, espionage, and the demonstrable applications of science, technology, and engineering.

With a few notable exceptions during the Cold War and its superpower space race, cosmonauts and astronauts (rather than the many unmanned satellites) dominated global attention and news headlines during the 1950s and 1960s, culminating with the 1969 Apollo moon landing. But the first fifteen years of the space race also saw the launch of hundreds of satellites—thousands, if you also count failures, balloons, soundings, and test instruments carried in the upper stages of the launching rockets themselves. This was the first huge wave of satellite growth and development.

The satellites of the 1950s and 1960s were mainly representative of either (and often, both) of two discursive and geopolitical worlds: the realm of global military power or the realm of what is sometimes called "big science." Not only had the superpowers competed for payload, orbit height, orbit duration, publicly demonstrable outer space experiments and techniques, and the moon with their human space missions, they had also deployed extensive satellite networks—in the jargon of the field, "satellite constellations." These networks were used for military communications; scientific experiments, observations, and measurements; electronic and photo intelligence (common acronyms for these surveillance and reconnaissance activities are ELEINT, PHOTOINT, and SIGINT); and, by the early 1970s, verification of mutual arms reduction treaties. Many satellite applications—such as weather observation, geodetics, or extraplanetary and solar research—had both military and scientific applications, and still others, such as PHOTOINT, eventually came to be recognized as having valuable nonmilitary applications for a wide range of scientific inquiries, scholarly research, and public-policy questions. The first publicly shared images of Earth from the perspective of the moon by Apollo 8 astronauts in 1968 globalized the image of a small blue-green orb alone in the universe, and stands as one of the most powerful images of the late twentieth century. In a more down-to-earth example, some of the first declassified satellite images of the desertification of the Aral Sea raised public awareness of long-term dangers to the global climate during the 1970s. Other photo releases led to changes in research fields such as cartography and archaeology, and to the formation of Geographic Information Systems (GIS), which later became a primary research infrastructure and database for a host of scientific disciplines.

As these satellite activities continued during the 1970s in the dual realms of global security and big science, the technology was also used in ways that significantly reshaped the global telecommunications and media landscape. The growing impetus of broadcast and telecommunications deregulation in many nations of the world spurred investments in new technologies such as satellites, while, for many other nations, the assistance, aid, and investment made by the United States and the USSR in conjunction with the 1960s growth of INTELSAT and INTERSPUTNIK meant that satellite distribution capability was arriving almost everywhere. If nothing else, virtually every nation now had a receiving dish or Earth station networked into its national television network for international program distribution and exchange. While somewhat forgotten in today's world of abundant undersea fiber optics, INTELSAT also meant that by 1970 international telephony was, at least in technical and infrastructural terms, a genuine service for every nation in the world. While in many cases access was limited by exorbitant costs, poor local telecommunications infrastructure, underdevelopment of fixed-line telephony, or political repression (most commonly, an assortment of these), INTELSAT and other satellite constellations expanded the international flow of telephony beyond extant conduits such as the well-established North Atlantic undersea cable systems and made full global telephony an everyday (albeit relatively expensive) worldwide reality for the first time.

During the 1970s, television growth in the industrialized world began its decisive movement beyond the standard model of one to three national television networks (sometimes augmented by independent television stations in large metropoles). This expansion, which included noncommercial or educational television stations as well as independents, culminated in the robust growth of cable television networks. Satellites were absolutely crucial to this expansion of the television landscape. The U.S. cable channel HBO began national distribution of HBO programming to local cable service providers in 1975. Less than a decade later, superstations such as WTBS and WGN had emerged, and new cable networks had also moved to satellite. By 1981, the big three U.S. commercial television networks—ABC, CBS, and NBC—had abandoned the AT&T coaxial cable system for national TV networking in favor of satellite distribution. Everyone was up in space, enjoying the new technologies and new economies of television network satellite distribution from the geosynchronous orbit. The U.S. case was not isolated, but it was exemplary. Going into the 1980s, satellite distribution by national television networks to their outlets and affiliates became the norm.

As deregulation, communication satellites, and television content expanded in the latter half of the 1970s—particularly in the Western Hemisphere and western Europe, but with growth evident around the world—more and more television and cable stations and networks turned to satellite technologies as a preferred infrastructural method of content delivery from the network to affiliates. The satellite-distributed television network signal was then rebroadcast over the traditional television-affiliate terrestrial tower or local cable provider's wire system to local markets and their viewers. At that time, few seemed initially to notice another growing trend—the tendency for some individual users and audience members to simply bypass the standard infrastructural chain of content delivery by procuring their own satellite dish and bringing the signal down to Earth directly into their own home or other locale.[5] Entry costs plummeted as the costs of satellite dishes and requisite receiving components continued to fall, making a satellite television package an increasingly attractive and affordable option for consumers.

Nations and regulators responded to this new phenomenon in different ways. Many nations chose to ease the path to home or private dish ownership and removed cumbersome regulations on the ownership of satellite dishes, and many other nations fitfully tried to block the path to that same ownership and stem the transnational flow of information via the home satellite dish. The tendency of television networks to cluster on the same satellites meant that if you could pick up one satellite television channel, you could get all the rest on that same satellite for no additional equipment costs. In the 1970s, few television satellite signals were routinely "scrambled" or encrypted by the networks, meaning direct satellite television viewers at that time generally did not face additional barriers of signal encryption or subscriber fees; the only new costs incurred were a one-time purchase of the dish and the receiving components. In our current age of Dish, DirecTV, and online media purchase services, it is all the more astounding that satellite television went through a "free" period without subscriber or program fees—but those days of free content were incredibly important for the first rush of home and individual satellite television growth, and removed a hurdle that could have discouraged many erstwhile home satellite-television pioneers.

Even more startling than the relative ease of technical access and no-cost programming was the fact that these do-it-yourself home satellite television tactics of the late 1970s were not confined within the borders of nations. Like broadcast and other wave-based telecommunication media prior to the satellite, the satellite signal, or footprint, as it came down to Earth cared not a whit for political borders or nation-state boundaries. The satellite footprint was an equal-opportunity provider. Citizenship mattered not. All that mattered were how many satellite footprints trod upon your own patch of the planet and could be captured by your dish. Thus for WGN Chicago and its new television superstation-on-the-satellite status, the "we are all citizens of one world without borders" techno-democracy of the satellite footprint meant, for example, Chicago sports teams became popular in Central American countries (such as Belize) where viewers, hotels, restaurants, cafés, and bars could access the WGN signal just as easily as farmers in Iowa, oil derricks in the Gulf of Mexico, or sports fans in Chicago. These manifestations of footprint democracy and global satellite access for all (or at least all who could get their hands on a dish, a receiver, and a monitor) were suddenly propagating all around the world. Footprints now came down to Earth everywhere. The big science of high entry costs and limited access to satellite technologies that marked the 1950s and 1960s was replaced in the 1970s and 1980s by low entry costs and abundant access to satellite technologies as a part of everyday life.

While this incredible change in the global means of communication took place via satellite, the realms of scientific research and global security that so significantly shaped the early decades of satellite growth continued to be active factors in satellite developments. Scientific and military satellite users did not sit dormant throughout the 1970s. Reconnaissance photo-intelligence satellites moved away from their first technology platforms—a film-based system that involved the sporadic return of exposed film (de-orbited canisters for U.S. satellites, or bringing down the entire satellite intact for the USSR)—to eventual real-time (or near-real-time) electronic image delivery. This had several implications, including the fact that hostile parties could no longer plan actions around the known times and observations of an adversary's satellite film retrieval. Formerly, once film had been retrieved, an adversary would know that another film retrieval would not likely take place for several days or more, meaning actions taken in the interim would not be discovered until days or weeks later.

Going in the other direction—not down to Earth with film and images, but as far away from Earth as possible—the 1970s also saw satellites such as the Pioneer and Voyager series, designed mainly to explore the outer planets of the solar system, but also built with the knowledge that these satellites, assuming continued functionality, would eventually leave our solar system for interstellar space. Pioneer and Voyager also captured global imagination, as they were equipped not only with communications components and scientific equipment but also with their chassis literally emblazoned with imagery and data about Earth and its humans for potential communication with extraterrestrial, intelligent-sentient beings. The potential for cognizance (or not) by theoretical extraterrestrials of human-rendered images of human figures, of the location of Earth as mapped by the nearest quasars, of the possibility for aliens to play back a recorded audio disc on one of the satellites, and for theorizing other human-alien communication aspects of these missions sent off debates that still continue on the nature, spirituality, and philosophy of human life in relation to the vast infinities of the cosmos. The human arts of representation not only entered outer space via the medium of the Pioneer and Voyager satellite series. In the early 1970s, some of the Apollo moon missions included art installations on the moon: a memorial to fallen astronauts and cosmonauts, and on another occasion, a small ceramic tile with work by Andy Warhol, Robert Rauschenberg, Claes Oldenburg, Forrest Myers, John Chamberlain, and David Norvos. The artistic tradition for satellites and space missions continued into the twenty-first century, when Damien Hirst designed the color calibration palette for the camera on the 2003 Mars satellite–ground explorer *Beagle 2*.

The 1980s also saw a vast expansion of new media in such communication technologies as cell phones, facsimile machines (faxes), videocassette recorders, personal computers, pagers, and early progenitors of what we now think of as convergent mobile devices such as Palms, Droids, and iPhones—PDAs or "personal digital assistants." Hints of competition entered the global satellite distribution industry with such entities as PANAMSAT, and the population of the planet was once again reminded of dual use, the partial discursive cleavage, and the semisecret relationship between security and outer space with new discussions of anti-satellite weapons and proto-projects such as the Strategic Defense Initiative. The collapse of Communism in Europe and the dissolution of the Soviet Union circa 1989–1991 brought tremendous changes into the world of satellites (as well as to many other worlds). But most of all, the transition to a post–Cold War era brought a huge expansion of satellite users, launchers, and service providers. The dissolution of the USSR created twenty-odd new nation-states with their own needs and agendas regarding telecommunications, information, and outer space. Since the Baikonur Cosmodrome, a crucial satellite launch site, was located in its national territory, the new state of Kazakhstan suddenly became as vital to the global satellite industry as Russia. Multinational corporations, particularly aerospace-defense industry behemoths, moved aggressively into the satellite industry. Lockheed Martin purchased INTERSPUTNIK, the old Soviet alternative constellation to INTELSAT. Loral had a turbulent experience in developing launch capabilities with China. The island nation of Tonga bought some old Soviet satellites, paid for their launch, and suddenly became a satellite "player" in the Pacific Rim.[6]

On the heels of the breakup of the USSR came the 1991 Persian Gulf War. SPOT, a French satellite remote sensing company (and at the time basically the only company in the world providing commercial high-resolution satellite imagery) joined the U.N. boycott of Iraq, and Saddam Hussein went "blind" as a result. American aerospace power overwhelmed Iraqi forces, and the rest of Arabia and the Middle East watched carefully, and subsequently invested heavily in satellite and telecommunications facilities. President Clinton, flush from the (illusory) Cold War peace dividend, helped lead the U.S. declassification and privatization of a host of satellite projects and image-data formerly restricted to government surveillance. The result was an explosion of new satellite service providers in areas such as remote sensing, GIS and cartography, and GPS and navigation. At the same time, the Clinton administration and the U.S. government wrestled with new wrinkles and manifestations of dual-use technologies and satellites, as the problematic uses of commercial launch sites in China for American satellites became a contentious political and security issue in 1996 with the LORAL case.

On the rapidly growing backbone of a World Wide Web with graphical user interfaces and exponentially expanding global bandwidth (some provided by satellites, more by ocean-floor fiber optics), satellite images and data fully entered the global tides of information flow and became mainstream commodities for consumers the world over. A huge host of satellite "products" beyond television programming was added to the everyday lives of users; GPS, satellite imagery, and telephony were all rapidly growing services directly available to the individual satellite consumer as the 1990s progressed. Not that television programs were ignored in the 1990s; the distribution of satellite television programs also exploded as Arabia and the Middle East, Africa, India, China, and the Pacific Rim all saw aggressive, competitive, and entertaining satellite television services emerge. These services were at times to the consternation of authoritarian governments accustomed to tight media control, or to the condemnation of religious leaders, or even to the typical anxieties of parents wondering what all these words and images on satellite television would "do" to their children. But these new media outlets were also met with enthusiasm and delight by millions of new users and viewers.

The twenty-first century arrived to global telecommunications and satellite media in a harmless threat, a hyped claim, a false alarm about global security known as "Y2K." It is likely true that a number of pre-Y2K actions in the late 1990s, such as updating obsolete equipment and software, the rewriting of operating codes for many computers, and other actions designed to mitigate the potential Y2K problem were largely successful in turning Y2K itself into a fizzle. Likewise, it is undoubtedly true that this "global upgrade phenomenon" for information and communications technologies in the run-up to Y2K accelerated the global pace of information and software upgrades and brought the global information society into greater technological synchronicity (thus, for example, contributing to new conditions for computer viral and denial-of-service attacks to be launched from global locales heretofore dormant as hacker havens).

In retrospect, the irony of Y2K is that the first decade of the twenty-first century, in tragic and in disturbing ways, turned out to be all about global security in its darkest and most phobic manifestations. The discourse of satellites and global security in the early twenty-first century took on new complexities, in part overdetermined by the Cold War past and in part newly informed by the technological limitations of satellite-based observation, measurement, and surveillance. Satellite technologies in the twenty-first century became part of a convergence of global media under a new rubric of global security. After the 9/11 attacks, and with the onset of the global war on terror, all media technologies—whether television newscasts, satellite images, or cell phone videos—were discursively recentered on global security. In what could be called (to borrow from theoretical physics) a sort of "unified field theory" of global media and global security, the discursive turn of global security in the twenty-first century subsumed satellites and all other media technologies into a quest for total information awareness, for endless surveillance, and for the decline of personal privacy concepts as a right for users of media and electronic communication technologies. The twenty-first-century discourse of global security is oddly symbiotic with a growing movement toward interoperability, compatibility, open-source programming, and other trends toward global harmonization of electronic media. When combined with a heretofore unprecedented global bandwidth and distribution-circulation capacity for electronic information, this discourse has significant potential to permanently alter future thoughts and actions regarding satellites and global media.

Critical Paths

As this brief sketch demonstrates, there are many fascinating paths for critical and historical investigation of satellite technologies. This book brings together a rich array of authors who detail various projects, sites, and concepts related to the satellite's development. Their contributions explore satellite applications including broadcasting, telecommunications, remote sensing, and global positioning. In the process, the authors delineate the players, locations, ideologies, and power constellations that are at once constitutive of the satellite's historical emergence and have formed in relation to its ongoing use around the world.

The book is organized into three sections: Concepts and Cartographies, Satellite Mediascapes, and Orbital Matters. In the first section, Concepts and Cartographies, authors explore some of the foundational concepts that have both fueled and been fueled by the emergence of satellite technologies. These concepts include legal definitions of outer space, notions of the local and the global, ideals of democratic communication, ocularcentrism and the quest for planetary omniscience, and the geopolitics of satellite footprints. While these chapters all delve into particular details, they also are concerned with issues that pertain to the "big picture" or "overview" of satellite technologies. The section is guided by a cartographic impulse, then, in that it sets out to generate a critical mapping of some of the principles and ideals on which satellite technologies and their uses are based.

The book's second section, Satellite Mediascapes, features chapters that focus on various power struggles and negotiations involving regulatory, geopolitical, and economic concerns that have occurred in the satellite broadcasting sector during the past forty years. In this section, authors examine satellite broadcasting services around the world, from the United States and Canada to Germany, the Middle East, and Africa. Contributors consider the notion of multiple geographies (local, national, regional, and global), the requirements of new media technology in transnational markets, and the friction that occurs when the vision of satellite broadcasting as a tool for utopian development rubs up against a vision of it as a lucrative multinational business venture.

In the third and final section, Orbital Matters, we turn to a consideration of satellite technologies in relation to themes of observation, materiality, and epistemology, exploring sites and systems that range from being little known or top secret to those that are highly public and have become media spectacles. Whether discussing mircosatellites, secret reconnaissance projects, or fallen satellites, the chapters in this section address crucial questions about the technology's relationship to the politics of secrecy, failure, spectacle, and truth-making. By bringing different satellite processes and histories into critical awareness, these chapters demonstrate how satellites in orbit *matter* in multiple ways—in their relationship to world history, in their production of global communication and visualities, in their extraction of and dependence upon material resources, in their relation to the environment, and in their capacity to expose state investments, strategies, and priorities.

It is worth taking a moment here to highlight a few points about the book's scope and structure. First, given the satellite's important role in remapping cultural, political, and economic boundaries from the Cold War era in the current moment of globalization, spatial motifs are used to structure the book's three major sections. This is not meant to suggest that the temporal is insignificant, but rather to emphasize the multiple sites of satellite study and spotlight the dynamic field of historical activity that extends from Earth's surface out through orbital space and beyond. By drawing attention to this domain, which might be generally referred to as the "satellite belt," several questions inevitably arise: Where are satellites developed and launched? Who owns most of the world's satellites? Which orbital slots are filled and why? What is the relationship between satellites in orbit and their footprints on Earth? Who are the players involved in satellite development, operation, and regulation? What happens to satellites when they fail or no longer function? Many of these questions will be addressed in this book, and the engaged reader will acquire a better understanding of the satellite's relationship to questions of history, power, space, and materiality.

Second, addressing such questions requires the study of satellites across multiple academic fields, using historical and critical approaches. *Down to Earth* brings together the work of scholars working across fields such as communication, geography, history, media studies, science and technology studies, and security studies. In so doing, the book provides a range of scholarly vantage points from which to consider satellite technologies, industries, and cultures. The terrain of satellite study is filled with complex layers that require interdisciplinary and international modes of engagement. While it is clear that satellite technology emerged in the midst of the Cold War and its development has been driven by global security, big science, and global capitalism, there are countless satellite sites to explore and histories to write. *Down to Earth* features research on developments and uses of satellites in Canada, China, Egypt, Germany, Kazakhstan, North Korea, South Africa, and the United States, and across Asia, Africa, Europe, North America, and the Middle East.

Finally, in addition to thinking about satellites across disciplines and in relation to various parts of the world, this book is intended to draw critical attention to a technology that is far away and physically removed from Earth and yet is fundamental to the lived conditions of our contemporary world. There is a need for further historical and critical investigation of the multiple sites, institutions, and infrastructures that satellites rely on, whether Earth stations, launching facilities, corporate headquarters, footprints, orbital paths, bandwidth, power sources, or regulatory agencies. While examining such sites throughout the book, contributors have made an effort to interweave macrolevel conceptual and institutional issues with microlevel, localized, everyday concerns. Bringing the satellite down to Earth, then, involves ongoing investigations of such issues as the development and ownership of satellites, regulation and allocation of the electromagnetic spectrum, orbital slot assignments, geopolitical agendas, risk management practices, and the digitization of satellite systems. It is our sincere hope that this book might not only provide readers with a more nuanced understanding of the history and present-day uses, considerations, and potential of satellites, but that it might also inform and serve as a catalyst for additional research into this topic, which has myriad global implications.

NOTES

1 For a listing, see "Satellite," accessed March 26, 2011, http://en.wikipedia.org/wiki/Satellite.

2 Trevor Paglen, *Blank Spots on the Map: The Dark Geography of the Pentagon's Secret World* (New York: Dutton, 2009), 97.

3 "List of Satellites in Geostationary Orbit," Satellite Signals website, updated March 1, 2011, accessed March 26, 2011, http://www.satsig.net/sslist.htm.

4 James Schwoch, *Global TV: New Media and the Cold War, 1946–69* (Urbana: University of Illinois Press, 2009).

5 For an interesting discussion of this practice, see Brian Springer's video, *Spin*, 1995, accessed March 26, 2011, http://video.google.com/videoplay?docid=-7344181953466797353#.

6 Anthony van Fossen, "Globalization, Stateless Capitalism, and the Political Economy of Tonga's Satellite Venture," *Pacific Studies* 22(2) (June 1999): 1–26.

CONCEPTS AND CARTOGRAPHIES

1

THE INVENTION OF AIR SPACE, OUTER SPACE, AND CYBERSPACE

JAMES HAY

This chapter offers a genealogy of three related discourses and programs about achieving, enacting, and managing communicative space: *air space*, *outer space*, and *cyberspace*. I first consider how something called "outer space" (a space of freedoms, an object of government and policy, and a space "settled," understood, and organized through a new regime of technologies fit for global communication) developed out of a pre–World War II conception of *air space* (having to do with both radio and flyover space, and hence a conception of space supporting and problematizing national sovereignty). I then suggest ways that these two historical conceptions of space and regimes of communicative space (*air-space* before World War II, and *outer space* during the Cold War) became a framework for imagining, requiring, and inventing something called "cyberspace."

In this chapter, I emphasize the ongoing problematization and regulation of extraterrestrial outer spaces since the late nineteenth century in order to suggest a historical connection between the reinvention of outer space and political modernization (the reinvention of liberal government). In this sense, I offer a way of thinking about the relation between the invention of communication, the invention of space (that solves or manages problems of communication), and the ongoing experiments aimed at "advancing" and re-inventing liberal government as a system invoking the virtue of communication as the most peaceful and civil way to exercise freedom and achieve security. While I focus on how this reinvention of communication, space, and government occurred in the West (and particularly from the United States), I also underscore how these ongoing programs of invention were rationalized as solving problems of global governance and of waging global peace through extraterrestrial communication. Furthermore, while I am interested in the practice and mentalities of invention, I also emphasize the experimentalism, failures, and insecurities surrounding the solutions for managing space, and in this way underscore the changing *regimes of truth* in governing and securing a "free world" and open skies, through communication technology.

This chapter makes a number of contributions to the discussion of satellite technologies of communication. First, it offers a history of the present, considering how the current ("neoliberal") governmental rationale about securing cyberspace has emerged out of a modern preoccupation with securing, regulating, and pacifying air space and outer space. Second, it addresses how understanding satellite technology depends on recognizing the historical and geographic relationship between communication and transportation technologies. Third, it rebuts technological determinist accounts of contemporary communication technology by emphasizing instead how the invention of communicative space has been predicated on the changing rationalities of liberal government—particularly from the United States. And fourth, it offers a theoretical and methodological alternative to accounts of modernity and globalization that see space as an epiphenomenon of communication networks, economies, and political government, and their modernity or modernization.

Modernity's Babel Complex

> It is by nature that one of the uses that God has given to the seas and rivers is that of opening up routes that communicate with every country in the world by navigation. And it is by police that we have made towns, public squares, and other places appropriate for this use, and that those of each town, province, and nation can communicate with all the others of every country by great highways.
> —*Jean Domat*, Le droit public

Although this chapter is mostly concerned with developments over the twentieth and twenty-first centuries, it recognizes the importance of an even longer history of communication, space, and government. Noteworthy research into this historical context includes James W. Carey's discussion of the relation between communication and transportation during the nineteenth century and how that relation temporally and spatially organized emerging nation-states such as the United States, and Armand Mattelart's explanation of "the invention of communication" as a swarming of scientific discourses and projects in eighteenth-century France preoccupied with the "health" of circulatory systems—from the human body to national territory— as unblocked, freely flowing arteries. Carey and Mattelart both call attention to the interdependence of communication and transportation, though Mattelart more fully acknowledges the relation between economic modernity (*laissez-faire*) and modern communication as transportation (*laissez-passer*), and between liberalism and the various "paths of reason" (the rationalities, technologies, and networks) on which communication's "invention" and modern application depended.[1]

The chapter's title refers to Mattelart's history of the invention of communication, but also examines the intersection between his history and Michel Foucault's writing about the emergence of liberalism (political modernity) through technologies of security and government. For Foucault, the birth of liberalism occurred through technologies of freedom that also operated as technologies of government. Whereas Mattelart points to the eighteenth-century French physiocrats in order to explain the "paths of reason" (the sciences of circulatory systems) that collectively represented communication as a relation between *laissez-faire* and *laissez-passer*, Foucault refers to physiocracy in order to consider how modern political economy developed as one of many "rationalities" and technologies of modern, liberal government.[2] Foucault's well-known studies of disciplinary power and a "security society" frequently examined the historical rationalization (separation and organization) of spaces such as asylums, hospitals, and prisons, but in his later career he occasionally discussed liberal government's emergence through a new regime of spatial technologies and rationalities designed to solve problems in managing distance and mobility.

Foucault's explanation of the relation between liberal government and modern "technics of space" is worth elaborating briefly to clarify some of the basic objectives of this chapter. He suggested that liberal government developed during a period when the old European conception of *police* or "police state"—whose model was the internal organization, communication, and monitoring of a town or polis—was giving way to the problem of governing *territory*.[3] Territorial government may have developed using the city as a residual model of government, that is, the "urbanization of territory" whereby streets radiating from a city center became analogous to networks of transport and communication radiating from state capitals. However, networks of transport, communication, and electrical grids (*technics* of space) increasingly solved the territorial problem of liberal government by becoming the practical rationalities for an extensive space and the space of the frontier. Second, these technics of space were indispensable to the emerging technics of liberal government because their modality and field of operation (territory, communication, and speed) were supposed to liberate—or rather, govern and secure—*through* the freedom of mobility and extensive space, through the maximization of *laissez-passer*. To *police* is to urbanize the practice and conception of government as an internal organization; to govern through freedom involved a new relation of space and power, specifically a recognition of the technologies for organizing and managing the extensiveness of

territoriality. Alongside Mattelart's account of the invention of communication, Foucault's account helps explain how the birth of modern communication developed as and through technologies of government—as both
free-flowing and well-*regulated* circulatory systems (the refinement of road,
bridge, and canal networks), which became integral to liberalism's technologies for making rational/scientific the means of achieving "fair-government"
and of continually overseeing "better," more effective ways for populations—
as free, independent, and mobile citizens—to govern themselves.

Foucault's account of liberalism's formation around these technics of
(extensive) space does not address two implications of his thesis that are central to this chapter: how the technics of space enabled and complicated the
security of territorial borders (the sovereignty and limits of territoriality)
while simultaneously enabling and complicating the *extension* of territorial
borders beyond or across the borders of nation-states, often through colo-
nialist projects launched by the Western bastions of liberal government. As
Andrew Barry has pointed out, the nineteenth-century space of liberal rule
that was made possible by the growth of networks of communication (such
as telegraphy), transportation (such as railroads), and electrical grids,
became instrumental to the security of borders and frontiers.[4] Governing the
border tested the practical capability of coordinating communication and
the rapid transport of people and supplies to and from borders, over increasingly vast distances. However, it also tested the limits of governing the extensiveness of communication and transportation networks—limits that called
forth a new regime of transnational government.

Implicit (but never elaborated on) in Barry's or Foucault's explanation is
how the freedoms ascribed to or supported by communication and transportation networks included the *transnationalization* of these networks. Although the transnational extension of communication and transportation
networks during the nineteenth century produced a space of international
circulation, the freedoms of transnational space posed particular problems
and contradictions for liberal government. For instance, the freedom of
movement and communication through extensive space made the border an
object of securitization even as the border needed to be designed for efficient
movement and communication through it with customs offices, passports,
and so on.[5] In this sense, the nation-state was only one scale of liberal government as extensive space and territory. Technics of national space, alongside the push for a *unified* international space of circulation, rapidly required
ancillary governmental institutions oriented specifically toward the transnational space of communication and transportation—an internationalization
of liberal governmentality in which the nation-states most committed to liberal government were most invested. As Mattelart has noted (echoing Carey's
points about telegraphy and the westward spread of religions in the United

States), these international networks of communication and transportation were both material and spiritual, rationalized by French philosopher Claude-Henri de Saint-Simon as the technical means of universal and association and brotherhood. The institutions that emerged to oversee international networks thus were preoccupied as much with assuring free movement across national borders as they were with a policing role—assuring the peacefulness and harmoniousness among nations dependent through the maintenance of the healthy, rational growth of the networks of universalization. From the 1850s through the 1910s, the number of interstate agreements and institutions establishing standards for transnational communication and transportation grew dramatically in Europe—there were 17 of these between 1850 and 1870, 20 between 1870 and 1880, 31 between 1880 and 1890, 61 between 1890 and 1900, and 108 between 1900 and 1910.[6] These institutions comprised an emergent regime of global governmentality, peace, and cooperation, one whose primary preoccupations were the health, stability, and normalization of international movements through communication and transportation networks.

Given that the modern governmental problem that Foucault identifies as territorial space was addressed in the second half of the nineteenth century through the proliferation and interdependencies of communication and transportation networks, these networks generally were able to connect populations and administrative agencies for whom communication was a *national* objective—a relation to the earth/territory (*terra/territorium*), as the location of birth and growth (*natio*). Not only did the speed and extensions of communication and transportation complicate this relation to *terra*, *territorium*, and *natio*, they required or called forth technologies for regulating both the breakdowns of peaceful coexistence/communication and of the freedoms of communication and exchange beyond national borders. The growth of international agencies of government and security targeted nothing short of the increasing potential for communication breakdown. They were both haunted by the specter of what could not be contained and the need to "advance" the apparatuses of global government—apparatuses that extended beyond native Earth and national territory and that "modernized" the Babel Complex.

The Freedom of the Air and the Law of the Air

The relation of freedom, government, and security that Foucault attributed to the nineteenth-century technics of space was decidedly *terrestrial*—their space of rule a *territorium*, the land surrounding a town or city. For Foucault, these technics of space, particularly by the late nineteenth century, became integral to a changing rationality of government that projected the town or city model of government—a model of streets radiating from a town's center—onto a more extensive, national space. Thus the old model of government became a template for grids of transport, electrical power, and communication radiating from capitals. In this sense, liberalism's rationality about the government of territory remained moored to the land example (the political territory as terrestrial) as much as it was committed to inventing technologies for overcoming distance and the terrestrial impediments that stood in the way of *laissez-passer*.

Until the twentieth century, the modern "Babel Complex" (the discord produced by the technology designed to extend communication transnationally) had not quite included the invention of extraterrestrial or air space as a problem or object of liberal government. Certainly there were scientific towers, such as Paris's Eiffel Tower, which Roland Barthes has described as born into a modern Babel Complex: if the Tower of Babel was a theological project used for communicating with God, the Eiffel Tower was a modern ascensional dream "almost immediately disengaged from the scientific considerations which had authorized its birth."[7] But as I intend to show in this section, even those towers became instrumental to the networks supporting a new rationality of liberal government preoccupied with, and acting on, air space.

Arguably the most challenging objective of emergent international government, and one of the most nettlesome challenges to the sovereignty of nation-states at the turn of the twentieth century, was the government of air space. In the first decade of the twentieth century, a rapidly intensifying preoccupation with air space led several key international (Western) conventions, committees, and treaties to formalize the first regime of air law. Through these legislative initiatives, the air or air space became a focal point of discourses about freedom and government, freedom of passage and sovereignty, aircraft and statecraft. If the "air," "heavens," or "ether" were premodern, "air space" was an object of rational, scientific calculation and governmental standardization, regulation, and policy.[8] To be free and open, the air needed to become air space that made rational its "fair" use. The nation-states most concerned about regulating the freedoms of air space were those most invested in the technologies of flight. They were also the bastions of liberal government.

Even though the rapid proliferation of meetings, pacts, and legislation about air space at the beginning of the twentieth century represented a formal and methodical consideration of air space, they often argued that the air (even in the "remotest" bastions and colonies of Western, liberal government) could not be rationally organized and managed, in part because the government of the earth/land provided little scientific or juridical precedent or certainty. Edouard Rollin, a representative at the 1906 meeting of the Institute of International Law in Ghent, argued that it was premature to formulate rules for the sky—a space that was as unknown and uncharted as the center of the African continent had been for Europeans fifty years earlier.[9] In 1911, British legal scholar Harold Hazeltine attributed the dangers of air space to the lack of "a firm basis in analogy," noting that whereas creating a buffer zone between national borders and international waters may have diminished the potential of surveillance or hostile threats, the increasing distance of aircraft above sovereign territory actually exacerbated the fear and threat of surveillance and/or hostile actions.[10]

Those regulating the skies above nation-states looked initially to maritime law, but the law of the air introduced a new problem: defining the vertical limits of sovereignty. From the earliest legal statements about air space as a space of freedoms and rules, the problem of calculating and standardizing the vertical limits of national and territorial sovereignty was grounded in laws concerning the protection of "private property" or privacy. The first seminal essay on air space in the twentieth century, M. Paul Fauchille's splendidly titled "Aerial Domain/Dominion/Property and the Juridical Regime of Aero-States," alluded to the violation or threat to private (terrestrial) property posed by balloons and aircraft.[11] Nascent in these discussions about the rights and security of property owners was another serious question: what constituted ethereal property? If air space could be owned and legislated using maritime law or the laws regarding land rights (that is, landowners could do anything above their property that they wanted), then deciding how to measure and standardize the vertical limits of sovereignty became crucial. There were various proposed solutions: ethereal property or sovereignty was "as high as humans can reach," presumably through technologies of transport); or, ethereal property should follow the topographic rise and fall of Earth, with a set height agreed upon by nation-states. Some argued that the Eiffel Tower should be designated as a basis for universal measurement. But terrestrial rights and laws were difficult to fix in the air: how close to a building or domicile would the pilot of a balloon or plane have to fly to commit an act of peeping, and thus of criminal trespass? Just before World War I, Hazeltine's *The Law of the Air* represented one solution: that the air is free unless air conduct threatens or infringes upon the rights, sovereignty, or security of private and public land owners.[12]

These writings, pacts, and programs about air space resulted in two related schools of thought: one, that air space was completely free ("free air"), and the other, that international formulas needed to be instituted to protect national sovereignty and private property. In "air space," the freedom of air was predicated on law and limits. Establishing a governmental formula for the vertical limits of sovereignty thus linked the "free air" and the freedoms exercised in the air as not only a matter of rules but of borders. Particularly in the West, air law and sovereignty (the exclusivity of territory marked by borders and the appropriate security apparatuses such as customs agents) thus involved applying an earlier reasoning about liberalism's space of rule, even as air space presented a new set of problems that called forth a new regime of governmental technologies. The sky, or more technically "air space," became a laboratory for testing the rules and limits of freedom necessary for national sovereignty and universal association. In this sense, air space became a medium for governing through peace and civility. Air space was a stage for international codes of conduct.

The invention of air space and the efforts to legislate the vertical limits of sovereignty were not simply about flyover space; they also pertained to the air space of radio signals and the growing synergy between technologies of transportation and communication that made air travel knowable, rational, safe, and governable. In that the birth or invention of air space involved recognizing limits (rules and borders) for the free and peaceful use of aircraft, these limits were supposed to make air space a space of *international communication* (in the sense propounded by Saint-Simon). In practice, the prospect of long-distance, precisely guided plane travel involved synergizing the technologies of air transportation and radio communication, "freeing" communication from the terrestrial technics and networks of space, and rationalizing air space for air travel and communication. By the 1920s, making the freedom of the air civil (that is, subject to civil codes and the rules of early international agreement) involved communication technology in very direct ways, because air vehicles (like ships) navigated with wireless forms of telegraphy and radio. Air travel and radio thus developed through one another by the 1920s. Their development in the first two decades of the twentieth century hastened various treaties and regulations pertaining to the freedom of radio- and flyover-air space, particularly in international air space. For instance, in 1908 an international agreement regarding the use of wireless telegraphy among ships and with land-based receivers became a template for the first rules regarding wireless radio in and from the skies—a pact administered by the fledgling International Office of Wireless Telegraphy and affecting twenty-seven countries, including the United States. By 1913, the year after the sinking of the *Titanic*, radio (or "wireless telegraphy") became a standard practice in Western shipping.

Radio communications both facilitated and problematized the government of a unified space of transnational circulation. By the early 1920s, radio operated through air space as part of the navigation of land and sea transport and also of aviation. Throughout the 1920s, as radio communications became integral to everyday life, they refined prior terrestrial communications' maintenance of a unified national space. However, during the 1920s and 1930s, in organizing nation-states as audiences and publics and linking nations with their colonies, radio communication also developed as a technology of

———————— Figure 1.1 Universal Pictures' 1930s logo—a rotating globe ————————
circumnavigated by a tiny airplane.

——— Figure 1.2 RKO's image ———
of an enormous radio tower
telegraphing sound waves
around the globe.

exporting (propagating and propagandizing) national cultures. And during wartime, radio served as a secretive and invisible way of gathering and conveying information from beyond national borders—making the foreign knowable, and thereby securing the homeland.

Some of the most well recognized iconography of the intersection between air space and Hollywood's emergent capitalization of global markets were RKO Pictures' branding of its films during the 1930s. RKO's image of an enormous radio tower telegraphing sound waves around the globe, like Universal Pictures' 1930s logo—a rotating globe circumnavigated by a tiny airplane—remain iconic to the present day (see Figure 1.1). The RKO logo in particular illustrated how profoundly the transformation of the radio towers (or the attachment of radio transmitting technology to skyscrapers) had stitched the nineteenth-century Babel Complex into a new, spatially "advanced" or extensive, rationality of air space and its liberal government (see Figure 1.2). The world as metropolis—as a new model of the city as a space of international communications and liberal government—had become (to use the title of RCA's new complex in New York City) a "radio city."[13]

Air space thus developed as a virtual/ideal space of absolute freedoms, against which and through which liberalism (as a governmental rationality) was constituted and "advanced" historically and geographically. Air space became one of the new frontiers or problems against which and through which the rules and sovereignty of free peoples and states could be calculated. Because the right to the air needed to be measured to be governed fairly and rationally, the project of such governance was as much about guaranteeing the security of borders and property as it was about the freedoms of public and private air transport to act in the skies above and across terrestrial borders. The early codes, programs, and institutions for assuring the civility of freedom in the air and the security and defense of public and private terrestrial sovereignty continued through the 1930s, as evidenced by the League of Nations' unsuccessful 1936 effort to mobilize international support for a "radiophonic nonagression" pact.

The international government of the air (rationalizing, civilizing, and pacifying the air as "air space") recognized the growing potential of air space as a threat to territorial sovereignty and as a new stage for warfare; controlling air space thus became both the objective and the means of waging peace and war. The origins of an "air force" (the exercise of power through and from air space) are not simply the outcome of military history and the inventions of technologies for combat but also of the technologies of international government oriented to both the freedom and pacification of the air in the late nineteenth and early twentieth centuries. The strategies of warfare in World War I and devastatingly in World War II not only developed through the communication and transportation technologies of controlling air space but also through the prior contradictions of exercising power through air space that was both "free" and "shared." Freeing and securing air space—governing through air space—became a reason for war, while the freedom of the air ("air power") became a new medium of military force. But controlling the air also became a stage for demonstrating the "finality" of war, or rather the continuation of war by other means. In the aftermath of World War II, the invention of a space beyond air space and the vertical limits of sovereignty (that is, inventing an "outer space") became the objective for governing the world through peace and communication.

The Invention of Outer Space–Governing the World through Communication and Peace

In his foreword to Andrew Haley's *Space Law and Government*, then vice president Lyndon Johnson noted that "the great new problems confronting civilization in the Age of Space require . . . that the principles of justice and order should be established in these early days of man's exploration of space."[14] More than rhetorical flourish, Johnson's emphasis on justice and order in outer space echoed the reasoning he had expressed since the late 1950s regarding the role of government in outer space. Johnson was one of the most outspoken supporters of bringing the space program under the *sponsorship*, rather than the direct control, of the federal government; this was partly a political tactic for rethinking the exclusively military and mostly clandestine early astronautical research, development, and exercises conducted during the Eisenhower administration. From the late 1950s, Johnson repeatedly justified the national space program as a means of establishing justice and order in and through outer space, of arranging outer space into a rational, liberal-democratic, and ethical sphere—as a "peaceful" endeavor rather than an exclusively or primarily military one.

Although one objective of accelerating and expanding a national space program was the imagined strategic military and geopolitical advantages doing so would provide the United States over Russia, the U.S. space program was not simply an epiphenomenon of the Cold War. The new world order that historians often associated with the Cold War had to do with an emerging way of understanding and governing space through many of the devices and technologies developed through the massive U.S. space program. The invention of outer space and its relation to the reinvention of liberal government following World War II developed through discourses and programs concerning the freedom and law of the air that had preoccupied Western nation-states and their early international agreements regarding transnational communication and transport since the nineteenth century.

In the first half of the twentieth century, the rules and sovereignty of free peoples were measured through and against the new frontier of air space, but there was a qualitative difference between regulating air space and outer space. "Outer space" became a new historical, geographic, and theatrical stage for shaping a discourse about rights and responsibilities, war and peace, security and risk—and thus for redefining the objectives of government and of national sovereignty on a global scale. Furthermore, as a new degree of space, a new state of freedoms, and a new object of government regulation, outer space became articulated to the political idea of a New Frontier. After World War II, and particularly by the late 1950s as astronautical launches became frequent, regulating *extra*terrestrial space became a key issue in rethinking government of and by nation-states. When Sputnik I began to orbit Earth in 1957, its legal status was unclear: what national laws did it violate, and if it violated no national laws, what international body governed its activity?

The seemingly unlimited, open, and unregulated nature of "outer space" during the 1950s revolutionized the reasoning about the security and insecurity of nation-states. This was particularly true of the United States and Russia, both of which viewed outer space as a zone for securing the nation against unfriendly overflight. At a time when outer space was still relatively unregulated, it posed new possibilities for expanding the limits of national sovereignty (in space and into the future) and for establishing a proprietary relation over bodies in outer space, while simultaneously posing new risks and requiring new knowledge and technology for managing these risks. The 1955 Open Skies Treaty (proposed by the United States and rejected by Russia) was an example of the mentality about liberalization and risk-management accompanying the space program. The pact proposed that flights by unarmed surveillance airplanes be permitted over participating nation-states and reasoned that overflights were necessary for promoting confidence, predictability, stability, and peace—a new international, geopolitical order exercised from above.

Making space knowable and manageable/controllable came to rely on the new relationship between outer-space transport and outer-space communication. Although radio became a widespread instrument for navigation and communication during the 1920s and 1930s in aerospace transport (as well as in ships and cars), radio and televisual instruments became indispensable to astronautical flight—integral to tracking, guidance, and information recovery via astronautical craft. In the late 1950s, Sputnik I (1957) and Vanguard I (1958), the first Russian and U.S. satellites, respectively, were designed to test and demonstrate the feasibility of radio transmission to and from outer space. And by 1962, the first U.S. attempt to land an unmanned craft on the moon relied on radio operations for most of its primary procedures: communicating and telemetering on the ground before and during launches; tracking, command, guidance, navigation, and telemetry during the flight; telemetry and televisual scanning after the expected landing; and the use of radiotelegraphy, radiotelephony, and data transfer throughout the operation—all at radically unprecedented distances. Collectively, these instruments attest to how quickly multiple and integrated forms of remote transmission and control became indispensable in developing the U.S. space program.

In 1962, the relation between space transportation and communication converged in Telstar I, an active-repeater communication satellite for global communication links that facilitated the first live, transcontinental telecast. Telstar represented a significant milestone in the formation of the U.S. space program and in the governmental rationality developing around this program for several reasons. First, Telstar represented the space program as a model for a new governmental rationality by breaking down the previous distinctions between military and civilian projects, and between public and private institutions. More than previous communication satellites, Telstar was designed and deployed as a cooperative venture between government and a consortium of the leaders of the U.S. communications industry—specifically, AT&T, RCA, ITT, and GTE. In 1962, an act of Congress permitted the formation of the Communication Satellite Corporation, capitalized by these companies and by the federal government. Second, Telstar was vital to publicizing the space program's capacity for global public relations and as a popular spectacle of national mobilization. Telstar celebrated the "tele-visuality" of the space program, which increasingly relied on telecasts from outer space to represent astronautical missions for television audiences in the United States and abroad. Third, Telstar marked the historical juncture when "broadcasting" as a national project began to be replaced by a new globalism facilitated by satellite telecasting. Telstar, in other words, redefined the objectives, capacities, and purview of television, and it celebrated a new *televisual stage* of globalism even though the telecasts emanated to and from North America and (via the Eurovision network) Western Europe. Through Telstar, this new stage of globalism became associated with, and virtually linked by, astronautical transport and satellite communication.

Unlike Russia, the United States moved quickly, through Telstar and subsequent telecasts of U.S. space missions, to shape a global imagination about outer space. Telstar also displayed the United States' accomplishment in outer space as a new standard of being modern and future-oriented; it represented the nation's new global telepresence in a world organized and exhibited via U.S. television production and distribution. The very first image and sound sent to outer space and back, and transmitted through the transnational network that Telstar launched, was an American flag pulsating to the U.S. national anthem in front of a futuristic, white, spherical radome at Andover, Maine. Later, as the satellite passed over the Northern Hemisphere, it transmitted a sequence of broadcasts representing various regions of the United States and Western Europe. Telstar's America linked new and familiar sites and sights (the Seattle World's Fair, Mount Rushmore, a Native American, John Kennedy in Washington, the Manhattan skyline, the Statue of Liberty) bundled and distributed as part of a new global map.

Telstar and contemporaneous astronautical exercises thus called into question an early-modern logic of mediation organizing the space of terrestrial flows, terrestrial distance, and the territorial sovereignty of nation-states. As part of a new political and cultural economy from outer space, and in imparting the sensation of simultaneity, Telstar muted questions and concerns about violating the vertical limits of sovereignty. Telstar telecasting depended upon the new linkage between astronautical transport and communication— a regime of transport and communication that David Harvey and others decades later would describe as the "time-space compression" of a late-modern (or, in Harvey's terms, "postmodern") globalism.[15] If, as Armand

Mattelart noted, communication was an idea or practice invented in the late eighteenth and nineteenth centuries as a condition of the idea of internationalism, then astronautical transport and communication were integral to a new stage of globalism—referred to in the late twentieth century as "globalization."

The globalization of Telstar and of subsequent communication satellites (such as the 1965 Early Bird satellite, a geo-stationary active repeater satellite that transmitted the first World Town Meeting) was not just about intercontinental simultaneity—or a new spatio-temporal relation between the living room and places thousands of miles away—or the emerging idea that outer space was the path to virtual space (for example, the telecast photo of Telstar by the Associated Press via Telstar for international distribution as a news story). Telstar made outer space available as a stage for a new, globalized representation of liberal governance, with the U.S. space program (a new kind of venture between the state and civilian institutions) as provider and broker of a globalized network.

The early spy and communication satellites, as well as manned space flights, occurred on a "third-dimensional" space that so exceeded the previous practices of flyovers as to transcend rule of nation-states. By the mid-1960s, the vertical limits of some satellites made them appear from Earth to be relatively stationary objects. Orbiting craft thus extended the limits of national sovereignty while claiming a new position—a third-dimensional space of freedom—where the laws and customs governing those borders did not necessarily apply. In certain respects, the enactment of the Space Act of 1958 was as much a response to Sputnik's ungovernability as it was a response to international consensus. Formalized in 1955 by nongovernmental associations of scientists, the Space Act provided that satellites—as exercises in science and communication, rather than strictly military exercises—did not constitute a breach of national sovereignty and thus were not a basis for war but were instead conditions for international peace. In other words, the Space Act was as much about a national insecurity as it was a statement about the changing conditions of freedom represented by the new dimension of space.

The proliferation of astronautical programs thus represented a new stage in the objectives of liberal government, particularly for the United States. The U.S. space program was rationalized as a technique for protecting and expanding a free society. Arguably, the massive scale of the space program indicated not just a resolve but a preoccupation about "open skies" and securing outer space as a new world stage for exercising freedoms, even as the United States and Russia were the only nations capable of operating in and from outer space, and even though protecting and expanding freedoms rested on new governmental technologies such as the space program. Astronautical space, as a space for a new convergence between communication and transportation, thus became a terrain where freedom and government could be reinvented, projected, and exercised. Liberalism, as governmental rationality, was redefined within the space metaphor and project. Outer space became the latest (and purportedly the ultimate) space of liberalization—a condition of reshaping the world through open skies. Outer space also became a new paradigm and a new object of liberal reform, while the space program supported a new paradigm and a new object of liberal government, and a new way of understanding the relation between *laissez-passer* and *laissez-faire.*

Reinventing Government and National Sovereignty in Cyberspace

As described here, history up to the mid-1960s emphasized how the ether became an object of reasoning about advancing liberal government, and how communication and transportation technology came to matter conjointly in programs for opening, securing, and governing an "outer space." This occurred as extraterrestrial space became an extension of transnational terrestrial networks and, as such, an objective in reinventing and advancing the global technology of liberal government. In essence, extraterrestrial space became a branch of global government. This history has current implications, particularly through the invention of "cyberspace," a term that gained currency during the 1990s through a neoliberal rationality about governing and securing the spaces of communication and transportation. In some respects, the invention of cyberspace acted on and repurposed the technical infrastructure and networks of air space and outer space. Just as the invention of air space and outer space occurred through various regimes and arrangements of transnational and global government, cyberspace also emerged as a prominent stage for advancing liberalism (historically and geographically), even as it problematized global, neoliberal government.

A thorough genealogy of cyberspace might actually begin as far back as the ancient Greeks, who used the word "*kybernetes*" to mean a steersman or governor, or to early Western uses of the term "cybernetic" to refer to the art or science of government. As early as 1948, Norbert Wiener adopted the same word in order to describe an entire field of theory about communication and control in animals and machines.[16] Although Wiener was particularly interested in the nineteenth-century scientific studies of control dynamics and feedback mechanisms (the "governor" being a device crucial to the invention of self-regulating machinery), even this branch of scientific invention and experiment was imbricated in nineteenth-century rationalities about liberal government, as when the French electro-physicist André-Marie Ampere adopted the term "*cybernetique*" to describe "the future science of government." The scientific discourse about the control dynamics of self-communicating and self-regulating machines provided the model for and technology of the modern liberal ideal of governing at a distance, thus linking self-governing machinery and self-governing social bodies, even if the term "cybernetics" would not gain significant traction until the second half of the twentieth century.[17]

During the 1990s, "cyberspace" typically referred to a "virtual" reality/space of computer-based communication and transport. This use of the term often placed considerable emphasis on cyberspace as the "end of geography"—an absolute ethereality, a space free of terrestrial impediments, a "no place." As Tiziana Terranova has pointed out, it is no coincidence that the view of cyberspace as a virtual or nonplace became a framework for a discourse about globalization as time-space compression: "Where the most common image of cyberspace used to be that of a virtual-reality environment characterized by direct interface and full immersion . . . , now the image is that of a common space of information flows in which the political and cultural stakes of globalization are played out."[18] Similarly, it is no coincidence that the emergence of a discourse about cyberspace as a new globalized space of communication and transport developed through a neoliberal rationality not just about free trade and global markets (a new economic liberalization in the narrow sense) but dedicated to privatizing and outsourcing public services and cultivating self-enterprising, self-empowered, and self-reliant citizens. The Clinton administration's projection of a national "information superhighway" became one of the instruments and objectives articulated through the administration's implementation of a National Partnership for Reinventing Government—a political rationality about the virtue of public-private partnerships in improving the workings of liberal government.[19] Less than three months into his presidency, on March 3, 1993, President Clinton told the nation, "Our goal is to make the entire federal government less expensive and more efficient, and to change the culture of our national bureaucracy away from complacency and entitlement toward initiative and empowerment."[20] Key to the modernization and makeover of government through private initiative and empowerment was the computer and information technologization that had begun to transform the banking and business sectors while also becoming a new target of investment and a model for an entrepreneurial economy—an "e-economy." President Clinton and Vice President Gore frequently compared the information superhighway to the transformations accompanying the creation of a federal highway system during the 1950s; thus cyberspace as "information superhighway" became central to an emergent liberal governmental rationality precisely at the time that outer space was becoming the "dormant frontier."

Although Vincent Mosco is right to point out that cyberspace was invented within a liberal rationality oriented toward a new stage of market freedoms, and although the reinvention of liberal government in and through cyberspace often was predicated upon a triumphalism about liberalism (and the space race) at the end of the Cold War, the "end of history" forecast by champions of a neoliberalism such as Frances Fukuyama (and emphasized by Mosco) was envisaged by those champions as a *return* to a purer state of freedoms but also as an acceleration of this state of freedom far beyond the limited communication networks of earlier periods.[21] In this sense, the invention of cyberspace as a time and space of government harkened back to air space and outer space as the most open and least impeded spaces of government. Air space and outer space became *objective correlatives* for a neoliberal reasoning about reinventing spatial states of freedom and government—states "predetermined to correspond to the preexisting idea in [their] living power . . . [that was] essential to the evolution of [their] proper end."[22]

This "back to the future" of liberal government was not exactly what William J. Mitchell referred to when he likened the electronic frontier of cyberspace to the Wild West in his influential (albeit much criticized) account of cyberspace as the end of geography ("a new land beyond the horizon").[23] However, Mitchell acknowledges (at least around the edges) the potential for risk and the need for securitization, noting that cyberspace would inevitably attract con artists and conquerors cut from the same cloth as those in the Wild West.

By the first decade of the twenty-first century, following but not simply a consequence of the World Trade Center attacks of 2001, the Bush administration's rearticulation of reinventing government played out partly along two fronts—one involving a profound deepening of federalism, privatization, outsourcing, and deregulation (the formula that less government equals greater freedom), and the other involving sweeping reforms for securitizing a "homeland" through a massive, decentralized administrative apparatus, the Department of Homeland Security. In no way were these two fronts incommensurate or contradictory; maximizing the localization, privatization, and personalization of government (the new "powers of freedom") opened up threats and instabilities for which a new regime of securitization could be rationalized.

One tool in this new regime of liberalization and securitization was the formulation of a National Strategy to Secure Cyberspace (NSSC)—a policy initiative that considered cyberspace to be a new problem of national and international security and thus a new problem of government.[24] Not only did the strategy adopt the expression "cyberspace" as a relatively new instrument and problem of government unfolding across various scales and through multiple agents (for example, as global and national government, public and private government, and the responsibility of self-governing citizens), it also cast cyberspace as a field of warfare and a virtual front in a war on terror being conducted on these various scales and by these multiple agents. This field of warfare occurred somewhere between peacetime and a perpetual state of alert. The NSSC articulated the security of a nation in cyberspace to a new stage of *partnership* between "the way that business is transacted [and the way] government operates."[25]

The NSSC proposed a concerted response to a set of threats that were both similar to and different from the ones that had been attributed to air space and outer space. At one point, it stated that "in the 1950s and 1960s, our nation became vulnerable to attacks from aircraft and missiles for the first time. The federal government responded by creating a national system to monitor our airspace. . . . Today, the nation's critical assets could be attacked in cyberspace. The United States now requires a different kind of national response system in order to detect potentially damaging activity in cyberspace."[26] In a policy document replete with visual illustrations, this statement followed an image of an individual (in shadowy silhouette) operating a personal computer in front of an enormous map of the world, with North America positioned between the individual's head and computer screen. The NSSC's map of cyberspace thus located the nation's insecure place within a world lacking limits and boundaries. If air space and outer space problematized the vertical limits of sovereignty, the NSSC's map of cyberspace accentuated the safety of a past "geographic isolation that helped protect the United States from direct physical invasion," noting that "in cyberspace, national boundaries have little meaning."[27] On a timeline that begins in 1995, a date that the NSSC attributed to the onset of an unsecured cyberspace, the strategy represented the security threats (and incidents handled) as spiking dramatically between 2000 and 2002, the first years of the George W. Bush administration.

By underscoring that cyberspace can best be secured by public-private partnerships and vigilant citizens, the NSSC cast cyberspace as a space for demonstrating not simply the *virtue* of but the *need* to accelerate a national strategy for reinventing government. The NSSC's mantra of public-private partnership and citizen empowerment suggested that its goal was nothing short of governing and securitization throughout society and that its insecurities about threats to cyberspace were comparable to its insecurities about a state that governs too much. As President Bush stated in his letter attached at the beginning of the NSSC, "Securing cyberspace is an extraordinarily difficult strategic challenge that requires a coordinated and focused effort from our entire society–the federal government, state and local government, the private sector, and the American people. . . . The cornerstone of American cyberspace security strategy is and will remain a public-private partnership."[28] And the authors of the strategy note later in the document that "our traditions of federalism and limited government require organizations outside the federal government take the lead in many [of] these efforts [to defend cyberspace]," noting that the Partnership for Critical Infrastructure Security has "played a unique role in facilitating private-sector contributions to the Strategy"[29] because "broad regulations mandating how all corporations must configure their information systems could divert more successful efforts by creating the *lowest common denominator approach* to cybersecurity."[30] The strategy's reasoning about the need for a flexible response that matches the threats and supposedly unregulated nature of cyberspace assumed that private "partnerships" (as crucial to reinventing government within a flexible economy) were more suited to the government of cyberspace than were older models of state regulation.

The political rationality of reinventing government informed the creation, design, and strategies of the Department of Homeland Security a few months after the NSSC was drafted. The Department of Homeland Security became the largest federal agency created since the immediate post–World War II years, even as it operated as a mechanism for coordinating numerous smaller agencies and initiatives such as ones overseeing cybersecurity and "public-private partnerships" for safety and security. In short, the department became a massive interagency network for assuring that the role of government in national security was privatized—dispersed throughout society, often through do-it-yourself resources for the self-defensive citizen-soldier[31] —even as its corporate partnerships operated as unregulated instruments under the Bush administration for spying on citizens' telephonic and Internet communications.

According to the NSSC, the healthy functioning of cyberspace that is the basis for national security is best achieved as much by reinventing government in the United States as through a new regime of global governmentality. Securing a nation in cyberspace involves multiple strategies of international cooperation—a term that ambiguously suggested reciprocity but also a global alignment with U.S. interests. Key initiatives included an alignment of international organizations and industry that would promote a global "culture of security," developing secure networks through the heightened engagement between U.S. industries and their foreign counterparts, promotion of a "safe cyber zone" across North American countries, increased real-time sharing of information from international surveillance networks ("watch and warning networks"), and "encouraging other nations to accede to the Council on European Convention on Cybercrime."[32] Nowhere, however, does the NSSC indicate a need for international agencies, such as those that proliferated in the West for the government of international communication and (subsequently) air space. Instead, the strategy emphasizes privatized, self-regulatory forms of securitization in and from nations. Although these objectives appear at the end of the document, perhaps suggesting that they are an afterthought to the strategy's emphasis on national security and sovereignty, they clearly affirm that the NSSC understands the strategies for policing and waging war in cyberspace as one front in a global war on terror. On this count, the government of cyberspace is oriented as much toward securitization and policing as the healthiness and peacefulness of a world in cyberspace.

It remains unclear how substantively a national strategy for reinventing government and a national strategy for securing cyberspace will be modified. The latter strategy became an immediate preoccupation of the Obama administration, which moved the institutional authority of cyberspace securitization from the Department of Homeland Security to the National Security Council, sanctioned the Pentagon's organization of a new offensive and defensive operational wing for "cyber-command," and announced the creation of a national "cyber-czar" (also cast more modestly and bureaucratically as the "cyber-coordinator"). In May 2009, President Obama provided a formal rationale for his administration's cybersecurity strategy, which declared the digital infrastructure to be a national asset upon which economic prosperity, public safety, and national security depended. Although this strategy sought to reassure U.S. citizens that it would correct the invasions of their individual privacy perpetrated under the Bush administration, it also stressed the need to ratchet up the comprehensiveness of a cybersecurity program—to make the program better suited to citizens' constant immersion in cyberspace.

Conclusion

The prior section's silence about satellites and extraterrestrial media may seem anathema to the primary sense that chapters in this book make of the expression "down to Earth." Of course there are important connections to be drawn between current satellite media and cyberspace's production, and government and securitization. It is my hope that the history charted here poses useful questions for future research about those connections, since addressing them lies beyond the scope of this chapter. That said, it is worth mentioning a few reasons for following the current line of analysis.

As Henri Lefebvre famously noted, space is not only produced, it is productive of future modes of production; it plays a role in making history.[33] This chapter has considered how "air space" and "outer space" were produced by a field of government that subsequently underpinned and was reinvented through cyberspace. In that sense, it outlines a way of thinking about how the government of cyberspace was not determined purely or primarily by technological invention and experimentation, unless one is willing to consider air space, outer space, and cyberspace (after Andrew Barry) as "technological zones" of government—government not just of spaces defined and demarcated by geographical and territorial boundaries but of "zones formed through the circulation of technical practices and devices."[34] Cyberspace has operated as a technological zone not only of government but also for re-inventing and "advancing" liberal government in the first decade of the twenty-first century. In this sense, we might weigh not only how the Obama administration rationalizes expanding the scope of cyber-securitization while correcting the invasions of privacy perpetrated under the Bush administration but also how making those corrections has involved the Department of Homeland Security's contemporaneous declaration that it would no longer authorize the use of satellite imaging for "spying" on U.S. citizens or coordinating police operations on them.

The technological zone of government that was at stake across the invention of air space, outer space, and cyberspace was international and, by the late twentieth century, global. The problem of maintaining this technological zone called forth successive agencies, standards, and regimes of international and global government. In this sense, the invention and government of cyberspace are nothing new. They have perpetuated, if not deepened, the late-nineteenth-century problem surrounding transnational communication and transportation—the "openness" not only of borders but of the less easily marked limits of air space and outer space, and the ongoing initiatives to rationalize and thus make peaceful and orderly the zones that are most difficult to govern because they represent the limits of national sovereignty. The invention and government of cyberspace (particularly the governmental problems of monitoring "foreign-ness" within a home territory) thus perpetuated or deepened a disposition of nineteenth-century liberal government.

However, the invention and government of cyberspace also have been shaped through and are productive of a new stage of liberal government wherein the virtue of open borders is matched, in and from the United States, by the creation of a national strategy (and subsequently a massive Department of Homeland Security) to secure and manage cyberspace—the ethereal *medium* for waging an ethereal "global war on terror." Securing cyberspace is predicated on the invocation of air space and outer space as historical templates of maximum freedom and maximum threat, as much as on a reasoning about having entered a new ("advanced") historical technological zone wherein government is not simply oriented above us (air space and outer space as a New or Final Frontier) but all around us, throughout life, in the unbearable lightness of being communicating and mobile subjects in the current technological zones of government. In this sense, cyberspace is the Babel Complex of the current (neo-)liberal governmentality, though a Babel Complex that turns on a new articulation of public and private government, and more than ever on the immanence (the "down-to-Earth-ness") of monitoring foreign-ness within the most routine and familiar enactments of citizenship.

NOTES

1 James W. Carey, "Technology & Ideology: The Case of the Telegraph," in *Communication as Culture* (Boston: Unwin-Hyman, 1989); Armand Mattelart, *The Invention of Communication* (Minneapolis: University Minnesota Press, 1996); see also James Hay, "Between Cultural Materialism & Spatial Materialism," in *Thinking with James Carey: Essays on Communications, Transportation, History*, ed. Jeremy Packer and Craig Robertson (New York: Peter Lang, 2006).

2 Michel Foucault, "Governmentality," in *Security, Territory, Population: Lectures at the College de France, 1977–78* (New York: Picador/Palgrave-Macmillan, 2009).

3 Michel Foucault, "Space, Knowledge, Power," in *Power: Essential Works, 1954–1984*, ed. James Faubion (New York: New Press, 2001); Foucault, "Governmentality."

4 Andrew Barry, "Lines of Communication & Spaces of Rule," in *Foucault & Political Reason*, ed. Andrew Barry, Thomas Osborne, and Nikolas Rose (Chicago: University of Chicago Press, 1996).

5 Craig Robertson, *The Passport in America* (Oxford: Oxford University Press, 2010).

6 Werner Sombart, quoted in Armand Mattelart, *Networking the World: 1794–2000* (Minneapolis: University of Minnesota Press, 2000), 7.

7 Roland Barthes, "The Eiffel Tower," in *The Eiffel Tower & Other Mythologies* (New York: Hill & Wang, 1979). I also draw the expression "Babel Complex" from Barthes's description of the Eiffel Tower, although I am deploying the term differently.

8 This distinction is implied in an early legal and policy doctrine; see Sir Henry Erle Richards, *Sovereignty over the Air* (Oxford: Clarendon Press, 1912), 5.

9 Edouard Rollin, cited in Harold D. Hazeltine, *The Law of the Air* (London: University of London Press, 1911), 8.

10 Hazeltine, *Law of the Air*.

11 M. Paul Fauchille, *La circulation aerienne et les droits des Etats en temps de paix* (Paris: Pedone, 1901/1910).

12 Hazeltine, *Law of the Air*, 80–81.

13 For a useful history on the relation of Radio City Music Hall to the vertical architectonics of New York City in the 1920s and 1930s, see Eric Gordon, "Invisible Empire of the Air," *Space & Culture* 8(3) (2005): 247–263.

14 Andrew Haley, *Space Law & Government* (New York: Appleton-Century-Crofts, 1963), vii.

15 David Harvey, *The Condition of Postmodernity* (New York: Blackwell, 1990).

16 Norbert Wiener, *Cybernetics* (Cambridge, Mass.: MIT Press, 1948).

17 James Clerk Maxwell, "On Governors," *Procedures of the Royal Society* 16 (London, 1868): 270–283, cited in Norbert Wiener, *Cybernetics*, 2nd ed. (Cambridge, Mass.: MIT Press, 1961), 11–12.

18 Tiziana Terranova, *Network Cultures: Politics for the Information Age* (London: Pluto Press, 2004), 42.

19 For greater historical context and detail on the National Partnership for Reinventing Government, see David Osborne and Ted Gabler, *Reinventing Govern-*

ment: How the Entrepreneurial Spirit Is Transforming the Public Sector (New York: Penguin, 1992); and Vincent Mosco, *The Digital Sublime: Myth, Power, and Cyberspace* (Cambridge, Mass.: MIT Press, 2004).

20 Transcript of President Clinton's address to the nation, accessed April 1, 2011, http://govinfo.library.unt.edu/npr/library/speeches/030393.html.

21 Mosco, *Digital Sublime*; Francis Fukuyama, *The End of History and the Last Man* (New York: Free Press, 1992).

22 Washington Allston, *Lectures on Art* (New York: Baker & Scribner, 1850).

23 William J. Mitchell, *City of Bits: Space, Place, and the Infobahn* (Cambridge, Mass.: MIT Press, 1995).

24 The National Strategy to Secure Cyberspace (2003) (hereafter cited as NSSC), accessed April 1, 2011, http://www.us-cert.gov/reading_room/cyberspace_strategy.pdf.

25 Ibid., iii.

26 Ibid., 19.

27 Ibid., 7.

28 Ibid., iii.

29 Ibid., xiii.

30 Ibid., 15.

31 James Hay and Mark Andrejevic, "Homeland Insecurities," *Cultural Studies* (July/September 2006): 331–348.

32 NSSC, 50–52.

33 Henri Lefebvre, *The Production of Space* (London: Blackwell, 1990).

34 Andrew Barry, *Political Machines* (New York: Continuum, 2001), 3.

2

DETHRONING THE VIEW
FROM ABOVE

TOWARD A CRITICAL SOCIAL ANALYSIS
OF SATELLITE OCULARCENTRISM

BARNEY WARF

A colleague of mine, working in a remote part of Amazonia, showed a local illiterate farmer there a satellite image of his property, explaining that it was taken by a machine floating so high up in the sky that it could not be seen. Incredulous, the farmer denied that such a thing was possible; it was, simply, beyond the horizons of possibility in his worldview. The enormous discrepancy between the views held by my colleague and the farmer illustrates that satellite images, far from constituting some "objective" vision of Earth, are always wrapped within and bounded by cultural understandings and assumptions.

One of the two primary functions of satellites is to "see" vast regions of Earth's surface to monitor land use and incorporate, process, and transmit visual data for remote sensing and weather forecasts. (The other primary function is communications.) Satellites may "see" in either a passive mode, with optical sensors that acquire radiation emitted by a sensed object, or active mode, using radar to "see" at night or through clouds. This chapter is concerned with the political and philosophical dimensions of satellite imagery, which are inextricably intertwined, particularly such imagery's relation to a mode of knowing called "ocularcentrism." Modern Western society has long held vision as the paramount sense called upon to produce knowledge, suggesting that seeing is synonymous with knowing. However, vision, as Martin Jay argues, is not simply a function of biology, but also a historically specific way of interpretation.[1] To visualize, to gain insight, to keep an eye on something, is to invoke a host of cultural and linguistic tools to make sense of reality. Yet while seeing and vision appear so natural, obvious, and undeserving of attention as to be taken for granted, satellite observations in fact are products of a long line of Western thought that privileges sight, manages it, and shapes it through a variety of cultural assumptions. In this light, satellites not only have profound economic and social impacts but epistemological ones as well. This dimension of the technology is rarely considered in the literature on this topic. Indeed, many readers of this essay may find the entire project rather puzzling, so entrenched is the domination of the visual within contemporary discourse. Rarely do the literatures on satellites and contemporary social theory intersect. The goal here is to bring these two bodies of thought into a creative tension with one another.

The chapter starts with a historical overview of ocularcentrism, how it came to be, and its changing forms as vision was repeatedly reconfigured in Western knowledge. Starting with the Renaissance, and the tsunami of intellectual change that it unleashed in philosophy, art, and cartography, the chapter maintains that ocularcentrism is not a natural, inevitable way of understanding the world, but a historically specific construct. Next, it turns to the ocularcentrism that dominates the use of satellites, focusing on three particularly important dimensions: the transmission of television imagery, the role of satellite data in challenging national sovereignty, and the panopticonic role of satellites in monitoring individuals and changing the contours of privacy. The conclusion points to alternative forms of analysis of satellite data that incorporate "ground truths," local knowledges, and the interests of the subjects of surveillance. Throughout, the goal is to motivate scholars of satellites to view them, and the images they produce, as more than simply technical phenomena, but as systems of knowledge irretrievably intertwined within changing relations of power, culture, and space.

Historicizing Ocularcentrism

The origins of ocularcentrism as a hegemonic mode of Western thought arguably lay with René Descartes, who argued persuasively that the only valid form of knowledge is the one that equates perspective with the abstract, disembodied, rational subject's mapping of space. Cartesian rationalism was predicated on the distinction between the inner reality of the mind and the outer reality of objects; the latter could rationally only be brought into the former through a neutral, disembodied gaze situated above space and time. With Descartes's *cogito*, vision and thought became funneled into a spectator's view of the world, one that rendered the emerging surfaces of modernity visible and measurable, and rendered the viewer without body or place. The multiple vantage points in art or literature common to the medieval worldview were displaced by a single, disembodied, omniscient, and panopticonic eye. Illumination was conceived to be a process of rationalization, of bringing the environment into consciousness through the modality of vision.

Simultaneously, a parallel transformation was under way in the visual arts.[2] The key discovery in this regard was the invention of linear perspective, first demonstrated by Filippo Brunelleschi in 1425, which involved the ability to represent three dimensions on a two-dimensional canvas. In 1435, Alberti and Toscanelli formulated the geometric rules of perspective that remained in place for the next 400 years. Thus, as explained by Robert Romanyshyn, "linear perspective vision was a fifthteenth-century [sic] artistic invention for representing three-dimensional depth on the two-dimensional canvas. It was a geometrization of vision which began as an invention and became a convention, a cultural habit of mind."[3] As Lewis Mumford noted, "perspective turned the symbolic relation of objects into a visual relation: the visual in turn became a quantitative relation. In the new picture of the world, size meant not human or divine importance, but distance."[4] Although perspectival paintings were only viewed by elites, these individuals were the key decision-makers of early modern society, and the diffusion of such paintings reflected and sustained a wider discourse of ocularcentrism. Perspective came to be a metaphor for the entire world of the Renaissance just as Florence came under the panopticonic gaze of the Medici aristocracy.[5] Paul Johnson takes this line of thought even farther, asserting that "the replacement of aspective art by perspective art was one of the greatest steps forward in human civilization."[6]

Ocularcentrism gradually came to infiltrate virtually every corner of the emerging modern worldview. Rather than the complex, convoluted visual and aural worlds central to the medieval world, Renaissance thought came to emphasize homogeneous, ordered visual fields—a taken-for-granted notion that profoundly influenced the subsequent trajectory of Western thought. As Denis Cosgrove observes, "modernity is distinguished by its concern with the human eye's physical capacity to register and to visualize materiality at every scale."[7] Derek Gregory likewise maintains that this particular knowledge/power configuration—a historically specific scopic regime—reproduced reality as a "world-as-exhibition"—that is, the world naked to the all-conquering vision of the modern explorer and scientist.[8] As Jay puts it, "it was this uniform, infinite, isotropic space that differentiated the dominant modern world view from its various predecessors."[9] In the process, "space was robbed of substantive meaningfulness to become an ordered, uniform system of abstract linear coordinates."[10] The ascendancy of ocularcentrism also initiated the long-standing Western practice of emphasizing the temporal over the spatial; thus Gearoid O'Tuathail argues that "the privileging of the sense of sight in systems of knowledge constructed around the idea of Cartesian perspectivalism promoted the simultaneous and synchronic over the historical and diachronic in the explanation and elaboration of knowledge."[11]

Not surprisingly, ocularcentrism also extended deeply into the exploding discipline of cartography. As J. Brian Harley and other critical cartographers have powerfully demonstrated, far from constituting a detached, objective viewpoint from nowhere (a view that reduces mapmaking to a *technical* process), mapmaking was (and is) a *social* process deeply wrapped up in the complex political dynamics of colonialism and political domination.[12] For example, the grid formed by latitude and longitude deployed by Europeans worldwide greatly facilitated the exchange networks of incipient capitalism, making global space smooth, fungible, and comprehensible by imposing order on an otherwise chaotic environment. The projection of Western power across the globe necessitated a Cartesian conceptualization of space as something that could easily be crossed. This function was performed well by the cartographic graticule of meridians and lines of latitude, which positioned the world's diverse locales into a single, unified, and coherent understanding designed by and for Europeans; this view also allowed places to be compared and normalized within an affirmation of a godlike view. Colonial mapping was thus not simply a tool for administration, but also a validation of Enlightenment science and a central part of the colonial spatial order: mapping offered both symbolic and practical mastery over space. Thus the discourses of space—in this case, maps—did far more than represent entities that existed before them; maps played an active role in *producing* that very geography. Spatial discourses, in short, are simultaneously reflective and constitutive of the reality they represent.

In 1839, the invention of photography, or "light writing," in the form of daguerreotypes ushered in a new age of representation that further legitimated the hegemony of the visual. In 1888, George Eastman's mass-produced Kodak camera made its appearance, and large numbers of people took to the art with gusto. The photograph became almost universally accepted as an accurate, unbiased, straightforward mirror of the world, one with the power to capture the fidelity of visual experience, to re-present the past faithfully.[13] Few technologies would validate the ocularcentrism of Enlightenment modernity so completely, producing what Jay describes as "a frozen, disincarnated gaze on a scene completely external to itself."[14] Modernist interpretations assumed photography to be unproblematic; as Catherine A. Lutz and Jane L. Collins write, "the photographer's intent, the photographic product, and the reader's experience were assumed to be one. For this reason, photographs, unlike other cultural texts, were held to be readable by even the simplest among us."[15] Thus photography was viewed as being a direct, unmediated reflection of objective reality, and the photographer's intent was held to be nonexistent or unimportant.

While the content of the photograph appeared self-evident, its context and meaning required interpretation. Almost immediately, the capacity of photographers to manipulate and retouch images began to undermine the taken-for-granted capacity of the camera to reflect reality objectively. Joel Snyder points out that "our vision is not formed within a rectangular boundary. . . . The photograph shows everything in sharp delineation from edge to edge, while our vision, because our eyes are foveate, is sharp only at its 'center.'"[16] Photography also formed an integral part of colonial regimes of administration: to photograph was not simply to record, but to control. John Tagg likens photographs to Foucault's panopticon, operating at the nexus of knowledge and power to control subjects by representing them in some ways and not others.[17]

Photography also had profound repercussions for the geographical imagination: with photography, the popularity of magazines concerned with travel and exploration soared. Travelers and explorers could easily record distant sites with unprecedented accuracy and detail, offering a means of remote visualization. As James R. Ryan states, "like steamships, railways, and telegraphs, photography seemed to dissolve the distance separating 'there' from 'here,' bringing new audiences face to face with distant realities."[18] Photography allowed the subject to roam wherever the camera went, inviting the viewer to identify with the camera lens, and thus was critical to the creation of late modern geographical imaginations. For explorers and anthropologists, the camera offered a means of bringing the world into Western eyes, extending the "world-as-exhibition" to new heights. According to Susan Schulten, photographs "translated distant lands and complicated scientific phenomena into easily discoverable realities."[19]

Ocularcentrism was taken to new heights, literally, with aviation, whose origins lay in the eighteenth-century experiments of the Montgolfier brothers. In 1856, the first aerial photographs were taken by Gaspard-Félix Tournachon, who ascended into the skies over Paris to view it from above, initiating a process of aerial photography that extended to the space age. In the twentieth century, regularized flights offered a panopticonic vision unmatched by rival forms of transportation. More than any other technology, the airplane allowed millions of individuals to view the world's surface from afar and appreciate its vast horizontality. Balloons, telescopes, cameras, and other such devices all represented mechanisms for achieving panoramic visions that supersede that of the body, but the airplane (unlike its terrestrial counterparts) offered passengers and pilots a "bird's-eye view"—an all-encompassing perspective that could purport to be objective and all-knowing.

All of these technologies, epistemological regimes, and ways of seeing laid the foundations for the cultural and political dynamics of satellite imagery today. By insinuating themselves deeply into the taken-for-granted matrix of presuppositions through which satellite data are gathered and interpreted, the historical legacy of ocularcentrism has been to legitimize some uses of satellite imagery and not others—a process that has profoundly shaped the legal and political dimensions that undergird the entire industry. Nowhere is this theme more evident than through an examination of satellite imagery and its relations to the scopic regime of the contemporary world.

Satellites and the Modernist Regime of Vision

In the long historical context of Western constructions of vision, satellites represent the latest chapter in a changing series of ways in which Earth, and the people who inhabit it, have been viewed. Whatever the historical era, the origins and impacts of satellites are felt keenly on the ground, and the images generated by this technology are social and political products that do more than simply reflect the people and places portrayed, but actively shape them. Frequently, satellite imagery has been framed in an understanding that privileges the skilled observer and minimizes the role of people and places being observed.

Satellites played a key role in the militarization of space during the Cold War.[20] The Soviet Union's launch of Sputnik in 1957 unleashed waves of panic in the United States, which temporarily viewed the USSR as a technological equal.[21] The first U.S. satellite, Explorer I, was put into orbit one year later. In 1960, the Central Intelligence Agency (CIA) established the National Reconnaissance Office to operate an emerging satellite espionage capacity.[22] Throughout the Cold War, satellites were instrumental in the discursive scripting of geographic space, its ideological construction within hegemonic modes of understanding shared by politicians, military planners, and the media.

After the Cold War, the gradual end of the monopoly position held by Intelsat and the rise of several national and private providers of satellite services complicated the industry's structure considerably.[23] Today, satellites are deployed by telecommunications companies, multinational corporations, financial institutions, and the global media to link far-flung operations, including international data transmissions, telephone networks, teleconferencing, and sales of television and radio programs. Because the producers of satellite technology, and most of the industry's users, are concentrated in Europe and North America, the production, transmission, and consumption of electronic discourses are inescapably intertwined with the Western domination of the global information infrastructure.

Large satellites capable of handling international traffic sit 35,700 km (22,300 miles) high in geostationary orbits. These are highly prized positions in space, because only in that narrow sliver of space do satellites and Earth travel at the same speed relative to each other, making the satellite a stable target for signals transmitted upward from Earth stations. From its vantage

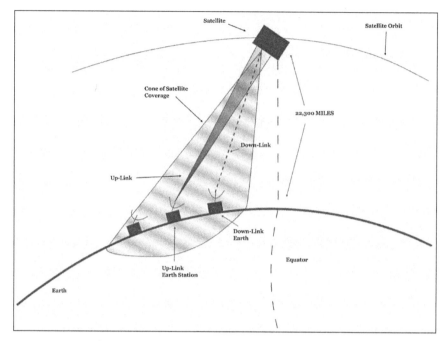

Figure 2.1 Earth stations, satellites, and footprints.

point, a broad-beam geostationary satellite can transmit to (or leave a "footprint" over) roughly 40 percent of Earth's surface (Figure 2.1); thus only three or four satellites are needed to provide global coverage. Because the cost of satellite transmission is not related to distance, the technology is commercially competitive in rural or low-density areas (for example, remote islands), where high marginal costs dissuade other types of providers such as fiber-optics firms.[24]

Unfortunately, given the hegemony of ocularcentrism, the uses to which these satellites have been put frequently fail to consider the perspectives of those being monitored. Remote sensing, for example, typically reduces the understanding of Earth's surface to physical processes; it is incapable of incorporating social processes into its images, although there is no inherent reason it cannot be sutured to conceptions of social process, change, and conflict. This line of thought builds on the emerging discipline of critical Geographical Information Systems (GIS).[25] This offshoot of critical cartography views GIS, like maps, as a social product with profoundly social origins and consequences. The same line of thinking can be extended to satellites, producing what might be labeled "critical satellite studies." Such an approach begins by embedding satellites and their images in relations of class, gender, and ethnicity, acknowledging in a Foucauldian sense that satellite imagery is a power/knowledge nexus.

From a feminist perspective, Karen T. Litfin maintains that satellites reinforce masculinist and positivist norms of an all-seeing detached observer, and unveils six assumptions that underlie taken-for-granted interpretations about the technology: that satellite remote sensing exhibits the inherent neutrality of science; that the science of satellites always leads to a rational, neutral public policy; that satellite knowledge always reduces uncertainty; that technological solutions (that is, satellites) always exist for social problems; that the global gaze proffered by satellites is useful, if not necessary, in addressing global predicaments; and that the planet is in dire need of being managed, preferably by those with access to satellite data. In each case, the assumptions are filled with inconsistencies, omissions, and ethical failures. She concludes that "the remote sensing project functions simultaneously as symptom, expression, and reinforcement of modernity's dream of knowledge as power." Central to this self-understanding is ocularcentrism, or as Litfin puts it, "the planetary gaze, relying on cameras collecting data at various wavelengths to inform us about the earth through color-coded computer simulations, is fundamentally a visual project."[26]

For example, pictures of Earth taken by the National Aeronautics and Space Administration (NASA) during the space program had important epistemological as well as political ramifications, and are ripe for deconstruction. As Cosgrove points out, for example, among the most significant impacts of the Apollo space missions was a new understanding of Earth, generating a vision of the world that lacked a clear center or periphery.[27] Far from comprising politically neutral representations, space photography legitimated and sustained a discourse of "one Earth" effectively embraced and encompassed by one nation, the United States. The Apollo images of Earth offered a perspective of humanity from the "outside," a view that reflected the globalization and worldwide ecumene that late modern capitalism had constructed. Doug Stewart, in a bit of hyperbole, claims that "the crowning achievement of space science, in fact, is probably neither the Moon landings nor the Jupiter flyby but rather the countless terrestrial uses, large and small, that people have found for Earth-orbiting satellites."[28] More recently, astronaut Thomas Jones affirmed that the Space Shuttle allows for a view of "the globe that fills the sky," a panopticonic perspective that never fails to awe.[29] Such lines of thought reflect and sustain the long-standing privileging of the visual, for global perspectives are visual rather than experiential. As Liftin notes, "the 'global view' afforded from the vantage point of space is certainly conducive to notions of 'global security,' but what might that mean in an unequal world?"[30]

Satellite imagery is deeply ocularcentric—as illustrated in the cases of television, national sovereignty, and individual privacy. In each case, the history of vision is reproduced in a way that minimizes the views of those whose lives are most affected by the technology. The point of these vignettes is to demonstrate that ocularcentrism is hardly some lofty, irrelevant philosophical speculation, but a set of practices deeply entwined with the dynamics of the global information economy.

Satellites and Television Space

Television has long been one of the central uses and missions of the satellite industry. The world's largest media companies rely heavily on geosynchronous communications satellites to provide a largely homogenous diet of television programs around the world. David Clark maintains that globalized satellite broadcasting of television is important in homogenizing the viewing options of consumers and enlarging markets for Western media firms:

> Irrespective of where they live, audiences around the world are fed a broadly similar diet of television. The same kind[s] of programmes are scheduled at the same times of the day. . . . Soap operas and quiz shows account for most of the daytime slots while children's programmes predominate in the early evening. These are followed by family viewing, the mid-evening news, drama, sport and adult television. The significance of this standard format is that it generates demand for particular types of programming, much of which is international in origin.[31]

Television, the first medium to stitch together the world as a collage of simultaneous sites and sounds divorced from their historical or geographical context, may be seen as a contemporary manifestation of ocularcentrism, with worldwide consequences. Paul C. Adams observed that "once people are able to acquire a television set, they use it with similar alacrity whether they live in Des Moines, Iowa, or Kragujevac, Yugoslavia."[32]

The medium has spawned a large, often contradictory set of interpretations. For optimists such as Marshall McLuhan, television, like other electronic media, formed the basis of the "global village," uniting disparate peoples through the power of the electronic message and destroying geographically based power imbalances.[33] For others, it formed a "vast wasteland"—a banal, anti-intellectual, and mentally debilitating world that robbed viewers of their critical powers of intellect and cultivated values of ephemerality and superficiality.[34] This process of standardization has important repercussions for local and national forms of consciousness and subjectivity, valorizing some forms of identity (predominantly Western ones) and devalorizing others. Arjun Appadurai views such phenomena as part of a global "mediascape" that interacts with other "scapes" to redefine the cultural geographies of global postmodernism.[35] Rob Shields asserts that electronic media eliminate the illusion of nearness or distance, dissolving subjects' sense of social and spatial proximity.[36]

Despite its origins during late modernity, television may be regarded as a distinctively postmodern medium by virtue of how it challenges modern emphases on linear rationality, contextual coherence, continuity of narrative, detached comprehensiveness, and objectivity.[37] Television offers a surreal field of vision that floods the viewer with massive volumes of unstructured information, a field in which local context is trivialized; the continuity of narratives is broken into incoherent, even random segments (for example, a murder followed by a fast-food advertisement); and the boundaries between fact and fiction are blurred to the point of nonexistence. Thus television is not simply a spectator in the creation of postmodern geographies, but an active participant, shaping the values and behaviors of billions of people worldwide. As Jean-François Lyotard emphasized, the proliferation of electronic information forms an integral part of the disintegration of centralized political and philosophical perspectives, the death of grand, sweeping metanarratives.[38] Jean Baudrillard suggests that this process coincides with the emergence of postmodern capitalism, in which images displace proximity as the source of discursive authority.[39]

Satellite Imagery and Challenges to National Sovereignty

Satellites, like the Internet, allow for an unprecedented, up-to-the-minute, twenty-four-hour-a-day flow of information, and a degree of transparency that one might rightly think is inherently good for the world. After all, Western thought since the Enlightenment has equated democratic governance with an informed electorate. Yet as Bernard L. Finel and Kristin M. Lord remind us, "transparency" is not synonymous with "democracy"; indeed, transparency in itself may not necessarily be beneficial.[40] By increasing transparency, satellite data may reduce pressures for conflict escalation, such as in resolving competing claims in border disputes. Conversely, satellite transparency may help to escalate conflicts by flooding decision-makers during a crisis with complex information, making it difficult to distinguish between meaningful and meaningless claims to their attention.

Since their inception, satellites have been central to concerns over "epistemic sovereignty," the fact that some nation-states know much more about the territory of their rivals than the observed states may know about themselves.[41] Led by the United States, purveyors of satellites insisted that national sovereignty stopped with the atmosphere, leading to a legal climate in which freedom of space reconnaissance prevailed. Generally, international law respects the freedom of outer space and preserves the right to acquire and disseminate satellite imagery without the consent of sensed states.[42] The unhampered flow of satellite traffic across national borders wreaks havoc with traditional notions of national sovereignty, an inevitable conflict that arises when signal footprints exceed the borders of a target country.[43] In the late 1960s, the rise of Direct Broadcast Satellites (DBS), which can transmit information directly to small receivers in the home, in particular made satellites a matter of political urgency for countries concerned about satellites' potential impacts. Because states cannot stop satellite imagery from crossing national borders, the technology poses a fundamental challenge to the Westphalian system of fixed, mutually exclusive borders of sovereign states that has formed a central part of the international political system since the seventeenth century. Television, one of the primary uses for satellite services, is vital in this regard: as Philippe Achilleas notes, "Television by satellite involves high political and legal stakes because of two underlying principles long considered to be antinomic: freedom of information and sovereignty."[44] Similarly, David Morley and Kevin Robins argue that "satellite broadcasting threatens to undermine the very basis of present policies for the policing of national space."[45]

Internationally, this issue has led to a schism between states that advocate the free exchange of information (generally those that produce the information transmitted by satellites) and those arguing on behalf of national sovereignty (generally those in the developing world). Frequently, impoverished states object to satellite broadcasts as a form of cultural imperialism, an issue that has been debated at length in the United Nations, including UNESCO and the Committee on the Peaceful Uses of Outer Space.[46] Some countries even allege that foreign satellite imagery constitutes a "weapon of mass destruction," representing invasions of their domestic markets that are essentially unregulated and that threaten national controls over information in much the same way that electronic funds transfer systems threaten national controls over money supplies.[47] States with monopolies over the mass media feel such challenges most acutely.

In 1977, the International Telecommunications Union addressed these concerns by introducing the national service principle, which required that satellite broadcasters minimize transmission into other countries' territories unless prior consent had been obtained. Such claims point to the relative ineffectiveness of international calls to limit cross-border flows of satellite imagery—attempts that are generally limited to the most extreme forms of violent propaganda (such as those used during the 1994 Rwandan genocide). Even when satellites are owned by national governments, such as Indonesia's Palapa or India's Insat systems, the contents of the programs transmitted are overwhelmingly Western in origin and orientation. As satellite imagery has become steadily more detailed and commercialized, the clash between commercial and national security concerns has become magnified, leading even some countries that were predisposed to openness to favor increasing restrictions.[48]

Some countries are straightforward in their attempts to maximize their control over international information flows. China and Malaysia, for example, outlaw the private ownership of satellite receivers, ostensibly on the grounds of protecting public morality or traditional culture, indicating a disjuncture between the interests of the state and civil society.[49] China and India expressly forbid foreign-owned satellite companies from broadcasting into their respective territories. States that seek to restrict imports of foreign media have typically found it impossible to assert national controls over global flows of information beamed from above, despite increasingly sophisticated "jamming" mechanisms. Such externalities point to the blurring of the distinction between foreign and domestic policy that communications technology has accelerated in a period of rapid globalization, and they illustrate the increasing "leakages" between the nation-state and the world system. Moreover, they point to the role of satellites, like the Internet, as potentially counter-panopticonic phenomena, and reveal their use to be a contested arena in which political positions are juxtaposed rather than given over to a single purpose.

Satellites versus Privacy?

As the resolution and accuracy of satellite imagery improved exponentially, their ability to identify ever-smaller objects rose accordingly. The visual power of satellites is typically expressed in terms of the length of the smallest feature that trained analysts can recognize. Although governments were highly reluctant to disclose the capacities of their surveillance systems during the Cold War, the post–Cold War climate of rampant commercialization has led providers of satellite imagery to tout their respective degrees of accuracy. Today's commercial satellites have attained levels of precision found previously only in military spy satellites. While the original 1972 Landsat satellite could recognize objects with a resolution of 80 m, the current generation of satellites, such as the Quickbird launched in 2005 by EarthWatch of Longmont, Colorado, can identify objects as small as 61 cm in length (Table 2.1). Obviously, the commercial applications and potential benefits of such imagery, such as assisting farmers in the application of pesticides or helping foresters identify particular species of trees, are enormous. Indeed, for this reason, the global commercial satellite image market enjoyed revenues greater than $2 billion annually in 2006.[50]

Not everyone is celebratory about satellites' mounting ability to monitor Earth in ever-greater detail, however. As satellite accuracy has improved, it has raised growing concerns about individual privacy and the potential abuse of the knowledge satellites produce. The very title of Robert K. Holz's influential book *The Surveillant Science: Remote Sensing of the Environment* acknowledged this issue.[51] In this light, surveillance technologies may be

———————— Table 2.1 Changing Maximum Resolution of Satellite Imagery ————————

Year	Satellite	Meters
1972	Landsat ERTS	80
1980	Discoverer	30
1985	ER2/AVIRIS	20
1992	SPOT	10
1999	KH-11	5
2001	KH-7	1
2005	Quickbird	6

Sources: Barney Warf, "Geopolitics of the Satellite Industry," *Tijdschrift voor Economische en Sociale Geografie* 98(3) (2007): 385–397; Marc P. Armstrong and Amy J. Ruggles, "Geographic Information Systems and Personal Privacy," *Cartographica* 40(4) (2005): 63–73.

used for the most sinister as well as the most liberating of purposes.[52] Perched high above Earth, satellites constitute what Mark Poster calls a "superpanopticon," "a system of surveillance without walls, windows, towers, or guards."[53] Remotely sensed data gathered from satellites, for example, can be cross-referenced with administrative records and data gathered from location-based services to reveal the identities of individuals, their characteristics, and behavior.[54]

Satellites are particularly crucial to this process because they are unobserved, are unchallengeable, and repeatedly take images of the same location at regular intervals, offering the opportunity to detect change. Police can use such images, for example, to estimate the size of crowds at political demonstrations and their rate of growth. Similarly, high-resolution thermal sensing "is eminently practicable from aircraft and terrestrial vantage points and has been widely used in energy audits and to deduce that cannabis cultivation is taking place in residences by sensing the waste heat produced by grow-lamps. In a spy-versus-spy escalation, growers are fighting back by increasing insulation and exhausting waste heat into sewer standpipes."[55] From relatively unobtrusive geodemographic marketing to radio-frequency identification (RFID) tags in clothing that track their wearers' movements to electronic anklets used for "correctional supervision" at home, satellites can be employed in a host of ways that render the panopticon not simply a metaphor but a lived reality. The combination of satellites, GIS, and video surveillance cameras has even led to concerns over geoslavery, a radically new form of human bondage characterized by location control via electronic tracking devices.[56] The fact that such data are commodified and traded raises severe doubts about the sanctity of individual privacy in the digital age.[57] Privacy, of course, is a highly fluid concept, with multiple historical and geographical meanings.

Under President George W. Bush's administration, the use of satellites to monitor terrorists as well as domestic activities reached unprecedented heights. A new office within the Department of Homeland Security, the National Applications Office, was established in 2007 to coordinate spy satellite information obtained from various agencies.[58] In addition to tracking terrorist movements, the agency is concerned with border control and law enforcement. The Patriot Act of 2001 dramatically enhanced the government's legal tools for wiretapping and surveillance, including surveillance by satellite. Critics and civil rights groups have raised strenuous concerns that domestic surveillance by satellites will also be deployed for espionage on various organizations that oppose the administration's policies, conjuring up images of "Big Brother in the sky." The Federation of American Scientists' Project on Government Secrecy, for example, has voiced concerns that the Bush administration's penchant for concealment could pave the way for a "surveillance state," in which the public domain is subject to continuous monitoring.[59]

Concluding Thoughts: Toward a Socially Responsible Use of Satellites

In light of these observations of how satellite images are social constructions, it is worth noting that they do not simply reflect the world's peoples and varied interests in all their unruly complexity; satellite images are simultaneously *constitutive* of those same relations. Because satellite imagery emanates from, and perpetuates, the long-standing tendency to equate meaningful knowledge with a "top-down" perspective, satellites have often been used as much for purposes of political surveillance as they have for addressing social and environmental predicaments. Such top-down applications have benefited users such as television companies and powerful states in the world system at the expense of those being watched, wreaked havoc with Westphalian notions of national sovereignty, and threatened individual privacy in unprecedented ways.

Once the ocularcentrism of satellites is made explicit, what alternatives exist? Is it possible to envision Earth in ways that are framed around less technocratic, more humane considerations? An emerging body of work points to perspectives that strive to be more empathetic in nature and to overcome the distance between viewer and viewed that is so readily sustained by ocularcentric discourses. Ecofeminists, postmodern development theory, nongovernmental organizations, critical social theorists, and practitioners of critical and participatory GIS have all conscientiously avoided the global gaze in favor of perspectives that take the local more seriously and solicit input from those who are monitored. The global diffusion of remote sensing expertise has generated a new arena of "satellite imagery activism," in which satellite data is used for non-panopticonic purposes.[60]

For example, as Litfin notes, "environmental advocacy groups and indigenous peoples in Southeast Asia, the Caribbean, the Amazon, and the Pacific Northwest are attempting to integrate their traditional knowledge into modern scientific methodologies through the use of satellite-generated data and mapping software."[61] Similarly, *People and Pixels*, the well-received volume published by the Committee on the Human Dimensions of Global Change, provided a series of analyses of how remote sensing data could shed light on social processes such as deforestation, as well as examples of how scientists using such information could constructively collaborate with activists working on poverty and environmental problems.[62] Finally, Rona A. Dennis and her colleagues studied the massive forest fires that swept Indonesia in the 1990s, utilizing participatory mapping that integrated geospatial techniques with local, interpretative accounts to reveal diverse causes in each region affected rather than one overlying causal source.[63]

As part of the global information infrastructure, satellites must be viewed in light of the profound disparities that exist between the world's wealthy and impoverished countries, as well as within them, and care must be taken to ensure that the use of satellites does not widen, rather than mitigate, this gulf.[64] Such a perspective begins with empathy and respect, rather than a desire to manage groups or communities, and, rather than attuning itself to an all-encompassing, panopticonic global gaze, focuses on the idiographic conditions of different locales.[65] Such a project would adopt explicitly emancipatory aims rather than a veneer of objectivity. For example, the nonprofit group WorldSpace, whose goal is to create "information affluence" in the developing world, launched several satellites with this ambition in mind, including the AfriStar satellite that provides radio to the entire continent of Africa, as discussed in chapter 10 of this book.[66]

Rather than engineering the understanding of peoples and places from above, a critical view of satellite engineering would incorporate ground truths derived from those below.[67] Land *cover* data, for example, would give way to land *use* data, with definitions of "wastelands" that are not simply based on views of vegetation from above, but that also incorporate farmers' perceptions. Unfortunately, existing remote sensing schemes leave little room for such feedback; as Wolfgang Hoeschele notes in his study of Kerala, India, "only once peasant knowledge has been transformed into government knowledge through the efforts of scientists is it regarded as valuable, while peasants are still regarded as ignorant."[68] Hoeschele concludes that "incorporating more peasant knowledge into the GIS-based analysis could be accomplished by adding several data layers about socially differentiated land use to the single layer of land cover derived from satellite data."[69] In short, it is misleading—even dangerous—to rely solely on satellite imagery abstracted from its social context. Similarly, Daniel Weiner and his colleagues constructed a GIS system incorporating satellite imagery in post-apartheid South Africa with the aim of fostering equitable agrarian land redistribution, a blend of top-down, bureaucratically driven data and bottom-up, indigenous knowledge.[70] Such efforts speak to the need for a critical satellite analysis sensitive to human concerns. As J. Douglas Porteus asked, "Do we have to be told that remote sensing . . . may result in a host of questions which can only be answered by its complement, intimate sensing, the microscopic approach performed at ground level?"[71]

NOTES

1 Martin Jay, *Downcast Eyes: The Denigration of Vision in Twentieth-Century French Thought* (Berkeley: University of California Press, 1994).

2 Denis E. Cosgrove, *Social Formation and Symbolic Landscape* (London: Croom Helm, 1984).

3 Robert Romanyshyn, "The Despotic Eye and Its Shadow: Media Image in the Age of Literacy," in *Modernity and the Hegemony of Vision*, ed. David Michael Levin (Berkeley: University of California Press, 1993), 349.

4 Lewis Mumford, *Technics and Civilization* (New York: Harcourt Brace, 1934), 20.

5 Samuel Y. Edgerton, *The Renaissance Rediscovery of Linear Perspective* (New York: Icon, 1975).

6 Paul Johnson, *The Renaissance: A Short History* (New York: Modern Library, 2002), 118.

7 Denis E. Cosgrove, ed., *Mappings* (London: Reaktion Books, 1999), 18.

8 Derek Gregory, *Geographical Imaginations* (Cambridge, Mass.: Blackwell, 1994).

9 Jay, *Downcast Eyes*, 57.

10 Ibid., 52.

11 Gearoid O'Tuathail, *Critical Geopolitics* (Minneapolis: University of Minnesota Press, 1996).

12 J. Brian Harley, "Deconstructing the Map," *Cartographica* 26(2) (1989): 1–20; J. Brian Harley, "Cartography, Ethics and Social Theory," *Cartographica* 27(2) (1990): 1–23; J. Brian Harley, *The New Nature of Maps: Essays in the History of Cartography* (Baltimore: Johns Hopkins University Press, 2002).

13 Susan Sontag, *On Photography* (New York: Farrar, Straus, and Giroux, 2007).

14 Jay, *Downcast Eyes*, 127.

15 Catherine A. Lutz and Jane L. Collins, *Reading National Geographic* (Chicago: University of Chicago Press, 1993), 28.

16 Joel Snyder, "Picturing Vision," *Critical Inquiry* 5 (1980): 505

17 John Tagg, *The Burden of Representation: Essays on Photographies and Histories* (Amherst: University of Massachusetts Press, 1988).

18 James R. Ryan, *Photography, Visual Revolutions, and Victorian Geography*, ed. David N. Livingstone and Charles W. J. Withers (Chicago: University of Chicago Press, 2005), 203.

19 Susan Schulten, *The Geographical Imagination in America, 1880–1950* (Chicago: University of Chicago Press, 2001), 171.

20 See, for example, Paul N. Edwards, *The Closed World: Computers and the Politics of Discourse in Cold War America* (Cambridge, Mass.: MIT Press, 1996); John Cloud, "Imaging the World in a Barrel: CORONA and the Clandestine Convergence of the Earth Sciences," *Social Studies of Science* 31 (2001): 231–251; and Jeffrey Richelson, *America's Secret Eyes in Space: The U.S. Keyhole Spy Program* (New York: Harper and Row, 1990).

21 Ryan Boyle, "A Red Moon over the Mall: The Sputnik Panic and Domestic America," *Journal of American Culture* 31 (2008): 373–382.

22 Mark Monmonier, *Spying with Maps: Surveillance Technologies and the Future of Privacy* (Chicago: University of Chicago Press, 2002).

23 Barney Warf, "Geopolitics of the Satellite Industry," *Tijdschrift voor Economische en Sociale Geografie* 98(3) (2007): 385–397.

24 Barney Warf, "International Competition between Satellite and Fiber Optic Carriers: A Geographic Perspective," *Professional Geographer* 58 (2006): 1–11.

25 See, for example, John Pickles, *Ground Truth: The Social Implications of Geographic Information Systems* (New York: Guilford, 1995); and Eric Sheppard, "Knowledge Production through GIS: Genealogy and Prospects," *Cartographica* 40(40) (2005): 5–21.

26 Karen T. Litfin, "The Gendered Eye in the Sky: A Feminist Perspective on Earth Observation Satellites," *Frontiers: A Journal of Women's Studies* 18(2) (1997): 39.

27 Denis Cosgrove, *Geography and Vision: Seeing, Imagining and Representing the World* (London: I. B. Tauris, 2008).

28 Doug Stewart, "Eyes in Orbit Keep Tabs on the World in Unexpected Ways," *Smithsonian* (December 1988): 70.

29 Thomas Jones, "A Globe That Fills the Sky: Geography from the Space Shuttle," *Geographical Review* 91 (2001): 252–261.

30 Litfin, *Gendered Eye*, 29.

31 David Clark, *Urban World/Global City* (London: Routledge, 1997), 126.

32 Paul C. Adams, "Television as Gathering Place," *Annals of the Association of American Geographers* 82 (1992): 118.

33 Marshall McLuhan, *The Gutenburg Galaxy: The Making of Typographic Man* (Toronto: University of Toronto Press, 1962).

34 Martin Esslin, *The Age of Television* (San Francisco: W. H. Freeman, 1982), 183. Esslin is echoing the sentiment Newton N. Minow expressed in "Television and the Public Interest," his address to the National Association of Broadcasters, Washington, D.C., May 9, 1961.

35 Arjun Appadurai, *Modernity at Large: Cultural Dimensions of Globalization* (Minneapolis: University of Minnesota Press, 1996).

36 Rob Shields, "A Truant Proximity: Presence and Absence in the Space of Modernity," *Environment and Planning D: Society and Space* 10 (1992): 181–198.

37 Romanyshyn, *Despotic Eye*.

38 Jean-François Lyotard, *The Postmodern Condition: A Report on Knowledge* (Minneapolis: University of Minnesota Press, 1984).

39 Jean Baudrillard, *The Illusion of the End* (Palo Alto, Calif.: Stanford University Press, 1994).

40 Bernard L. Finel and Kristin M. Lord, "The Surprising Logic of Transparency," *International Studies Quarterly* 43 (1999): 315–339.

41 Rosemary J. Coombe, "Authorial Cartographies: Mapping Proprietary Borders in a Less-Than-Brave New World," *Stanford Law Review* 48 (1996): 1357–1366.

42 Ram Jakhu, "International Law Covering the Acquisition and Dissemination of Satellite Imagery," in *Commercial Satellite Imagery and United Nations Peacekeeping: A View from Above*, ed. James F. Keeley and Robert Neil Heubert (London: Ashgate, 2004), 11–30.

43 M. Samwilu Mwaffisi, "Direct Broadcast Satellites and National Sovereignty: Can Developing Nations Control Their Airwaves?" *African Media Review* 5(1) (1991), accessed March 13, 2011, http://archive.lib.msu.edu/DMC/African%20Journals/ pdfs/africa%20media%20review/vol5no1/jamr005001008.pdf.

44 Philippe Achilleas, "Globalization and Commercialization of Satellite Broadcasting: Current Issues," *Space Policy* 18 (2002): 37.

45 David Morley and Kevin Robins, *Spaces of Identity: Global Media, Electronic Landscapes and Cultural Boundaries* (London: Routledge, 1995), 43.

46 Jon T. Powell, "Direct Broadcast Satellites: The Conceptual Convergence of the Free Flow of Information and National Sovereignty," *California Western International Law Journal* 6(2) (1975): 1–40; Kathryn M. Queeney, *Direct Broadcast Satellites and the United Nations* (Amsterdam: Kluwer Law International, 1978).

47 Colby C. Nuttall, "Defining International Satellite Communications as Weapons of Mass Destruction: The First Step in a Compromise between National Sovereignty and the Free Flow of Ideas," *Houston Journal of International Law* 27(2) (2005): 389–428.

48 Jakhu, "International Law."

49 Amos Owen Thomas, "National Sovereignty in an Age of Transnational Television," in *Television, Regulation and Civil Society in Asia*, ed. Philip Kitley (London: Routledge, 2003), 249–260.

50 John Yaukey, "Satellites Raise Privacy Questions," *Inside Technology*, accessed March 13, 2011, http://www.usatoday.com/tech/columnist/ccyau005.htm.

51 Robert K. Holz, *The Surveillant Science: Remote Sensing of the Environment* (Boston: Houghton Mifflin, 1973).

52 Stephen Graham, "Spaces of Surveillant Simulation: New Technologies, Digital Representations, and Material Geographies," *Environment and Planning D: Society and Space* 16 (1998): 83–504.

53 Mark Poster, *The Mode of Information* (Chicago: University of Chicago Press, 1990), 121.

54 Marc P. Armstrong and Amy J. Ruggles, "Geographic Information Systems and Personal Privacy," *Cartographica* 40(4) (2005): 63–73.

55 Ibid., 65.

56 Jerome E. Dobson and Peter F. Fisher, "Geoslavery," *IEEE Technology and Society Magazine* 22(1) (2003): 47–52; Jerome E. Dobson and Peter F. Fisher, "The Panopticon's Changing Geography," *Geographical Review* 97 (2007): 307–323; Peter F. Fisher and Jerome E. Dobson, "Who Knows Where You Are, and Who Should, in the Era of Mobile Geography?" *Geography* 88(4) (2003): 331–337.

57 Michael R. Curry, "The Digital Individual and the Private Realm," *Annals of the Association of American Geographers* 87 (1997): 681–699.

58 Eric Schmitt, "Liberties Advocates Fear Abuse of Satellite Images," *New York Times*, August 17, 2007, accessed March 13, 2011, http://www.nytimes.com/2007/08/17/us/17spy.html.

59 For more on government secrecy, see http://www.fas.org/sgp/.

60 John C. Baker and Ray A. Williamson, "Satellite Imagery Activism: Sharpening the Focus on Tropical Deforestation," *Singapore Journal of Tropical Geography* 27 (2006): 4–14.

61 Litfin, *Gendered Eye*, 41.

62 Committee on the Human Dimensions of Global Change, *People and Pixels: Linking Remote Sensing and Social Science* (Washington, D.C.: National Academies Press, 1998).

63 Rona A. Dennis et al., "Fire, People and Pixels: Linking Social Science and Remote Sensing to Understand Underlying Causes and Impacts of Fires in Indonesia," *Human Ecology* 33(4) (2005): 465–504.

64 Linda Main, "The Global Information Infrastructure: Empowerment or Imperialism?" *Third World Quarterly* 22 (2001): 83–97.

65 Yaakov Jerome Garb, "Perspective or Escape? Ecofeminist Musings of Contemporary Earth Imagery," in *Reweaving the World: The Emergence of Ecofeminism*, ed. Irene Diamond and Gloria Orenstein (San Francisco: Sierra Club, 1990).

66 Noah A. Samara, "Information Affluence for the Developing World: The Vision and Work of WorldSpace," *Development Practice* 4 (1999): 479–482.

67 Wolfgang Hoeschele, "Geographic Social Engineering and Social Ground Truth in Attappadi, Kerala State, India," *Annals of the Association of American Geographers* 90(2) (2000): 293–321.

68 Ibid., 298.

69 Ibid., 311.

70 Daniel Weiner, Timothy Warner, Trevor M. Harris, and Richard M. Levin, "Apartheid Representations in a Digital Landscape: GIS, Remote Sensing, and Local Knowledge in Kiepersol, South Africa," *Cartography and Geographic Information Systems* 22 (1995): 30–44.

71 J. Douglas Porteus, "Intimate Sensing," *Area* 18 (1986): 251.

3

THE GEOSTATIONARY ORBIT

A CRITICAL LEGAL GEOGRAPHY OF SPACE'S MOST VALUABLE REAL ESTATE

CHRISTY COLLIS

This chapter begins 22,236 miles (35,786 km) above Earth's equator, where a satellite drifts eastward at 6,897 miles (11,100 km) per hour. The satellite receives information from Earth and bounces it back. The satellite is an average one: about 12.5 feet (3.8 m) high, and, with its solar panel "wings" extended, about 85 feet (26 m) wide. It weighs 3,807 pounds (1,727 kg), including its fuel, which it will use to maintain its precise orbital position over the course of its operational lifespan of about 15 years.[1] Two aspects of this satellite make it particularly important, neither of which has to do with the satellite itself. Its importance rests instead on its location—its geography. At this precise height over the equator, the satellite moves at exactly the same speed as the point on the planet beneath it: it forever stays in the sky above a single fixed point on Earth. Second, because it is above the equator, the satellite can "see" 42 percent of Earth's surface at once, from 81°N to 81°S: what is called its "terrestrial footprint" is larger than that which could be achieved by a satellite in any other orbit around Earth.[2] As such, it is a particularly powerful communications tool: the receiving stations on Earth below it do not need to be adjusted or calibrated, because the satellite never moves from its position above them, and its data can be broadcast to 40 percent of the Earth's surface at once. What makes this satellite so powerful, and so valuable, is that it is located in the Geostationary Earth Orbit (GEO):[3] the single orbital belt, 22,236 miles above the equator and a relatively miniscule 18.6 miles (30 km) wide, in which satellites orbit at the same speed as the ground below them. Because of its special properties, the GEO is Space's most valuable position:[4] with a satellite in GEO, a communications provider does not have to pay the massive costs associated with maintaining several satellites to provide full-time coverage, or construct multiple Earth stations or moving receivers. With only three satellites in GEO, a communications provider can cover almost the entire planet. For satellites, which currently carry much of the world's communication data as well as its navigation and meteorological information, the GEO is *the* place to be. But, as the above citations of the GEO's size indicate, the GEO is not infinite: satellites have to be positioned apart from each other so that they do not interfere with each other's transmissions; they are, to borrow Siegfried Weissner's phrasing, strung along the GEO's thin belt "like pearls on a string."[5] Only so many pearls can fit on a string, particularly when they have to be spaced at prescribed intervals. This chapter addresses two key questions about the valuable GEO: who, if anyone, owns it; and what kind of a cultural space is it?

The chapter is grounded in two theoretical approaches: cultural geography and critical legal geography. The chapter is framed by the cultural geographical concept of "spatiality," a term that signals the multiple and dynamic nature of geographical space. As spatial theorists such as Henri Lefebvre assert, a space is never simply physical; rather, any space is always a jostling composite of material, imagined, and practiced geographies.[6] The ways in which cultures perceive, represent, and legislate that space are as constitutive of its identity—its spatiality—as the physical topography of the ground itself. Critical legal geography, the second theoretical field in which this chapter is situated, derives from cultural geography's focus on the cultural construction of spatiality. In *Law, Space and the Geographies of Power*, Nicholas Blomley asserts that analyses of territorial law largely neglect the spatial dimension of their investigations; rather than seeing the law as a force that produces specific kinds of spaces, they tend to position space as a neutral, universally legible entity neatly governed by the equally neutral "external variable" of territorial law.[7] "In the hegemonic conception of the law," W. Wesley Pue similarly argues, "the entire world is transmuted into one vast isotropic surface" on which law simply acts.[8] But as the emerging field of critical legal geography demonstrates, law is not a neutral organizer of space, but is instead a powerful cultural technology of spatial production. "Rather than seeking to bridge the gap between law and space, the argument here is that there is no gap to bridge," Blomley explains.[9] Or as David Delaney states, legal debates are "episodes in the social production of space."[10] International territorial law, in other words, *makes* space, and does not simply govern it. Drawing on these tenets of critical legal geography, as well as on the Lefebvrian concept of multipartite spatiality, this chapter does two things. First, it extends the field of critical legal geography into Space, a domain with which the field has yet to substantially engage. Second, it demonstrates that the legal spatiality of the GEO is both complex and contested, and argues that it is crucial that humanities scholars understand this dynamic legal space on which Earth's communications systems rely.

Thinking carefully and critically about the legal geography of the GEO is important, and increasingly urgent. One aspect of this importance is entirely practical—most of our communications, meteorological, and navigational systems depend on satellites in the GEO. In fact, it is no understatement to say that global communication and navigation now depend on the GEO. As Barney Warf notes, "satellites and earth stations comprise a critical, often overlooked, part of the global telecommunications infrastructure." Castells's notion of a "space of flows," Warf continues, "would be impossible without the skein of earth stations and orbital platforms that lie at the heart of the [satellite] industry."[11] In an article on the astropolitical environment, Fraser MacDonald similarly notes that "our lives already extend to the Outer-Earth in ways that we entirely take for granted."[12] Lisa Parks describes satellites as "moving persistently through orbit, structuring the global imaginary, the socioeconomic order, and the tissue of everyday experience across the planet."[13] Accordingly, the satellite industry—one that is largely centered in the desirable GEO—has become an increasingly powerful component of global economies: satellite world revenues in 2004 were US$103 billion, and were predicted to exceed US$158 billion by 2010.[14] Understanding the GEO, and in particular its legal geography, is thus critical to understanding how the very infrastructure of world communication works.

Understanding the legal geography of the GEO is also of particular importance to cultural theorists. As the previous paragraph indicates, for media and communication scholars—or for anyone with an interest in communication—understanding the legal geography of the GEO is fundamental: the GEO is *the* physical and legal space that allows contemporary communication practices to exist. For cultural theorists concerned with the creation of social spaces, the ways in which ideological forces produce the social and material world, or the ways in which cultures interact with and shape their environments, the GEO is of profound salience. Yet disturbingly little scholarship on Space exists inside what can roughly be called the humanities. Despite the demonstrated ability of humanities scholars—cultural geographers, critical legal geographers, media and communication scholars, and cultural studies scholars in particular—to understand and explain cultural, historical, and political phenomena, when it comes to Space, there is a curious, critical silence. As Parks notes, "Despite the global significance of satellite technologies, cultural theorists have been relatively silent about their ramifications."[15] MacDonald makes a similar argument, noting that for many humanities scholars, Space seems too absurd, odd, and abstract a subject with which to engage: this assumption, however, is a direct result of a lack of understanding of the centrality of Space (and in particular the GEO) to everyday life.[16] My experience as a Space cultural theorist demonstrates that this lack of understanding at times leads to condemnation of Space scholarship: when people are starving on Earth, I've been scolded, how can you morally justify sitting around thinking about Outer Space? Yet as this chapter, and this book, signal, Space is imbricated into our lives, our social organization, our cultures, and the power politics of the world. As MacDonald notes, "what is at stake—politically and geopolitically—in the contemporary struggle over outer space is too serious to pass without critical comment," and is too serious to be left to engineering, scientific, and legal scholars alone.[17] "In cultural theory the satellite has been missing in action, lying at the threshold of everyday visibility and critical attention," Parks notes.[18] It is time, then, for cultural theorists to extend their analytical skills, their attention, and their distinct, critical perspectives beyond the surface of Earth.

This is not a chapter about what kind of a space I think the GEO should be, or how I think it should be created as a legal geography. Similarly, this chapter is not a recondite philosophical argument about the nature of spatiality. Instead, this chapter provides an anatomy of the legal geography of the GEO, a spatial history of this valuable and contested site on which we now rely.[19] It is expository rather than argumentative for one key reason: few humanities and social science scholars are aware of the existence of the GEO itself, let alone its complex cultural history and constitution. Before debates about the GEO can be initiated, and before humanities scholars can lend their critical thoughts and insights to the struggle for the GEO, the GEO first needs to be understood and anatomized, which is the purpose of this chapter.

1957–1967: "A Lawless Environment"

It is a common assumption that when the Soviet Sputnik satellite launched as part of the 1957 International Geophysical Year blasted into orbit, it entered an entirely lawless environment, a space that simply did not exist as a legal geography.[20] To some extent, this is true—laws specifically pertaining to Space did not exist in 1957—yet Space was not entirely lawless. From as early as 450 B.C., Roman law asserted that the airspace over crops belonged to the crops' owners, and that if a neighbor's tree intruded into this airspace and impeded the crop's growth, the crop's owner had the right to remove the overhanging branches. Legal space, in other words, was not solely terrestrial; it reached up into the air and extended the legal geography of possession beyond Earth's surface.[21] In 1587, English law similarly constructed airspace as a legal geography, adapting the Roman legal principle of *cujus est solum ejus est usque ad coelum* (he who owns the land owns up into the sky) so that private property owners owned the airspace superjacent to their land.[22] The legal geography of the air became a particularly pressing issue during World

War I, as aerial bombing emerged as a key aspect of war for the first time. The Paris Peace Conference of 1919 formalized the legal status of the air, giving each state "complete and exclusive sovereignty over the air space above its territory," and for the first time airspace became the legal possession of states.[23] Any state that wanted to fly into the airspace of another state would have to seek permission first, thus recognizing airspace as a legal possession. Airspace as property, it was assumed, extended upward from the coastal boundaries of the subjacent state: the legal geography of state possession reached indefinitely into the sky.

But as far as Space was concerned, this legal geography was entirely abstract until 1957, when Sputnik, the first satellite, achieved Earth orbit. Sputnik overflew various states, including the United States, but no state protested that its legal airspace had been violated by the satellite.[24] At international law, a state territorial possession is created not only by physical acts (such as the planting of a flag, the visiting of unvisited land, or the construction of a government base) but also by other states' recognition of it.[25] When the Soviet Union's Sputnik traveled through Space without official protest, Space became an entirely new legal geography—a space beyond state possession, a vast space owned by no one at all. It was not yet clear what kind of a legal geography Space was, but one thing became clear: Space was not a state possession. The fact of satellites' physical presence in Space meant that what had hitherto been an abstract and theoretical legal geography suddenly became an occupied one. A legal geography that accommodated this development was clearly required. In 1961 and 1963, the General Assembly of the United Nations (U.N.) passed two resolutions on Space, suggesting in 1961 that "the exploration and use of outer space and celestial bodies shall be carried out in accordance with international law" and in 1963 that Space was not subject to national appropriation.[26] Although these resolutions were nonbinding, and although there were not yet any specific international laws for Space, they marked a beginning of the legal geography of Space: the laws of Earth, according to the U.N., encompassed Space; Space was, albeit vaguely, a legal geography. The United States, seeing this legal geography of nonpossession as an opportunity to keep Space open for its own spy satellites, and seeing it as a way to avoid a costly extraterrestrial war, quickly endorsed this new spatiality.[27] Yet already in the two U.N. resolutions, a deep contradiction regarding the legal geography of Space arose: customary law, which the first resolution suggests should be applied to Space, allows states to claim territory; indeed, the management of states' territorial possessions is one of the cornerstones of international law.[28] Yet the second resolution suggested that Space was unavailable to states—that it was a new kind of legal geography, akin only to the high seas and the unclaimed sector of Antarctica.[29] At this point, the fundamental bifurcation in the legal geography of Space (a split that continues to characterize the GEO today) emerged: was Space just another area to enfold into Earth's standard legal geographies, particularly those that create Earth as a series of state territorial possessions, or was it a different kind of space altogether, a shared space in which Earth's standard legal geographies did not apply? Was the legal production of Space simply a matter of extending Earth's legal geographies upward, or was it a matter of creating an entirely new kind of legal geography?

Throughout the 1960s an increasing number of satellites achieved orbit, entering this unstable legal space. The first satellite to achieve geostationary orbit was NASA's Syncom 3 (measuring 28 by 15.35 inches [71 by 39 cm]), in 1964; as of that date, the GEO became an occupied space. The bar-fridge-size Syncom 3 broadcast the 1964 Olympics in Japan to the United States, beaming the first television content ever across the Pacific.[30] Not only did Syncom 3 integrate East Asia into the expanding global media and communication system, it also proved the value of the GEO: unlike earlier satellites, Syncom 3 never disappeared from "sight" of its Earth receivers back in the United States.[31] In 1965, the U.S. corporation COMSAT's Early Bird, the first privately owned satellite, entered the GEO, and with this, the legal geography of the GEO was suddenly no longer a matter of state-state competition, adding a complication to existing legal frameworks. Early Bird began to highlight the economic value of the GEO: it could carry 240 telephone calls, almost ten times the maximum available on the transatlantic analog line.[32]

As more satellites entered the GEO, a second fundamental split in its legal geography emerged. The 1961 U.N. resolution stated that satellite communications should be made available to all states. In other words, the GEO should be readily available to any state that wanted to use it, rather than monopolized by the states or corporations with the economic and technological power to get there first. But as the numbers of satellites increased, and with them the volume of radio signals on Earth and in Space, concern emerged that satellites' radio waves would interfere with existing radio signals on Earth, and with each other's signals in Space. In 1963, the U.N.'s International Telecommunication Union (ITU), which is in charge of allocating and organizing international radio and communication frequencies, held its first Space conference, or World Administrative Radio Conference (WARC). The 1963 Space-WARC allocated specific and limited radio frequencies and frequency locations to Space in general, and to the GEO in particular, to ensure that satellite transmissions from the GEO did not interfere with Earth signals or with transmissions from other GEO satellites.[33] This seems logical and reasonable, but the legal geography it established for the GEO was far from universally accepted.

Under the ITU process, to secure a GEO position (or "slot") and its associated radio frequency, a satellite's owner must register first with the ITU, which then checks that the slot positioning is acceptable in relation to other satellites, and then assigns that slot and its frequency to the satellite.[34] Developing states argued that this effectively meant that states with economic and technological power could and would help themselves to the GEO in a first-come, first-served rush of satellites, and that states with less money would effectively find themselves locked out of the GEO.[35] They argued that a different kind of legal geography, one that favored developing states, was required. This second division between visions of the GEO's legal geography is based in two contradictory ontologies of law. In the civil law approach, law creates geographies based on specific principles; in this vision, legal geographies are produced first, shaping and governing all future activity within them. This chapter uses the term "a priori" for this legal ontology and its resultant geographies. In the other ontological approach, one based in the Anglo-American tradition of common law, legal geographies are generated in response to activities and practical issues as they happen; they are based on cases and activities rather than on abstract principles.[36] This chapter uses the term "a posteriori" for this legal ontology and its resultant geographies.[37] These are fundamental questions about the relationship between law and geography, and about the purpose of legal geography. Michel Foucault notes that places never simply "are"; rather, places are always constructed as "places for" certain people or activities.[38] Creating "places for" is precisely the role of legal geography: for *what* and for *whom* the GEO was constructed was the question.

In 1965, Space in general and the GEO in particular became a legal geography of a posteriori common law: the developed world's satellites that were beginning to populate it created GEO as a *res nullius*, or land owned by no one, available to whomever could get there first.[39] As such, by 1967 the GEO was a space for the developed world, and particularly for the United States; at the end of 1966, only the United States and the Soviet Union had launched satellites, and all eleven of the satellites in the GEO were American.[40] With the ITU's first-come, first-served approach to GEO slot and frequency allocation, and Space's apparent lack of ownership, it seemed as if the GEO was becoming a familiar type of legal geography: one in which the wealthiest and most powerful states (and in particular the United States) rushed in and claimed "squatters' rights" to the most valuable real estate, the "geostationary gold."[41] As Warf notes, access to satellites, and in particular to satellites in the GEO, "reinforces . . . terrestrial power-geometries of states in the world-system."[42] MacDonald reinforces this point, writing, "Space exploration, then, from its earliest origins to the present day, has been about familiar terrestrial and ideological struggles here on Earth."[43] The geo-power politics of Earth had begun to incorporate and inform the GEO.

1967–1975: Space Transformed

The legal geography of Space—and of its most valuable real estate, the GEO—did not remain static, however: in 1967 it was radically transformed. In the 1960s, numerous new states emerged as a result of global decolonization, many of them determined that legal geographies should no longer reflect the power and property imbalances of the colonial era. This movement, referred to as the New International Economic Order, was led by developing states, particularly in the U.N., whose one-state/one-vote system granted them structural force.[44] One way in which this new order could be achieved, they argued, was the legal creation of specific spaces as the Common Heritage of Mankind, now known as the Common Heritage Principle (CHP).[45] In 1967, Arvid Parvo, the Maltese ambassador to the U.N., famously proposed that the high seas should become a CHP; in the same year, the Argentine ambassador, Aldo Cocca, similarly proposed to the U.N.'s Committee on the Peaceful Uses of Outer Space that Space should be created as a legal geography based on CHP principles.[46] In the vision of legal geography of the CHP, areas such as Space, the high seas, and Antarctica should not be available to state possession, should be jointly managed by all states, should be used for peaceful purposes only, and most controversially any economic benefits from their exploitation should be shared equitably among all states.[47] This was a radical new legal geography, one that proposed to supplant the dominance of the legal geography that had defined Earth since the 1648 Peace of Westphalia: that is, the partitioning of Earth into states' possessions.[48] It also proposed to transform Space and the GEO from their then-current legal status as a *res nullius* to a variety of *res communis*. In short, the developing states proposed to transform the legal geography of Space radically and entirely.[49]

In 1967, these proposals partially succeeded, and Space became an entirely new legal geography. In that year, the Treaty on Principles Governing the Activities of States in the Exploration and Use of Outer Space, Including the Moon and Other Celestial Bodies (or, more succinctly, the Outer Space Treaty, or OST) was adopted by the U.N. General Assembly to legally transform Space. The United States, the USSR, and the United Kingdom immediately ratified the treaty, giving it significant international power. The OST produced Space—including the GEO—as a space of international parity and justice, stipulating in Article I that "the exploration and use of outer space shall be carried out for the benefit and in the interests of all countries irrespective of their degree of economic or scientific development." No longer was the GEO simply another *res nullius* space "out there" awaiting national claimants; after the OST, the GEO became a shared international space—a *res communis*—in which state claims, weapons testing, and scientific secrecy were banned.[50] Eilene Galloway suggests that the fact that the Soviets beat the United States into Space had a profound impact on the ensuing legal geography of Space: suddenly aware that the USSR had the potential to conduct Space warfare and to potentially occupy prime GEO slots, the United States quickly agreed that Space should be a space of peace and international cooperation.[51] Article IV of the OST bans military activity from Space, making Space "exclusively for peaceful purposes"; Article I also makes all of Space open to "free use" and exploration. Article II of the OST is the most powerful in terms of spatial construction: according to Article II, "Outer Space, including the Moon and other celestial bodies, is not subject to national appropriation by claim of sovereignty, by means of use or occupation or any other means." With Article II, Space became a legal geography of nonpossession, a *res communis*. Although the exact height at which sovereign airspace ends and Space begins remained under debate, with Article II, Earth's legal geographies of state sovereignty were given a ceiling. While developed states lauded the OST's prevention of Space warfare, developing states celebrated its new legal geographical regime: the OST, as India argued, would prevent "extraterrestrial colonialism."[52] State territorial sovereignty and *res nullius* were confined to Earth; the GEO had become a unique legal geography of international peace and sharing.

However, while the OST was a legally binding treaty (at least for the states that ratified it), it did not instantly solve the problems of the GEO. In fact, it appeared to exacerbate them. If the GEO was available for "free use" by anyone, developed states and organizations argued that they were entirely within their rights to continue to populate it as quickly and as thoroughly as they wanted: it was, after all, a legal geography of open access.[53] In the early 1970s, India and Indonesia began to plan satellite systems for the GEO slots above their terrestrial territories, but developed states and the U.S.-dominated international satellite consortium INTELSAT refused to adjust their satellites to accommodate them.[54] Although the OST prevented state claims to possession of the GEO, it apparently did not prevent the "squatters' rights" a posteriori spatiality of the ITU. Because the ITU is a regulatory body lacking enforcement powers, it could not compel the existing satellites' owners to accommodate India and Indonesia.[55]

The question of the legal geography of the GEO was thus back on the agenda at the 1971 WARC, which passed Resolution Spa 2-1, declaring that ITU slot allocations "should not provide any permanent priority for any individual country . . . and should not present an obstacle to the establishment of space systems by other countries."[56] But this aspirational spatiality did not override the a posteriori legal geography of the GEO: satellites have a limited operational lifespan, so stating that they were not allowed to permanently occupy slots was not particularly powerful. The a posteriori spatiality of the GEO emerged largely unchanged from the 1971 WARC, but not unchallenged: France, the United Kingdom, and Latin America all argued that a new legal geography for the GEO was required. In 1973, the ITU declared the GEO a "limited natural resource," akin to Amazonian rain forests or the aquifers of the Sahel. With this, the GEO was for the first time legally differentiated from Space in general; it became a finite new legal geography within the infinity of Space. No longer was the GEO simply a space to be managed equitably; it was now a "resource" necessary to human life, and a limited resource at that. If all states, as the 1971 WARC resolved, should be able to establish space systems in the GEO, and if the GEO was officially limited, then a posteriori spatiality was no longer legitimate. Although the a posteriori spatiality remained in place in 1974, it was beginning to erode.[57] Once again, the legal geography of the GEO became unstable.

1975–1982: Bogotá and the Moon

Until 1975, the general argument of the developing states was that some kind of a priori system should characterize the legal geography of the GEO so that the "limited resource" would be available to all states, regardless of their current level of technological capability. In 1975, however, frustrated that this legal geography was not evolving at the same speed at which developed states were putting new satellites into the GEO, Colombia made a radical move and asserted sovereignty over a section of the GEO. In 1976, at the thirty-first U.N. General Assembly, Panama did the same; and in 1976, Colombia, Brazil, Congo, Ecuador, Indonesia, Kenya, Uganda, and Zaire together created and signed the Bogotá Declaration, in which they collectively claimed sovereignty of the sections of the GEO superjacent to their terrestrial territories. This was a radical move because it was the first claim to Space sovereignty: to date, the GEO remains the only area of Space that has been subject to formal state claim. It was also a radical moment in the legal geography of the GEO. According to the Bogotá Declaration, the GEO was not part of Space, and was therefore not a part of the legal geography of the OST. Although it seems audacious, excising the GEO from Space had a legal basis: in 1975—and indeed still today—there was no legal agreement as to exactly where airspace ended and Space began.[58] Bogotá simply asserted that Space began beyond the GEO.[59] The Bogotá Declaration applied the legal principle of contiguity (the same principle through which states claim sovereignty to areas of the continental shelf off their shores) to the GEO, firmly placing the GEO within the Earth's legal geography of state sovereignty.[60] The declaration argued that the signatories' national appropriation of the GEO was hardly radical or novel: "under the name of a so-called non-national appropriation," they argued, "what was actually developed was technological partition of the orbit, which is simply a national appropriation" by the developed states, and in particular by the United States.[61] The GEO was subject to de facto national claims based on the a posteriori legal geography in place. Maintaining their proposed legal geography of the GEO, the eight Bogotá signatories, as well as Gabon, refused to sign off on satellites assigned at WARC 1977 and 1979 into "their" slots. The Bogotá Declaration treated existing satellites as earlier European empires had treated the residents of "new" territories they "discovered" and claimed: it stated that "equatorial countries do not condone the existing satellites or the positions they occupy on their segments of the Geostationary Orbit nor does the existence of said satellites or use of the segment unless expressly authorized by the state exercising sovereignty over this segment."[62] With the Bogotá Declaration, the legal geography of the GEO became yet more complex and contested, and the geopolitical nature of its contours became more visible.

Almost all states and space organizations rejected the Bogotá Declaration outright. Some developing states argued that it would effectively lock them out of access, the United Kingdom stated that the GEO was a part of Space and therefore unavailable for claim under the OST, and Belgium insisted that the equatorial states were unable to "effectively occupy" their GEO claims and were therefore unable to "perfect" those claims at law.[63] The majority of counterarguments rested on the GEO being a part of Space, even in the absence of a clear legal definition of Space. The legal production of geography is particularly visible in these exchanges: not only were they concerned with what kind of a space the GEO should be, they also sought to establish what and where, exactly, Space was. The Bogotá Declaration was particularly productive in its stimulation of new debates about the legal geography of the GEO. The GEO, responses to Bogotá affirmed, was part of Space, and not part of Earth's legal geographies of possession. While Bogotá did not succeed in transforming the GEO into the property of equatorial states, it did firmly situate developing states on the agenda of GEO spatial considerations: that developing states should have equitable access was now largely accepted; that transforming the GEO into Earth-bound sovereign territory was the way to accomplish this was refused.

Throughout the 1970s, increasing numbers of satellites claimed slots, particularly as international organizations emerged. These included INTELSAT, Intersputnik (established in 1974 to service the USSR, Eastern Bloc states, Iraq, and Syria), Immarsat (established in 1973 as a nonprofit cooperative of twenty-eight countries that provides maritime satellite services), Eutelsat (a European organization established in 1977 to service European communication, Eutelsat launched its first satellite in 1983), and Arabsat (established in 1976 to service members of the Arab League of states, Arabsat launched its first satellite in 1986). In competition with these large, state-based organizations, in 1972 the U.S. Federal Communications Commission (FCC) proclaimed its "open skies" policy, "allowing any party to apply for orbital slots, effectively initiating the era of privately-owned domestic commercial satellites."[64]

While physically the GEO continued to fill—it remained a *res communis* governed by a posteriori spatiality—legally, it assumed yet another character in 1979. The 1979 U.N. Agreement Governing the Activities of States on the Moon and Other Celestial Bodies (the Moon Treaty) attempted to transform Space again: it constructed Space not only as *res communis* but also as a CHP area.[65] Unlike the earlier OST, which created Space as a space beyond state sovereignty, the Moon Treaty created Space as a very specific form of economic geography. The Moon Treaty stipulated that any wealth derived from Space had to be shared equally with Earth's less-developed, poorer states, thus preventing Space from becoming a site for increased economic power for already-wealthy states: developing nations should benefit equally from any proceeds of Space exploration, particularly from resource extraction.[66] The Moon Treaty transformed the vague *res communis* of the OST into a codified economic geography. No space-faring state has ever signed the Moon Treaty, rendering it largely impotent. Space, they assert, may be a geography of spatial sharing, but not of revenue sharing.

In this way, in the 1970s the proposed spatialities of the GEO swung from one extreme of legal geography to the other—from state sovereignty to a CHP economic geography of nonsovereignty and enforced revenue sharing. Yet while these a priori geographies hovered over the GEO, seeking traction, the growing numbers of satellites entering the GEO maintained its status as an a posteriori legal geography, dominated by the developed world.

1980–1988: Space WARCs and a Combined Legal Geography

In 1979, the ITU convened the first General WARC since 1959. Its aim was to finalize, once and for all, the allocation of all frequencies, particularly those associated with specific slots in the GEO. Marked inequality continued to govern the legal geography of the GEO: 90 percent of the world's radio spectrum was, at the time, allocated to countries with only 10 percent of the world's population; of the seventy-four satellites in the GEO in 1982, only four belonged to developing states.[67] The 1979 WARC resolved that all states have equal rights to the limited natural resource of the GEO, but failed to effect this aspirational moral geography into the more powerful domain of established legal geography.[68] The 1982 Plenipotentiary Conference of the ITU injected a similar resolution, stating that "the special needs of developing countries and the geographical situation of particular countries" should be taken into account in GEO frequency/slot allocation procedures, a resolution that remained similarly aspirational to the 1979 proposal.[69] How these ideals might be activated into law was the agenda for the 1985 and 1988 "ITU WARC on the Use of the Geostationary Satellite Orbit and the Planning of Space Services Using It." Arguments were heated, and ontological: the developed world's proponents of the a posteriori legal geography characterized the ITU's consideration of the developing world as an egregious "intrusion of political pressures into an ostensibly objective, technical engineering issue."[70] Reserving specific slots for developing states, they argued, ran counter to the OST, which stipulated that no area of Space could be appropriated.[71] Developing states, on the other hand, continued to argue that the GEO, like other non-sovereign spaces such as the deep seabed and Antarctica, should be an a priori domain in which economic and power inequalities were balanced. Debate revolved around semantics: did the "equitable access" of the 1973 convention mean "equal access" to the GEO, or did it mean in proportion to expenditure?[72] In terms of legal geography, WARC 85/88 generated a compromise.

The 1988 Implementation Session of the WARC finally generated a new, combined legal geography of the GEO. It guaranteed every state one predetermined arc (PDA) in the GEO's Fixed Satellite Service expansion bands.[73] In other words, every state gained the right of occupancy to a PDA of the GEO closest to its territory. For the first time, the GEO became an a priori geography, clearly tethered to the terrestrial geographies of Earth's states. But it was only a very limited a priori geography. As Jannat C. Thompson points out, the Fixed Satellite Service expansion bands legislated into PDAs constituted only 1 percent of the Space spectrum, and further, the ruling did not apply to most telecommunication slots.[74] Existing satellites were grandfathered, or exempted from the rule.[75] Additionally, although each state gained a PDA, the WARC regulations allowed that PDA to be shifted without that state's consent in order to accommodate other satellites.[76] In order to maximize use of the GEO, "additional use provisions" allowed anyone to place a satellite into another state's PDA, but only for a maximum of fifteen years.[77] The 1980s WARCs also confirmed the GEO as a space dominated by states, rather than by private organizations: it stipulated that private satellites had to provide service for any state whose PDA they occupied, and assigned PDAs only to states.[78] While some commentators argue that the WARC changes of the 1980s did not go far enough in reinventing the GEO as an equitable space, others argued with equal vigor that the changes would retard GEO development and use. Despite these entrenched oppositional positions about the ontological nature of the GEO's legal geography, with the 1980s WARCs, the GEO became a space of combined a priori and a posteriori spatialities; the GEO had become a new legal geography.

1988–1995: Tonga and the GEO

Responding swiftly to this new spatiality, in 1988 the tiny Pacific nation of Tonga rocked the legal geography of the GEO. In 1987, Mats Nilson, a U.S.

satellite entrepreneur, approached Tonga's King Taufa'ahua Tupou IV with an idea for the PDA of the GEO Tonga had received as a result of the 1980s WARCs. Tonga's PDA comprised the last remaining sixteen slots that link Asia, the Pacific, and the United States. Given that the market for telecommunications traffic between the United States and Asia totaled US$2.5 billion in 1989, and was growing at 21 percent per year, these slots were particularly valuable.[79] Nilson proposed that he and Tonga form a national satellite organization, Tongasat, which would claim all of these slots, and then lease them to anyone who could pay for them at US$2 million per slot per year.[80] From Nilson's perspective, Tonga was an ideal partner for him in this plan. Tonga was not a member of INTELSAT, so was not constrained by INTELSAT's regulation that blocks member states from launching their own satellites; Tonga had no national communications regulatory board to slow down satellite plans; and Tonga now had the legal basis to claim its PDA in a valuable area of the GEO. The plan also appealed to Tonga: it would provide Tonga with the telecommunication infrastructure it lacked; it would give Tongans access to global telecommunications and thereby "create or reinforce a notion of their country as one part of the world community," and it would increase the national budget by 20 percent.[81] Thus, between 1988 and 1990, the newly formed Tongasat—in which Nilson had a 20 percent stake—formally filed claim with the ITU for the last sixteen slots in the Pacific sector of the GEO.

Tonga's move was in keeping with the legal geography of the GEO established by the 1980s WARCs: it mobilized the a priori spatiality according states the right to priority of claims in their assigned PDAs, as well as the a posteriori spatiality of first-claimed, first served. But this was the first time a developing nation had claimed the GEO for largely commercial, rather than simply domestic telecommunications, purposes. INTELSAT reacted furiously. INTELSAT dominated the GEO: most states claimed slots in their PDAs on behalf of INTELSAT, which had the finances to populate them with satellites that would then provide telecommunications to the claiming state. Jonathan I. Ezor observes that INTELSAT had become complacent about its dominance of the GEO: not imagining that developing states such as Tonga might attempt to configure the GEO for their own, competitive, purposes, INTELSAT had not claimed the remaining Pacific slots, simply assuming that "it had a lock" on them.[82] Seeing its monopoly eroded, INTELSAT rushed to the ITU, arguing that the Tongan GEO claims broke with the legal geography of the GEO by transforming PDAs into property.[83] INTELSAT stated that Tonga was setting a dangerous precedent for "claim-staking" and "hoarding" of GEO slots by developing states, as well as creating "financial speculation in the GEO" in contravention of the "spirit" of the new legal geography of the GEO.[84] While U.S. satellite corporation Columbia Communications demanded that the FCC deny Earth reception to any satellites in Tonga's PDA, INTELSAT went further and insisted that the ITU must change its regulations at the 1992 WARC, and therefore again change the legal geography of the GEO itself.[85]

INTELSAT's challenges, and Tonga's actions, revolved around interpretations of the nature of the GEO's legal geography: who, and what, was the GEO for? For Tonga, the OST and the 1980s WARCs produced the GEO as a place for developing states to gain some degree of economic competitiveness with developed economies, a space for the creation of global economic parity. For INTELSAT, the OST and the a posteriori aspects of the WARCs produced the GEO as a place for the continuation and further development of the existing economic order, a place for Earth's major powers. In adjudicating this dispute, the ITU would be creating the legal geography of the GEO. If it sided with Tonga, it could be creating the GEO as a form of private property, which ran counter to the geography established by the OST.[86] In siding with INTELSAT, the ITU would override its own legal geography of compromise, one that made the GEO a site for equalizing the global economy for developing states. It would mean admitting that PDAs could be overridden by states or organizations with superior Space capacities and bargaining power: it would mean transforming the GEO into a solely a posteriori legal geography. Once again, the ITU created a compromise geography: Tonga was granted six of its claimed sixteen slots, and was allowed to auction these off as planned; it dropped its claim to the other ten. In 1991, then, the GEO became a legal geography of a priori/a posteriori compromise that also comprised some private property—insofar as states could auction or lease "their" PDAs.

Conclusion: A Tangled Geography—the GEO 1993–2008

The GEO remains a space of conflict, compromise, and instability, all results of its complex and evolving legal geography. Although fiber optics now carry some of the communication data once solely carried by satellites, the GEO is under increasing pressure, particularly as the markets for satellite TV, telephony, and broadband grow.[87] As Ram Jakhu notes, "The existing spacefaring nations are increasing the number of their satellites. At the same time, more nations want to launch and own their own satellites."[88] Further, militaries around the world rely heavily on satellites for communication, remote sensing, and navigation, putting substantial pressure on the GEO: although warfare in Space is prohibited by the OST, the GEO is highly militarized.[89] As Peter I. Galace states, "the wars in Iraq and Afghanistan are eating up massive satellite bandwidth to support coalition military operations . . . military use will generate 46 percent of all satellite service revenues from 2002 to 2007."[90]

As this chapter has indicated, the capitalization of Space, and debates about whether Space should be incorporated into Earth's organizing structures—particularly capitalism and state territorial possession—have shaped the legal geographies of the GEO for more than forty years, and will continue to do so. The case for creating the GEO as a legal geography outside of capitalism and outside of state possession has been made, and debated vigorously, for decades. Understanding the history of these debates, and of the GEO, is key to understanding the ways in which capitalism and state expansionism have and, more important, have not shaped the legal geography of Space.

The current legal geography of the GEO is unlikely to hold. As demand for limited GEO slots increases, the ITU's ability to maintain its legal geography of compromise is diminishing. The ITU's legal geography faces two key problems. First, the ITU is now inundated by "paper satellites": because the right to a slot is determined by registration with the ITU rather than by the existence of an actual satellite, states with no capacity for launching satellites are preemptively claiming slots they may never occupy. Of 1,300 applications for slots submitted to the ITU in the early 2000s, 1,200 were for paper satellites. Processing this glut of claims means that the ITU now needs three years just to catch up with existing claims filed.[91] The ITU is attempting to deal with this by charging a filing fee (thus transforming the GEO into a form of purchasable property) and by insisting that a state must occupy any slot it has been granted within seven years. Second, it is increasingly apparent that the ITU might have been able to play a major role in the creation of the legal geography of the GEO to date, but it has no power to enforce that geography.[92] In 1993, for example, Indonesia complicated the Tongasat situation further by simply launching its PALAPA-B1 satellite into one of the slots claimed by Tonga. It did this without registering with the ITU, arguing that the GEO was an a posteriori *res nullius* rather than an a priori space and that Tonga's claims to the GEO were illegitimate.[93] The ITU could do nothing: only the collapse of the Indonesian satellite project resolved the conflict over this slot. Similarly, when Japan and Tonga's occupation of Vietnam's PDA blocked Vietnam's VINASAT-1's entry, the ITU could not compel them to shift.[94] Many private space development organizations and legal scholars now argue that the OST only applies to states, and that Space is therefore available to claim by nonstate actors.[95] "It is impossible to deny," argues Andrew T. Park, "that space is the next strategic frontier, both militarily and economically speaking. Unfortunately, the creators of the current legal regime for space failed to foresee the rate at which these advancements would take place."[96] In other words, the legal geography of the GEO must, again, change.

More than 22,000 miles above Earth, a legal geography upon which the global communications infrastructure depends remains riven and unstable, characterized by power struggles and ideological challenges. In this, it is no different from any of Earth's historical resource frontiers: creating the GEO's legal geography is little different from when, at the dawn of European imperialism, legal scholars debated who and what the lands of the "new world" were for. Creating the legal geography of the GEO both reflects and will shape terrestrial power relations, communication infrastructures, and global economics far into the future. It is well past time for humanities and social science scholars to attend to this geography critically and thoroughly. As MacDonald argues, "a critical geography of Space is long overdue."[97] When it comes to Space's most valuable real estate, the GEO, this need is particularly urgent.

NOTES

This chapter was first published in David Bell and Martin Parker, eds., *Space Travel and Culture: From Apollo to Space Tourism* (Keele, U.K.: Wiley-Blackwell, 2009), 47–65.

1 "What Is a Satellite?" Boeing Satellite Systems, accessed March 13, 2011, http://www.sia.org/industry_overview/sat101.pdf.

2 T. S. Kelso, "Basics of the Geostationary Orbit," *Satellite Times* 4(7) (1996): 76.

3 The Geostationary Earth Orbit (GEO) is also sometimes referred to as the GSO and sometimes as the Clarke Orbit, after Arthur C. Clarke, who proposed its existence in 194. The GEO should not be confused with geosynchronous orbits, which do not remain in place over a single terrestrial point. Satellites in geosynchronous orbit circle Earth once a day, but unlike satellites in the GEO, they do not remain in place over a single terrestrial point. Satellites in geosynchronous orbits will pass north and south of the equator in the course of a day; satellites in the GEO will remain directly above the equator at all times. Maintaining coverage of a specific terrestrial area using satellites in a geosynchronous orbit requires two expensive solutions: terrestrial Earth stations that can move in order to track the mobile satellite, and multiple satellites so that one satellite is always over the terrestrial area. More than one satellite means multiple launch costs, multiple satellite purchases, and multiple satellites to maintain. Geosynchronous orbits are thus less valuable as real estate than the GEO. See also Kelso, "Basics," 77; and Marvin S. Soroos, "The Commons in the Sky: The Radio Spectrum and Geosynchronous Orbit as Issues in Global Policy," *International Organization* 36(3) (1982): 667.

4 To differentiate between the physical place, Space, and the general concept of space, the former will be capitalized in this chapter.

5 Siegfried Weissner, "The Public Order of the Geostationary Orbit: Blueprints for the Future," *Yale Journal of World Public Order* 9(2) (1983): 225.

6 Henri Lefebvre, *The Production of Space*, trans. Donald Nicholson-Smith (London: Blackwell, 1991).

7 Nicholas Blomley, *Law, Space and the Geographies of Power* (New York: Guilford, 1994), 28.

8 W. Wesley Pue, "Wrestling with Law: (Geographical) Specificity versus (Legal) Abstraction," *Urban Geography* 11(6) (1990): 568.

9 Blomley, *Law, Space*, 37.

10 David Delaney, "Running with the Land: Legal-Historical Imagination and the Spaces of Modernity," *Journal of Historical Geography* 27(4) (2001): 494.

11 Barney Warf, "Geopolitics of the Satellite Industry," *Tijdschrift voor Economische en Sociale Geografie* 98(3) (2007): 385–397.

12 Fraser MacDonald, "Anti-*Astropolitik*: Outer Space and the Orbit of Geography," *Progress in Human Geography* 31(5) (2007): 594.

13 Lisa Parks, *Cultures in Orbit: Satellites and the Televisual* (Durham, N.C.: Duke University Press, 2005), 7.

14 Ram Jakhu, "Legal Issues of Satellite Telecommunications, the Geostationary Orbit, and Space Debris," *Astropolitics: The International Journal of Space Politics and Policy* 5 (2007): 176.

15 Parks, *Cultures*, 5.

16 MacDonald, "Anti-*Astropolitik*," 610.

17 Ibid., 593.

18 Parks, *Cultures*, 7.

19 For more on spatial history in particular, see Paul Carter, *The Road to Botany Bay: An Essay in Spatial History* (London: Faber, 1987).

20 Henri Wassenbergh, *Principles of Outer Space Law in Hindsight* (The Hague: Martinus Nijhoff, 1991), 15.

21 Herbert David Klein, "Cujus est solum ejus est . . . quousque tandem?" *Journal of Air Law and Commerce* 26 (1959): 237–254.

22 F. B. Schick, "Space Law and Space Politics," *International and Comparative Law Quarterly* 10(4) (1961): 681.

23 S. Latchford, "The Bearing of International Air Navigation Conventions on the Use of Outer Space," *American Journal of International Law* 53(2) (1959): 403.

24 Oliver J. Lissitzyn, "Some Legal Implications of the U-2 and RB-47 Incidents," *American Journal of International Law* 56(1) (1962): 138.

25 Gillian Triggs, *International Law and Australian Sovereignty in Antarctica* (Sydney: Legal Books, 1986), 33.

26 The Declaration of Legal Principles Governing the Activities of States in the Exploration and Use of Outer Space. Resolution 1721A (XVI) was adopted by the United Nations on December 20, 1961; Resolution 1962 (XVIII) was adopted two years later, on December 13, 1963. See also Bin Cheng, *Studies in International Space Law* (Oxford: Clarendon, 1997), 71, 85.

27 Christopher Stone, "Orbital Strike Constellations: The Future of Space Supremacy and National Defense," *Space Review* (2006), accessed March 13, 2011, http://www.thespacereview.com/article/628/1.

28 Robert Yewdall Jennings, *The Acquisition of Territory in International Law* (Manchester, England: Manchester University Press, 1963), 2.

29 M. J. Peterson, "The Use of Analogies in Developing Outer Space Law," *International Organization* 51(2) (1997): 245–274.

30 Nandasiri Jasentuliyana, *International Space Law and the United Nations* (The Hague: Kluwer Law International, 1999), 281.

31 Parks, *Cultures*, 21.

32 Mark R. Chartrand, *Satellite Communications for the Non-Specialist*, SPIE Press Monograph PM128 (Bellingham, Wash.: SPIE Publications, 2004), 6.

33 Once approved by member states, the results of WARCs have the legal status of international treaties.

34 Soroos, "Commons," 229.

35 Jasentuliyana, *International Space*, 288.

36 Ibid., 292.

37 This chapter also uses these terms because they are the terms generally used in legal discussions of GEO spatiality.

38 Michel Foucault, *Power/Knowledge: Selected Interviews and Other Writings, 1972–1977* (Brighton, England: Harvester, 1980).

39 At international law, *terra nullius* is defined as land belonging to no one, and is thus available for possession through "discovery," claiming, and "effective occupation." "*Res nullius*" is the term used when this legal geographical term is applied to Space, as technically, Space is not "land."

40 United Nations Office for Outer Space Affairs, "Online Index of Objects Launched into Outer Space," (2008), accessed March 13, 2011, http://www.unoosa.org/oosa/osoindex.html.

41 Soroos, "Commons," 666; Barney Warf, "International Competition between Satellite and Fiber Optic Carriers: A Geographic Perspective," *Professional Geographer* 58(1) (2006): 1–11.

42 Warf, "International Competition," 385.

43 MacDonald, "Anti-*Astropolitik*," 597.

44 Robert F. Gorman, *Great Debates at the United Nations: An Encyclopedia of Fifty Key Issues, 1945–2000* (London: Greenwood, 2001), 16.

45 "Common Heritage Principle" is now used (rather than "Common Heritage of Mankind") to reflect the legal status of the idea, and the existence, of women.

46 Parvo's bid was largely successful. In the 1982 U.N. Law of the Sea Convention, the deep sea–bed was declared a CHP area; under the convention, revenues earned from deep-sea mineral extraction must be shared with developing countries through the International Seabed Authority. Christopher C. Joyner, "Legal Implications of the Concept of the Common Heritage of Mankind," *International and Comparative Law Quarterly* 35(1) (1986): 196.

47 Jasentuliyana, *International Space*, 139.

48 Gearóid Ó Tuathail (Gerard Toal), *Critical Geopolitics: The Politics of Writing Global Space* (Minneapolis: University of Minnesota Press, 1996), 4.

49 For more on the CHP, see also Bernard P. Herber, "The Common Heritage Principle: Antarctica and the Developing Nations," *American Journal of Economics and Sociology* 50(4) (1991): 392–406; and Joyner, "Legal Implications."

50 United Nations Office for Outer Space Affairs, "Treaty on Principles Governing the Activities of States in the Exploration and Use of Outer Space, including the Moon and Other Celestial Bodies (1967)," accessed March 13, 2011, http://www.unoosa.org/oosa/en/SpaceLaw/outerspt.html.

51 Eilene Galloway, "Organizing the United States Government for Outer Space: 1957–1958," in *Reconsidering Sputnik: Forty Years since the Soviet Satellite*, ed. Roger D. Lanius, John M. Logsdon, and Robert W. Smith (London: Routledge, 2000), 309–326.

52 Jasentuliyana, *International Space*, 142.

53 Susan Cahill, "Give Me My Space: Implications for Permitting National Appropriation of the Geostationary Orbit," *Wisconsin International Law Journal* 19 (2000–2001): 236.

54 Jasentuliyana, *International Space*, 152. As discussed in Warf, "Geopolitics," 390, from 1964 to 2001 the U.S.-based INTELSAT was a cooperative international satellite organization. States can become members/shareholders of INTELSAT; voting power is determined by share ownership. In 2007, the United States owned nearly 25 percent of INTELSAT shares. In 2001, INTELSAT became a private company. INTELSAT dominates the satellite industry and the GEO, although this dominance is being eroded by the increasing emergence of national satellite systems.

55 Jakhu, "Legal Issues," 181.

56 Jannat C. Thompson, "Space for Rent: The International Telecommunications Union, Space Law, and Orbit/Spectrum Leasing," *Journal of Air Law and Commerce* 62 (1996): 279–311. Resolution Spa 2-1 was adopted as Article 33 at the 1973 ITU Plenipotentiary Conference.

57 Ibid., 293.

58 Alexandra Harris and Ray Harris, "The Need for Air Space and Outer Space Demarcation," *Space Policy* 22 (2006): 3–7.

59 Weissner, "Public Order," 287.

60 Triggs, *International Law*, 90; and Jasentuliyana, *International Space*, 153.

61 Quoted in Weissner, "Public Order," 238.

62 Ibid., 237.

63 Stephen Gorove, "The Geostationary Orbit: Issues of Law and Policy," *American Journal of International Law* 73 (1979): 452.

64 Warf, "Geopolitics," 392.

65 Brandon C. Gruner, "A New Hope for International Space Law: Incorporating Nineteenth Century First Possession Principles into the 1967 Space Treaty for the Colonization of Space in the Twentieth Century," *Seton Hall Law Review* 35 (2004): 327.

66 Herber, "Common Heritage," 393.

67 Weissner, "Public Order," 289.

68 Andrew T. Park, "Incremental Steps for Achieving Space Security: The Need for a New Way of Thinking to Enhance the Legal Regime for Space," *Houston Journal of International Law* 28(3) (2006): 877; and B. L. Waite and F. Rowan, "International Communications Law, Part Two: Satellite Regulation and the Space WARC," *International Law* 20 (1986): 357.

69 Thompson, "Space for Rent," 289.

70 Waite and Rowan, "International Communications," 356.

71 Ibid., 363.

72 Thompson, "Space for Rent," 294.

73 Waite and Rowan, "International Communications," 364.

74 Thompson, "Space for Rent," 295.

75 Cahill, "Give Me," 233.

76 Thompson, "Space for Rent," 296.

77 Jasentuliyana, *International Space*, 75.

78 Cahill, "Give Me," 234.

79 Edmund L. Andrews, "Tiny Tonga Seeks Satellite Empire in Space," *New York Times*, August 28, 1990, accessed March 13, 2011, http://www.nytimes.com/1990/08/28/business/tiny-tonga-seeks-satellite-empire-in-space.html.

80 Harvey J. Levin, "Trading Orbit Spectrum Assignments in the Space Satellite Industry," *American Economic Review* 81(2) (1991): 42.

81 Jonathan Ira Ezor, "Costs Overhead: Tonga's Claiming of Sixteen Geostationary Orbital Sites and the Implications for U.S. Space Policy," *Law and Policy in International Business* 24 (1992–1993): 919; Andrews, "Tiny Tonga."

82 Ezor, "Costs Overhead," 939.

83 Cahill, "Give Me," 244.

84 Thompson, "Space for Rent," 281.

85 Ezor, "Costs Overhead," 927.

86 Ibid., 935.

87 Warf, "International Competition."

88 Jakhu, "Legal Issues," 178.

89 Park, "Incremental Steps."

90 Peter I. Galace, "Asia's Satellite Industry: Winning by the Numbers," *Satmagazine .com* 4(3) (2006): 20, accessed March 14, 2011, http://www.satmagazine.com/june2006/june2006.pdf.

91 Jakhu, "Legal Issues," 182.

92 Park, "Incremental Steps," 871.

93 Thompson, "Space for Rent," 282.

94 VINASAT-1 finally launched in 2008.

95 See, for example, David Everett Marko, "A Kinder, Gentler Moon Treaty: A Critical Review of the Moon Treaty and a Proposed Alternative," *Journal of Natural Resources and Environmental Law* 8 (1992–1993): 293–346; Glenn H. Reynolds, "Space Property Rights: An Activist's Approach," *To the Stars* (September/October 1998): 19–21; and Gladys E. Wiles, "The Man on the Moon Makes Room for Neighbors: An Analysis of the Existence of Property Rights on the Moon under a Condominium-Type Ownership Theory," *International Review of Law, Computers and Technology* 12(3) (1998): 513–534.

96 Park, "Incremental Steps," 878.

97 MacDonald, "Anti-*Astropolitik*," 593.

4

"Freedom to Communicate"

Ideology and the Global in the Iridium Satellite Venture

MARTIN COLLINS

To wander onto the terrain of the 1990s global is to invite disorientation. Its media expressions and literature seem a jumble of outlooks—of promotion and critique, of declamations of control and unruly realities, and of totalizing visions and their limitations in an ever locally grounded world. For a taste of these jostling perspectives, consider these two nearly contemporaneous quotes:[1]

> *Freedom to communicate, anytime, anywhere*
> For the first time, Iridium shrinks the size of instant, reliable, truly worldwide communication to fit comfortably in the palm of your hand. And with a single telephone number, it follows you from isolated regions to international capitals, across borders, oceans, time zones. . . . [S]imply stated, there is nothing like Iridium. And for someone like you—who sees the world as one—there will be nothing in your way.

> The half-century since the end of World War II has been a period of unprecedented American hegemony over the rest of the planet. The confident mobility and the implicit threat that go with an aerial perspective have helped give a face to that hegemony. . . . [T]he United States has demanded, as a sort of natural right, that its citizens and media be able to pass unhindered across the borders of nations and continents. For fifty years, the assumed mobility of the view from above has been a virtually unavoidable component in a sort of unconscious popular cosmopolitanism, a set of expectations about the openness and submissiveness of the world that are shared widely even among Americans who never leave their country.

The first quote highlights the confident entitlement of business-class travelers living in a capitalist world tailored to their needs. Techno-enthusiasm and a "master of the universe" vibe seem to promise smooth transit across the global stage. Yet hints of disorder come through in the acknowledgment of "isolated regions." Unpredictability and risk seem to shadow the exhilaration of global motion "for those with nothing in their way." In "seeing the world as one," the text implies an alternate world that is not-one, of stratification between haves and have-nots, of a reality of locally grounded differences and opposition.

The second quote lays out a classic and germane critique, suggesting the post–World War II lines of power that have helped make that business-class vision seem natural. The handy organizing lens of hegemony resolves the churning of the global into neat familiar patterns of dominance and accommodation, of center and periphery.[2] But such critique itself was part of the intellectual field that composed the global. In particular, the transcendent ideals of the Enlightenment coexisted with the realization that such ideals were not "above history" but an accompaniment to the particularities of pre–World War II European expansionism.[3] Iridium, the historical example at the center of this essay, offers good empirical meat for the hegemonic assessment and for its limitations.

Iridium was a transnational business venture conceived by Motorola, a Fortune 500 corporation, which received encouragement from a range of U.S. governmental entities.[4] Its core idea, first sketched in 1987, involved placing a network of sixty-six satellites in low Earth orbit to provide cellular telephony service over the entire surface of the planet.[5] The project achieved a number of "firsts"—in technology, in manufacturing, in business organization, and in having the market rather than the state undertake a vast space project. Its planet-embracing technological envelope (a communications first) made the global an *actuality* rather than an evocative metaphor—an important way in which the 1990s differed from earlier eras of the global.[6] There was now congruence in extent between human action and Enlightenment ideals.

Iridium epitomized the period's entanglement of global action and ideals. In the post-1989, post–Cold War years, the U.S.-led venture drew in more than a dozen investors representing a diverse sampling of countries and corporations around the world. These included, notably, state-derived investments from former Cold War adversary Russia, the Peoples Republic of China, and India, as well as seed money from companies in Taiwan, Japan, South Korea, Thailand, Germany, Italy, Saudi Arabia, Venezuela, and Brazil—each of which then placed a director on the new company's board. Through a single project, the consortium planned to create a fully global technical infrastructure and a fully global market. It did: during the 1990s, the system was designed, funded, placed into space, and, in late 1998, began commercial operation. Its status as a unique corporate-national-global entity was codified by the International Telecommunications Union (an arm of the United Nations responsible for regulating transnational communications), which assigned the planetary, boundary-transcending Iridium its own telephone country code. At a cost of approximately US$7 billion (the largest privately financed technology project ever undertaken), Iridium stood as a symbol of an unfolding future in which neoliberal ideals brought technology, corporations, markets, states, and international politics into new, beneficial alignment. As such, Iridium was one of the darlings of a 1990s media often smitten with the idea of business entrepreneurship as the vanguard of a new post–Cold War order.[7]

The disorientation of the global, and its resistance to simple explanation, arises from its multiple and heterogeneous interactions, particularly as Western constructs push into tension and confrontation with non-Western modes of life.[8] In the post–Cold War period, this engagement takes place in a specific frame of ideology and practice: neoliberalism—a belief in markets as the preeminent mechanism for stimulating economic creativity, promoting individual freedom and self-determination, and achieving progressive social transformation. The market, rather than the state, exemplified the quest for realizing Enlightenment universals, and served as the ideal for the regulation of the social sphere (rather than the economic sphere). Neoliberalism secured the social standing of capitalism, but with a twist. It legitimated classic modes of dominance *and* the claim of their possible reconstitution into structures that enhanced the capacity of individual and community self-determination.[9] Iridium's moment in the 1990s, thus, was not that of imperialist enterprises in the late nineteenth or early twentieth century; different value structures were in play, ones that at least gave a rhetorical nod to postcolonial independence movements and the agency of non-Western actors. Institutions such as the United Nations gave substance (if only partial) to the transformation of pre–World War II patterns of imperialism. These political and institutional shifts, in turn, shaped and helped legitimate the rise of neoliberalism in the post-1960s period; both outlooks, in somewhat different ways, drew on the transnational cultural power of universalism.

The deeper question that has preoccupied postcolonial and globalization studies is whether Western leverage (primarily) operated with hegemonic gusto (the first half of the neoliberal equation), or whether countervailing interests forced some kind of co-construction, or mutual agency, in which individuals and communities shaped the global (the assertion of progressive benefits in the second half of the equation).[10] And embedded in this duality is another aspect of the global. One might see the shifting, increasing flow of commodities and media signs as a prominent form of *diffuse* power, a field of daily experience sometimes aligned with but often distinct from the purposive interests of *all* political actors.[11] Or, said slightly differently, and adapting Daniel Bell's argument in *The Cultural Contradictions of Capitalism*, culture (as a system of values, practices, and signs) might both serve and muddle the capitalist enterprise.[12] Agency and explanation in the global are not just a problem of applying a hegemonic analytic or of uncovering mutual agency and coproduction, but also of this amplified condition of semiotic experience—what Frederic Jameson has called a "second nature."[13]

From its inception, Iridium confronted as a foundational problem such intersecting meanings of the global—and the fact that as a nexus for corporate, state, and media actions, the venture was itself part of the very process of creating those meanings. This circumstance permeated the enterprise: it was evident in the design of its communication system, work practices on the manufacturing floor, corporate efforts to shape the identities of Motorola and Iridium employees, the financial and political organization of the project (especially, in its involvement of many non-U.S. and European investors), corporate lobbying to create favorable national and international trade and regulatory regimes, relations with the U.S. military, and as ideology and image. And when, on occasion, Motorola and Iridium had to define their view of the world explicitly, they confronted an undeniable reality: that via this venture a major U.S. corporation was looking to extend its already multinational business interests into new realms, but in a landscape that both facilitated prior methods of power and reshaped them. The fluid nature of the global allowed this reality to be framed in different ways, with different emphases. Image, ideology, and business goals could be configured to make near-polar opposite messages possible and plausible—sincere identifications with Enlightenment-oriented values *and* pointed ideological communications, reminiscent of earlier imperialist outlooks, both came out of the project.

This essay explores these expressions of ideology and their connections to the global. The focus is on a selective but informative slice of Iridium's cultural presence: its treatment in the U.S. media, particularly as seen through two outlets, *Wired* magazine (an icon of the 1990s), and the company's own in-house publication, initially called *Iridium Today* and then *Roam*. This account thus attempts to enhance our understanding of those Western actors—corporations and their nation-state partners—so central to the 1990s global and which even in their prominence have not yet been examined in detail.

Framing the Neoliberal Global

The history of the intellectual crosscurrents of the last decades of the twentieth century still needs to be written—to understand how neoliberalism rose as an ideology and achieved a prominent position in Western culture, and to detail those communities of discussion, institutions, individuals, and media patterns that gave it life.[14] One aspect of the ideological turn to markets has been sketched by Daniel Yergin and Joseph Stanislaw in their work, *The Commanding Heights*.[15] Yergin and Stanislaw offer an account of how market ideology gained traction in the politics of the United States and Europe through a network of academic advocates (particularly via the University of Chicago), conservative think tanks, and long-standing probusiness publications such as *Fortune* and the *Wall Street Journal*. International connections also were essential to this movement, as U.S.-based academics assisted international organizations and advised on how to invigorate the economies of developing countries. The turn to the market, thus, was not just about how to organize the U.S. economy, but an argument for reconfiguring the global landscape.

But market talk, despite its increased prominence in U.S. electoral politics over the 1970s and 1980s, was largely a conversation and an ideological contest among elites—impassioned, but largely abstract. Fred Turner's *From Counterculture to Cyberculture* suggests how 1960s countercultural thought, which placed a high value on romantic individualism and small-scale communitarianism, converged with elite market talk to enrich neoliberalism's intellectual ambit and social resonance.[16] The vehicle for this convergence was personal computing—or, more specifically, it was the genealogy of this technology, its strong sociological connection to countercultural figures such as Stewart Brand, and to countercultural networks and enclaves, such as the WELL (the Whole Earth 'Lectronic Link), the *Whole Earth Catalog*, and liberal-oriented universities. This romantic turn, paradoxically, allowed the market to stand also for the nonmarket—of individuals and groups pursuing interests that transcended the grind of capitalism. In this context, all high-tech as covered by traditional media seemed to benefit from a default view that it contributed to the social good, at least at first blush. Military-generated technologies, big-business developments, and Silicon Valley start-ups all seemed variations on a fuzzy, neoliberal progressive theme.

The fall of the Berlin Wall in November 1989 raised this outlook to a new level of prominence, clearing the Soviet Union and its state-centered economic practices from the field of international competition. The moment was not lost on neoliberal exponents, one of the foremost of whom was Francis Fukuyama. Published in late 1989, his essay "The End of History?" gained near-instantaneous recognition as the signature statement of the triumphant standing of market-centered ideology.[17] This reception reflected the cresting wave of the neoliberal in the media. During the 1980s, major news organs, think tanks, and a vastly expanded business press had created and fed an appetite for the exploits of CEOs, corporations, iconoclast innovators and marketeers, wealth-making, the next great thing, and the future as an unfolding terrain awaiting its techno-political possibilities. The end of the Cold War only amplified this condition.

Motorola gave Iridium its public unveiling in June 1990, seven months after the richly symbolic Berlin event cascaded through the transnational mediasphere. From its debut, Iridium's meaning was shaped by the cultural, ideological, and historical shifts and events of the 1970s and 1980s, accelerated by the disintegration of the Soviet empire. The tangle of neoliberal positions all found a toehold in Iridium, making the project a handy symbol for the post–Cold War moment. Its enamoring qualities abounded, with forward-leaning attitude as important as its practical goals and possibilities.

Knowingly and unknowingly, Iridium's leadership tapped into the prevailing cultural currents. The rollout featured four press events held simultaneously in London, Melbourne, Beijing, and New York City—a nod to the project's geographic scope and the realities of generating interest in key financial, media, and political circles. The New York City event was the focal point. The renowned Hayden Planetarium played host, adding a historical echo to the new venture; in 1951 the Hayden hosted the Symposium on Space Travel, a first-of-its-kind event that helped galvanize public interest in space exploration six years before the launch of Sputnik.

Adding to the "what to make of it" factor was Motorola's political and public standing in 1990. It had emerged from the 1980s as a poster child for how American firms might reinvent themselves in the face of intense global competition, particularly from Japan.[18] In 1988, the company received the first Malcolm Baldrige National Quality Award, established in 1987 by Congress and each year presented by the sitting president to encourage extraordinary accomplishment in improving manufacturing techniques and quality control. Linked to this achievement was the company's striking success in building and benefiting from the fledgling cellular telephone market—its "bat-wing" logo cell phones had become an early icon of this consumer and cultural phenomenon.[19] Motorola's high standing in elite political circles and with many consumers only added to the stakes and geopolitical issues that circled around the Iridium announcement.

Designed to attract worldwide attention, the geospatially distributed four-events-at-once display garnered an enthusiastic media response—even if accounts did not quite know how to focus their narratives and settle on defining metaphors. More than 1,400 newspapers carried the announcement, often on the front page. The New York Times ran its account on page A1: "Science Fiction Nears Reality: Pocket Phone for Global Calls."[20] In good pop culture fashion, the announcement also found its way into a Johnny Carson monologue and a Batman comic strip. The Beijing event received substantial coverage in China, running on the evening news. Approximately 250 million Chinese viewers heard parts of the Motorola press release and saw dubbed portions of a promotional video depicting how the satellite communications system would work.[21]

In linking science fiction visions (with a dash of techno-religion) to Iridium, the New York Times came closest to capturing the range of meanings embedded in the project, tying together themes of the global, of relentless innovation, and of the centrality of markets and corporate-driven leadership: "The small and portable telephone that can be used anywhere on Earth has been a staple of science fiction and a Holy Grail of telephone engineers for several decades. Today Motorola Inc. will become the first company in the world to announce plans to build and operate such a phone system . . . and could allow the user to make and receive calls from the North Pole to Antarctica."[22]

Motorola itself sought to sustain this type of narrative. Soon after the public rollout, it embarked on an advertising campaign and made an explicit connection between the company's international reputation in communications and the new venture: "Our experience in committing to new ideas gives us the conviction to act, filling the needs of a fast-moving world. . . . [O]ur satellite-based Iridium system [is] intended to bring personal communications to every square inch of the earth."[23] In early 1992, the *New York Times* returned to Iridium, noting that the project "continues to engender both awe and skepticism . . . awe, because only a company of Motorola's standing . . . would hazard so vast a project, a constellation of 77 satellites arranged in Copernican complexity. Skepticism, because even Motorola might not solve the daunting financial and technical problems of building an airborne AT&T and then finding customers for it." To emphasize the venture's forward-lookingness, the author quoted Iridium's chairman, Robert Kinzie: "This is not just a phone; it is a vision."[24]

The Global as Seen through *Wired* and *Roam*

No publication captured and gave expression—boldly and in broad explanatory sweep—to this assemblage of commitments, values, and excitements more than *Wired* magazine. Founded in 1993, the publication was the creation of old counterculture hands, including Louis Rosseto, who claimed that 1960s icon *Rolling Stone* magazine served as a template.[25] *Wired* made explicit neoliberalism's fundamental appeal: that the cocktail of markets, technology, and individualism fed on deep, compelling emotions of individual and group identity-making and of a sense that those in the know (*Wired* readers and their social kin) were fashioning a new historical era. They were pioneers, tilling a new land with new tools, making a new culture.

In its first year, the magazine turned its attention to Iridium, publishing a feature article by Joe Flowers titled simply "Iridium." The first paragraph showed the *Wired* narrative style, a hip fusion of technical and business doings with questions of meaning for the social and personal: "As big business goes, it doesn't get any bigger. Imagine a sixty-six satellite system of such stupendous ambition that you can phone anyone, anywhere on the planet. . . . Players: Motorola, the big Japanese electronic companies, dozens of local PTTs [postal, telegraph, and telephone services], an alphabet soup of national and international regulatory bodies. Here's the story of a dance that tells us a lot about who we are, what we expect, and how we deal with change at the end of the millennium."[26]

Then with a passing pop culture reference to Dick Tracy's wrist phone, the author, in a careful blend of tongue-in-cheekness and acceptance of the cultural assumptions of his *Wired* audience, laid out a view of life in the global:

> But worse than that, a cellular phone can't easily leave town. Mine won't work at all in that big nothing on the drive to Las Vegas. . . . If I go to Europe, the phone won't work at all. . . . No, I'm with Tracy—I want a real phone, something I can toss in the pocket of my genuine Banana Republic photojournalist's vest and take anywhere.
>
> I can feel your frustration. . . . Have faith. You have not been forgotten. The big boys are working on it. Give them another five years, and your troubles will be over. You'll be connected, always and everywhere, clear channel, error-corrected, voice- and data-capable, page-able, locate-able, and encrypted—all with one phone number, no matter where you are. Ask and ye shall receive.[27]

This telling captured one thread of countercultural neoliberalism—a certain self-indulgence and a tendency to see issues of political economy as a distant sporting contest. The *Wired* style gave great weight to revealing the structural and power dynamics of the global, but deferred any judgment or critique. The author neatly synopsized Iridium's transnational political maneuverings, involving the project's numerous *dramatis personae*: "What's really going on is something between a minuet and a World Wrestling Federation Monster Mash. . . . The dance is political and corporate . . . the business decisions push the politics; the politics mold the technology, around and around it goes."[28] What made a story like Iridium interesting was that all this churning involving the "big boys," in net effect, coincided (at least superficially) with countercultural support for techno-political transformations that seemed to facilitate individual empowerment. Missing were those 1960s countercultural questions that asked about modes of production and how the networks of power that enabled them shaped the political field. That type of critique would erupt, from other quarters, as global trade regimes such as the World Trade Organization (WTO) drew deeper scrutiny.[29]

In the several years after the publication of this *Wired* piece, media coverage (primarily newspapers and magazines) predominantly focused on discrete developments in Iridium—investments, regulatory hurdles cleared, top management changes, agreements with China and Russia to carry Iridium satellites into orbit (in the post–Cold War moment, a much-noted aspect of the project), progress in spacecraft manufacture, and comparative takes on the project in relation to its primary competitor, Globalstar. In covering these developments, accounts often raised the question of whether the expanding geographical reach of ground-based cellular networks was undermining the business rationale for Iridium. This skepticism, again, reflected that universal human question: "Can the hero overcome travails to fulfill the quest and reach the goal?" But this plucking at the hero's cape became muted as Iridium began a historic and impressive run of rocket launches beginning in May 1997 and continuing into early 1998, creating in less than a year the biggest, most complex satellite constellation ever put into space. The *New York Times* 1990 headline "Science Fiction Nears Reality" received its bookend complement in early 1998 in the clip "Iridium Satellites Close to Girdling the Globe."[30] The market had achieved a feat well beyond anything attempted in the decades-long history of state-sponsored space communications activity.

Alongside this more traditional reporting on Iridium, the venture also integrated itself into the unfolding developments of neoliberal politics and culture. Responding to the dramatic spread of the Web and its burgeoning spectrum of international users, the company created a corporate website. This was a static, description-tilted presentation, but made especially concrete the striking array of international corporate and state investors in the project. More in active conversation with the cultural moment was an in-house corporate magazine, also established in 1996, first called *Iridium Today*, and then renamed *Roam* in early 1998. In the business-enthusiastic 1990s the corporate magazine was an expanding genre. And for good reason: it filled a hybrid space in which a corporation could attempt to appeal to potential customers, well-heeled travelers, and the media—thus, in one literary stroke, reaching out to shape its base of consumers and introduce perceptions and symbols that might gain circulation in wider flows of the media. *Wired* took note of this phenomenon and dubbed those corporate technology publications "gadget gazettes."[31] It described *Roam* as a cross between *Conde Nast Traveler* and *Forbes*.

Love of the gadget as gadget was part of the magazine's slant, especially in its first years. But as the change of name to *Roam* suggests, the magazine also attended to readers' desires to link a "thing" to a complex of lifestyle interests and preferences. This was part of an attitude toward consumption that the *Harvard Business Review* dubbed the "experience economy"—consumers increasingly tended to see purchases as opportunities to create a narrative about their own lives.[32] As *Wired* insightfully noted, for Iridium this taste for techno-fetishism combined with a salient fact: "In fact, if anything, *Roam* tries to generate demand where supply is as yet unrealized. It is . . . mostly a tease. . . . The success of gadget gazettes is that they do what objects cannot: they put the product into a context, an environment that shows the product being used."[33]

But, of course, not all consumers across the global stage were equal; only those who could "see the world as one" might participate in such acts of imagination. Iridium played to this in a number of ways, including emphasizing the theme of wanderers and seekers who traversed the international landscape in search of meaning and experiences—achieved through intimate connection with choices in consumption. In this vein, the company sponsored the Iridium Adventure Series, which included the Eco-Challenge in Morocco, the World Championship Nippon Cup, and the Dakar Rally, tapping into the global-leaning consciousness of environmentalists and extreme sports enthusiasts.[34] The company also served as one of the sponsors for the Iditarod dog-sled race. To appeal to another social niche, the firm convinced Neiman Marcus to carry the phone in its prestigious Christmas catalog, known as *The Book*. A Neiman Marcus spokesperson explained, the "offering is consistent with our desire to be associated with only the top products, services, and companies. Our customers are among the busiest most well-traveled in the world. They need to be in touch, no matter where they are or what they are doing."[35]

While such breezy promotions of narrative-centered consumption obscured underlying politics, they reflected a fascination with and self-awareness of the making of the global—executed tepidly, but still there. As one significant exemplification, *Roam* covered the middle registers of globalization, a range of activity that did not typically surface in the major media. One type of social site instrumental in connecting the developed and developing world over issues of regulation and investment was the international trade show. Not atypical of the phenomena and its value to Iridium was this event summarized in the magazine:

> Film has Cannes, fashion has Paris, and telecommunications, of course, has Geneva. . . . Telecom 95 confirmed that the telecommunications industry is booming indeed. Nearly 200,000 people, including some 400 government ministers, 2,143 journalists, and thousands of others, descended on Geneva for the 10-day event. Nearly 800 exhibits from 46 countries were featured. The large ones cost millions and featured lavish displays, including laser shows, glass elevators, and waterfalls. For the uninitiated, it was difficult to see beyond the dazzle. One industry veteran was overheard debating the merits of underwriting construction of a hospital instead of pouring funds into an exhibit booth that would be ogled and then dismantled eight days later.

Without any sense of dissonance, the review continued that "Telecom 95 was distinguished by its international themes, including an opening ceremony speech delivered by South African President Nelson Mandela. His speech was an eloquent plea for including the developing world in the global information infrastructure."[36] During the mid-1990s, Iridium chairman Robert

Kinzie traveled to and spoke at dozens of these conclaves in every region of the world.[37]

These international extravaganzas revealed a deep tension in the day-to-day process of constructing the global. Corporate and consumerist excesses intermingled with nation-straddling issues of wealth disparity and political equity. Government and international regulatory officials mixed with CEOs and salespeople, developed and developing world convening under the same big top. In pointing to the contrast of the "ogled" exhibition booth and the need for hospitals, Iridium marked the fault line it walked. It needed the on-the-ground goodwill and political support of those in the developing nations and sought to use its leverage in national and international forums to adapt local and global structures to the company's benefit. This balancing act was particularly noticeable in Iridium's several-year relationship with the International Telecommunication Union (ITU). Their interaction began soon after the Iridium rollout announcement in 1990. The central issue was how a United Nations entity—one that gave special weight to advancing the interests of developing countries—might adapt to the neoliberal agenda. By 1996, as Iridium's manufacturing and launch preparations were well along, that question had been answered: completely. After all, Iridium, as a prime instance of the merger of first world and other-world actors, seemed a confirmation of the neoliberal's possibilities. The ITU's transformation helped legitimate this outlook as a way to enhance economic well-being transnationally and still keep a posture of attentive help to developing countries' needs.

The "gadget gazette" was there to share this convergence of interests with its readers through an interview with Pekka Tarjanne, secretary general of the ITU: "My favorite quote on the global information infrastructure . . . is that it is not global unless it is really global. . . . We know that today, some two-thirds of mankind are outside of the telecommunications network. . . . But I think it is good to speak about globality, because that ensures that we look at the globe as our market . . . and I have said that on the global level, the ITU's role should be to make sure that there is global regulation whenever it is needed, but only when it is needed. Sort of a minimalistic principle."[38] The secretary general, a representative of the nonmarket public interest on the international stage, casually invoked neoliberalism's most basic tenet—"that we look at the globe as our market"—a view perfectly in keeping with that of Iridium's.

Major media outlets in the United States reported on Iridium in snapshots, both with respect to it activities and their meaning. Iridium's in-house publication captured a richer, if obviously slanted, view of the venture's alignment with globalization's main lines of power. Again, though, Iridium found its most adept scribe in *Wired*—with its odd 1960s-cum-1990s counterculture interest in the big picture but muted concern for the real-world implications of the politics it so insightfully delineated. In October 1998, on the eve of the venture's launch of service, a much-anticipated moment of vindication and triumph for Motorola and its start-up, *Wired* returned to Iridium in a prominent feature story, "The United Nations of Iridium."[39] In the years since 1993, when the magazine first assessed the venture, the neoliberal outlook had gained credence, symbolized, for instance, in the establishment in 1996 of the WTO and exemplified by the seeming march-of-progress advance of communications technologies, particularly the exponential growth of the Web. In elite politics, Iridium had a ready connection with these developments. In 1996, President Bill Clinton toured an Iridium manufacturing facility in New Hampshire, the event a microcosm of the interconnections among the global, technology, and politics. The complex of associations was more richly evident in the staging of Iridium's "cut the ribbon" inauguration of service in fall 1998, covered by *Roam*:

> *From the Rose Garden*: At midnight Greenwich Mean Time on Sunday, November 1, Iridium became the first provider of global mobile-telephone services. Helping us launch this new era of global communication was U.S. Vice President Al Gore, who placed the first official Iridium phone call on the preceding Friday afternoon. Standing in the White House Rose Garden, the Vice President used an Iridium phone to call Gilbert M. Grosvenor, chairman of the National Geographic Society and a great grandson of Alexander Graham Bell. To mark the occasion, the Vice President greeted Grosvenor with the historic words that Grosvenor's great-grandfather had spoken to Tom Watson, his assistant, when they completed the world's first telephone call in 1876: "Watson, come here, I want to see you."[40]

Gore's pop culture association with the Internet only added to the symbolism of the moment, in which the global present was linked with, then distinguished from, the past.

Wired's 1998 article began to connect all these dots of the global. The article began with "it's a bird, it's a plane, it's the world's first pan-national corporation," but then gathered its analytic stride:

> The real importance of Iridium . . . is [as] the world's first pan-national corporation, a global partnership created, from Day One, without control by any one country. It takes that emblem of twentieth-century capitalism, the multinational company, and kicks it into the next millennium. When a Coca Cola, Siemens, or Ford expanded overseas, ultimate control remained at—and profits were repatriated—home. Iridium's core identity is defined by its transcendence of national borders, a structure that is particularly post–Cold War. It's a harbinger of what ever-less-restricted global free trade can bring about. . . . Iridium may well serve as a first model of the twenty-first century corporation.[41]

Giuseppe Morganti, CEO of Iridium Italia (a separate business entity within Iridium that provided services to most of Europe), drew the contrast between an imperial past and the neoliberal present: "Multinationals are comparable to the idea of national colonialism, where cultures are places to be conquered. Iridium is something that starts as a global entity."[42] The tail end of that analogy was left benignly unexamined: what kind of power structure was the pan-national global, and how might one characterize collaboration and absorption, in distinction from colonial control?

On the surface, it looked different and had the edginess and energy of a new social experiment: "Iridium's partners are assigned territories to manage on their own, forming separate companies. Fifteen of these operations, with names like Iridium Italia and Iridium China, have been started—each independent, each with its own CEO and governing board. Four times a year, twenty-eight Iridium board members from seventeen countries gather to coordinate overall business decisions. They meet around the world, shuttling among Moscow, London, Kyoto, Rio de Janeiro, and Rome . . . resembling a United Nations in miniature."[43] This description of organization and process was largely correct, but did not get at whether all the partners were equal and, especially, did not examine Motorola's special standing as initiator, largest investor, and critical provider of the billions of dollars worth of system hardware and software.[44]

Citing Arthur C. Clarke as a promoter of global comity, the author quietly introduced a key aspect of Iridium's utopian and global resonance: it was about outer space, our use and control of it, the possibilities it held for transforming life on terra firma; to get above Earth was to enhance our ability to comprehend and change ourselves. The very act of getting above Earth, to live and see through this perspective, made natural new forms of organization and production, the very thing Iridium represented. As with its sociotechnical cousin, the Web, Iridium stood for the frontier, that social space that allowed pioneers to energetically converse with and reconstitute the culture of the status quo. And in the neoliberal moment, it was capitalism that opened further the channels of engagement between the frontier and the life of the status quo.

Morganti, "an affable philosopher-businessman," touched on one organizing motif of the global—given intensified, practical meaning by Iridium—namely, the erasure of distance, upending a long-standing predicate of the human condition.[45] "This is the first civil application of the global village. This is a historic event. From the prehistoric period, from creation, it is the first time that mankind can overcome any problem of distance." As a coda to the article, David Bennahum linked this philosophic claim back to the corporation-inflected present: "Whether Iridium succeeds or fails in matching the expectations in its business plan, it has changed the world's perception of the inaccessibility of space and led the way toward creating corporations ever more disassociated from national identity and geography."[46] This kind of talk and the *Wired* article represented the highest tide in assessing Iridium's meaning. Within several months, the venture crashed from a transcendent exemplar of change to the ignominy of bankruptcy court. The U.S. Department of Defense appreciated the system's special capabilities and in 2000 helped push for Iridium's sale to new investors (renamed Iridium Satellite). As of mid-2009, the satellite system is still in operation. In its new guise, the U.S. government has been one of the company's biggest customers, particularly via use by the military during the wars in Afghanistan and Iraq.

Conclusion

What might this partial tour of the Iridium landscape tell us? Missing from this account is a full sense of how the multiple, heterogeneous interactions of the global—the contact and coexistence of powerful and less powerful actors— worked, especially as to how the non-Western states and corporations in Iridium saw and participated in the project. That story still needs to be elaborated. But, at least in Iridium, cohabitation in the close quarters of a major business venture did seem to unsettle lines of power and make the neoliberal a zone of experimentation. In the constellation of interests that promoted it, neoliberalism seemed a recrudescence of older and not-so-old instantiations of hegemony. Its robust rhetoric featured much sleight-of-hand (including Iridium's use of a classic trope such as "Freedom to Communicate") in promoting the market as a providential engine of social and economic good. But the 1990s, it might be argued, yielded a distinct cultural formation. The ability to act over the totality of the planet (technologically and via the market), the circulation of Enlightenment ideals (whether as a cover for power or a resource for countering oppression), and the interests of many locals came into more frequent confrontation, with more consequence, making the global a testing ground for new configurations of practice and meaning.

NOTES

1 Top quote from "Freedom to Communicate," advertising brochure, Motorola 1998, unfoldered, box 8, Iridium Papers, NASM (hereafter cited as IP); bottom quote from Bruce Robbins, *Feeling Global: Internationalism in Distress* (New York: New York University Press, 1999), 4.

2 A classic and strong statement of the global as hegemony is that of Pierre Bourdieu and Loïc Wacquant: "Cultural imperialism rests on the power to universalize particularisms linked to a singular historical tradition by causing them to be misrecognized as such." See Pierre Bourdieu and Loïc Wacquant, "On the Cunning of Imperial Reason," *Theory, Culture, Society* 16 (1999): 41–58.

3 The ideas of decentering and historicizing the West, for example, are central to Dipesh Chakrabarty's work and the discussion it has stimulated. See Dipesh Chakrabarty, *Provincializing Europe: Postcolonial Thought and Historical Difference*

(Princeton, N.J.: Princeton University Press, 2000). Too, Chakrabarty's and Robbins's analyses stand as markers of how academic critique itself had become globalized. This point (in this and other contexts) has stimulated scholars to probe how knowledge production in the academy can serve to reinforce dominant power structures. See, as one perspective, David Harvey, "Cosmopolitanism and the Banality of Geographical Evils," *Public Culture* 12(2) (2000): 529–564. The complex intersection of globalism and the academy is engaged most thoroughly in the work of Arif Dirlik. See, for example, Arif Dirlik, Vinay Bahl, and Peter Gran, eds., *History after the Three Worlds: Post-Eurocentric Historiographies* (Lanham, Md.: Rowman & Littlefield, 2000).

4 For a more detailed account of Iridium, see Martin Collins, "One World, One Telephone: One Look at the Making of a Global Age," *History and Technology* 21 (2005): 301–324.

5 The original design idea called for seventy-seven satellites, which inspired the name Iridium—in the periodic table, the element iridium has the atomic number 77. A redesign of the system in 1992 resulted in the reduced number of satellites.

6 As the space age developed, the deployment of satellite systems for monitoring or communicating around Earth had been approaching such a state. In the civilian (nonmilitary government) and commercial sectors, communications and meteorological and other remote sensing systems provided either near-global or global coverage on a periodic (nonconstant) basis. The same probably was true of military and intelligence satellite systems—with one important exception. The military-produced Global Positioning System, a constellation of twenty-four satellites, was in development just prior to Iridium's inauguration and became fully operational in 1995.

7 On the role of the media in promoting the enthusiasm for markets, see Thomas Frank, *One Market Under God: Extreme Capitalism, Market Populism, and the End of Economic Democracy* (New York: Doubleday, 2000).

8 Of course, such transcultural intermingling has a long history. See James Clifford, *Routes: Travel and Translation in the Late Twentieth Century* (Cambridge, Mass.: Harvard University Press, 1997). The issue for the recent period is arriving at useful analytic concepts. Descriptive-cum-analytic metaphors such as "hybridity" and "circulation" are indicative of the effort to gain intellectual purchase on this problem.

9 For a useful synopsis of neoliberalism and its relation to earlier ideologies of the market, see Paul Treanor, "Neoliberalism: Origins, Theory, Definition," accessed March 15, 2011, http://web.inter.nl.net/users/Paul.Treanor/neoliberalism.html; David Harvey, *A Brief History of Neoliberalism* (New York: Oxford University Press, 2005); and Robert J. Antonio, "The Cultural Construction of Neoliberal Globalization," in *The Blackwell Companion to Globalization* (Oxford: Blackwell, 2007), 67–83.

10 The intersecting literature of postcolonial studies and globalization is now vast. Works that capture the methodological problems with thoughtful nuance include: Arjun Appadurai, *Modernity at Large: Cultural Dimensions of Globalization* (Minneapolis: University of Minnesota Press, 1996); Frederick Cooper, *Colonialism in Question: Theory, Knowledge, History* (Berkeley: University of California Press, 2005); Anna Lowenhaupt Tsing, *Friction: An Ethnography of Global Connection* (Princeton, N.J.: Princeton University Press, 2005); Robert Ferguson, *The Media in Question* (London: Arnold, 2004); Ulf Hannerz, *Transnational Connections: Culture, People, Places* (London: Routledge, 1996); Fredric Jameson and Masao Miyoshi, eds., *The Cultures of Globalization* (Durham, N.C.: Duke University Press,

1998); Rob Kroes, *If You've Seen One, You've Seen the Mall: Europeans and American Mass Culture* (Urbana: University of Illinois Press, 1996); Charles S. Maier, *Among Empires: American Ascendancy and Its Predecessors* (Cambridge, Mass.: Harvard University Press, 2006); Edward W. Said, *Culture and Imperialism* (New York: Vintage Books, 1994); Mark Poster, *Information Please: Culture and Politics in the Age of Digital Machines* (Durham, N.C.: Duke University Press, 2006); and Saskia Sassen, *Globalization and Its Discontents* (New York: New Press, 1998).

11 On the connection of Earth applications satellites to this issue, see Lisa Parks, *Cultures in Orbit: Satellites and the Televisual* (Durham, N.C.: Duke University Press, 2005).

12 Daniel Bell, *The Cultural Contradictions of Capitalism*, 20th ed. (New York: Basic Books, 1996).

13 Frederic Jameson, Pierre Bourdieu, Jean Baudrillard, and Jean-François Lyotard are the main reference points for this analytic outlook. On Jameson, see Fredric Jameson, *Postmodernism, Or, the Cultural Logic of Late Capitalism* (Durham, N.C.: Duke University Press, 1991); and Jameson and Miyoshi, *Cultures*. For more on Jameson's importance to this discussion and his centrality to the related issue of postmodernity as a descriptor of the postwar condition, see Perry Anderson, *The Origins of Postmodernity* (London: Verso, 1998). Bourdieu's signature work is Pierre Bourdieu, *The Logic of Practice* (Palo Alto, Calif.: Stanford University Press, 1990). On Baudrillard, see Jean Baudrillard, *Selected Writings* (Palo Alto, Calif.: Stanford University Press, 2001); the introduction by Mark Poster provides useful insight on the arc of Baudrillard's thinking. He began publishing on these issues in 1968 and continued through his death in 2007. Lyotard's writings have been equally seminal; see, as his best-known example, Jean-François Lyotard, *The Postmodern Condition: A Report on Knowledge* (Minneapolis: University of Minnesota Press, 1984).

14 For an intellectual history of capitalism, see Howard Brick, *Transcending Capitalism: Visions of a New Society in Modern American Thought* (Ithaca, N.Y.: Cornell University Press, 2006).

15 Daniel Yergin and Joseph Stanislaw, *The Commanding Heights: The Battle Between Government and the Marketplace That Is Remaking the Modern World* (New York: Simon & Schuster, 1998).

16 Fred Turner, *From Counterculture to Cyberculture: Stewart Brand, the Whole Earth Network, and the Rise of Digital Utopianism* (Chicago: University of Chicago Press, 2006).

17 Francis Fukuyama, "The End of History?" *National Interest* 16 (1989): 3–18. His views were developed further in Francis Fukuyama, *The End of History and the Last Man* (New York: Free Press, 1992).

18 For an overview of Motorola's business activity and standing in the 1980s and 1990s, see Dan Steinbock, *Wireless Horizon: Strategy and Competition in the World Wide Mobile Marketplace* (New York: Amacom, 2003).

19 When President George H. W. Bush campaigned for reelection in 1992, he made a point of visiting a Motorola plant in Schaumburg, Illinois, home of the company, to make a pitch for his Agenda for American Renewal, a response to a then-weak economy. In his speech at the plant, Bush, referring to the award and the company's national and international business triumphs, offered, "If we use this as a microcosm of our country, they're [Motorola workers] writing the future for our whole country, the future for the United States of America. What you are doing is the perfect putdown for the professional pessimists, the doomsayers, some of whom say we can not compete in a changing world. And you've taken the challenges of this new world, and you have done what America has always done—reinvented them as opportunities for yourselves, for your families, and for every single American." Memorandum from Roni Haggart to GRO staff, September 28, 1992, unfoldered, Box 6, IP, NASM.

20 Keith Bradshear, "Science Fiction Nears Reality: Pocket Phone for Global Calls," *New York Times*, June 26, 1990.

21 These events in China are related in Mark Gercenstein, Oral History Interview, Iridium Oral History Project, NASM. Gercenstein was Iridium's representative in China at the time of the press announcement.

22 Bradshear, "Science Fiction."

23 *Wall Street Journal*, November 19, 1991.

24 Anthony Ramirez, "The Ultimate Portable-Phone Plan," *New York Times*, March 18, 1991.

25 For a useful overview of *Wired*, see Thomas Streeter, "The Moment of Wired," *Critical Inquiry* 31 (2005): 755–779.

26 Joe Flowers, "Iridium," *Wired* 1.05 (November 1993), accessed March 15, 2011, http://www.wired.com/wired/archive/1.05/iridium.html.

27 Ibid.

28 Ibid.

29 For a critique of and resistance to such developments, see, as an example, Naomi Klein, *Fences and Windows: Dispatches from the Front Lines of the Globalization Debate* (New York: Picador USA, 2002).

30 "Iridium Satellites Close to Girdling the Globe," *New York Times*, April 16, 1998.

31 Tom Dowe, "The Gadget Gazettes," *Wired* (1998), accessed March 15, 2011, http://www.wired.com/culture/lifestyle/news/1998/08/14585.

32 Joseph Pine and James H. Gilmore, "Welcome to the Experience Economy," *Harvard Business Review* (July–August 1998): 97–105.

33 Ibid.

34 See *Roam* 4(4) (1998): 6.

35 Ibid.

36 *Iridium Today* 2(2) (1996): 4.

37 See, as an example, Iridium's president Robert Kinzie's travel and meeting schedules in folder "Speeches 1997," Box 11, IP, NASM.

38 *Iridium Today* 2(3) (1996): 29.

39 David S. Bennahum, "The United Nations of Iridium," *Wired* 6.10 (October 1998): 134–138, 194–201, accessed March 15, 2011, http://www.wired.com/wired/archive/6.10/iridium.html?pg=1.

40 *Roam* 5(1) (1998): 4.

41 Bennahum, "United Nations."

42 Ibid.

43 Ibid.

44 On the complex patterns of hegemonic control and resistance, see Charles Bright and Michael Geyer, "Regimes of World Order: Global Integration and the Production of Difference in Twentieth-Century World History," in *Interactions: Transregional Perspectives on World History*, ed. Jerry H. Bentley, Renate Bridenthal, and Anand A. Yang (Honolulu: University of Hawai'i Press, 2005), 202–238.

45 This motif has been central to the globalization literature; as a classic expression, see Roland Robertson, *Globalization: Social Theory and Global Culture* (London: Sage, 1992).

46 Bennahum, "United Nations."

5

THE NAVSTAR GLOBAL POSITIONING SYSTEM

FROM MILITARY TOOL TO GLOBAL UTILITY

RICK W. STURDEVANT

The NAVSTAR Global Positioning System (GPS), which enabled users to determine their precise location in three dimensions and time within billionths of a second, evolved from concept to operational system in slightly more than two decades.[1] Colonel Bradford Parkinson, U.S. Air Force (USAF) director of the newly formed GPS Joint Program Office in 1973, directed his team to synthesize a design from several competing programs.[2] The USAF launched the first operational GPS satellite in 1989 and declared a twenty-four satellite constellation fully operational in April 1995.[3]

The Department of Defense (DoD) needed GPS primarily to deliver weapons on target, but recognized the system's nonmilitary potential. To withhold full accuracy from enemies but provide GPS service to civilian users, the USAF designed the system with a protective feature called "selective availability" (SA) that, when used, gave U.S. and allied forces significantly more precise signals than other users received. In September 1983, after Korean Airline Flight 007 went astray and was shot down by Soviet fighters, President Ronald Reagan assured the world that once GPS became operational, the coarser signal that could have saved the Korean flight would remain continually and universally accessible at no cost. As GPS approached operational status in the early 1990s, civilian and commercial users, who already had ten times as many GPS receivers as the military, mounted an increasingly vocal campaign for unrestricted access to the more precise signals. Finally, in May 2000 President Bill Clinton acknowledged the global utility of GPS and directed immediate discontinuation of the SA feature.[4]

With increasing demand for accuracy, users found ways to augment GPS. For small areas, those included pseudolites (ground-based transmitters that could be configured to emit GPS-like signals) and differential GPS (DGPS), which required a high-quality GPS "reference receiver" at a surveyed location. The Wide Area Augmentation System (WAAS), one type of Wide Area DGPS, used reference receivers at multiple monitor stations and a master-control hub to achieve similar results over a larger region.[5]

Other countries and regions developed their own GPS augmentation systems. Russia had its own Global Orbiting Navigation Satellite System (GLONASS). China launched its first Beidou prototype navigation satellite in October 2000. Europe followed suit with GIOVE-A in December 2005, the first step toward an operational system called Galileo. Even as GPS's global utility manifested itself, integrating all the systems into a single Global Navigation Satellite System (GNSS) remained possible. Participants in the International GNSS Service (IGS), a voluntary federation of more than 200 agencies worldwide, contemplated just such a possibility at meetings like the IGS Analysis Center Workshop in Miami, Florida, in June 2008.[6] Since a full accounting of GPS's impact would require hundreds of pages, what follows is merely illustrative.

Military Applications

The first use of GPS-guided weapons occurred in Operation Desert Storm, the first Gulf War, in 1991. During the initial hours of the air campaign, bombers delivered AGM-86C Conventional Air-Launched Cruise Missiles, each containing GPS loosely coupled with an existing inertial navigation system, and achieved several exact hits. During the 1990s, the U.S. military converted fundamentally "dumb" bombs into GPS-guided Joint Direct Attack Munitions (JDAMs), high-precision ordnance capable of destroying multiple targets in a single sortie at any time of day or night, in any kind of weather. During Operation Allied Force in 1999, U.S. B-2 "Spirit" stealth bombers delivered 650 JDAMs against Serbia with devastating accuracy. Munitions manufacturers strove to put GPS into smaller projectiles, and, in 2008, army artillery in Afghanistan first fired a 155 mm GPS-guided howitzer round during combat. These new projectiles offered greater lethality and lower collateral damage.[7]

In 2004, U.S. Marines in Iraq witnessed the first combat use of a GPS-assisted parafoil, a key component of the Joint Precision Airdrop System (JPADS). The first operational use of JPADS in Southwest Asia supplied ammunition and water to troops in Afghanistan in 2006. By enabling accurate delivery of cargo pallets from release altitudes above 25,000 feet, these systems protected aircraft and crews from inexpensive surface-to-air missiles.[8] Meanwhile, paratroopers obtained GPS-equipped helmets. French forces had the Operational Paratroopers Navigation System in 2005, and U.S. Marines contracted for helmet-mounted ParaNav units in 2008.[9]

A rapidly expanding family of remote-controlled systems relied on GPS. Unpiloted aerial vehicles equipped with GPS flew over Bosnia and Kosovo during the late 1990s. In 2001–2002, Predator drones carrying Hellfire missiles attacked Taliban and Al Qaeda forces in Afghanistan. The GPS-equipped REMUS autonomous underwater vehicle scouted for mines in the port of Umm Qasr in 2003. By 2005, the DoD contracted for GPS-enabled robotic vehicles to extract and evacuate wounded soldiers from the battlefield.[10]

The military's most pervasive GPS use involved tracking forces in real time. During Operation Desert Storm, GPS aided large troop formations positioning and maneuvering in relatively featureless desert terrain. By 2005, U.S. and allied forces in Kosovo, Afghanistan, and Iraq relied on more than 10,000 Blue Force Tracking devices. The enhanced situational awareness allowed battlefield commanders to plan and coordinate maneuvers with unprecedented precision and saved American lives by reducing the so-called fog of battle. In December 2008, the U.S. Army increased by US$3.2 million its contract with Comtech Telecommunications Corporation for Movement Tracking System equipment, and the North Atlantic Treaty Organization (NATO) ordered additional equipment worth US$7.3 million from Globecomm Systems for NATO's GPS-Based Force Tracking System.[11]

Civil and Commercial Applications

When GPS receiver technology and cellular phones became more affordable during the mid-1990s, industry analysts and entrepreneurs perceived an emerging multibillion-dollar market, especially for location-based services to nonmilitary users. A 2002 survey of 20,000 U.S. households revealed strong consumer interest in GPS. Applications spanned a colorful spectrum, across which curious observers could see both distinct and mottled hues.[12]

MASS TRANSIT AND FLEET MANAGEMENT. Governments and businesses used GPS technology for managing vehicle fleets more efficiently. Benefits ranged from labor savings and scheduling efficiency to better bus-stop placement and enhanced passenger safety. Cities like Denver, Colorado, and Dallas, Texas, installed automatic vehicle location (AVL) technology in buses to keep riders informed of projected arrival times, assist mobility-impaired riders, and expedite response time in emergency situations. Meanwhile, Netherlands-based Büchner Transport, a company whose fifty trucks delivered house plants, fruits, and vegetables under controlled temperatures throughout Europe, began equipping its entire fleet with AVL after experimenting with it on two vehicles from December 1991 to June 1992.[13]

Before long, GPS devices infiltrated fleet management and measurably improved that sector's business practices. Between 1988 and 2004, San Diego–based Qualcomm, the world's largest AVL supplier, installed its system on more than half a million commercial vehicles belonging to more than 1,500 trucking firms. Rocky Mountain Tracking, Inc., of Fort Collins, Colorado, estimated in May 2006 that equipping an average-size fleet of ten vehicles with its GPS devices for monitoring speeds, routes, and idle time resulted in monthly savings of more than US$6,000. Phil Nolan, Islip community supervisor on Long Island, New York, put GPS devices in 614 official vehicles to deter inappropriate usage and thereby saved almost 14,000 gallons of gasoline over a three-month period compared to the same period a year earlier. When Denver, Colorado, equipped seventy-six official vehicles with GPS in 2007, drivers traveled 5,000 fewer miles than in the same period the previous year. After using GPS units in its service vehicles for seven years, Orkin, Inc., the second-largest pest-control company in the United States, discovered a 40 percent drop in its automobile accident rate. One professionally conducted survey in 2008 reported a 19.2 percent reduction in miles traveled after companies began using GPS-enabled services. That reduction meant less fuel consumption and less environmentally harmful emissions.[14]

Within the fleet management arena, emergency-response efforts benefited immensely from GPS. Saving time often meant saving lives or property. A handful of Pennsylvania townships used dashboard-mounted GPS units in ambulances to plot the shortest, best routes when responding to calls in unfamiliar areas. When the Space Shuttle *Columbia* broke apart over Texas during reentry on February 1, 2003, emergency responders relied on GPS and geographic information system (GIS) capabilities to conduct their search for debris and map the precise location of whatever they found. Firefighters in North Las Vegas, Nevada, depended on GPS-enabled GIS software to respond more efficiently to calls. In 2008, the conclusion of Europe's HARMLESS project demonstrated the possibility of using space-based navigation to help save some of the 2,000 European lives lost annually because of difficulty in pinpointing and rapidly reaching the location of victims after an emergency call.[15]

A survey by Aberdeen Group of more than 300 fleet management professionals in September and October 2007 revealed "sizeable improvements" as a result of implementing GPS-enabled and other location-based services in their field operations. On average, costs for fuel, maintenance, and overtime dropped roughly 13 percent. Service response times improved by almost 24 percent, and miles traveled dropped by 19.2 percent. Both fleet and workforce utilization rose more than 25 percent. Aberdeen Group concluded, "While these improvements equate to varying amounts for firms of different sizes and across different industries, these do amount to substantial enhancements or savings."[16]

In 2008, 25 percent of demand-response services and 15 percent of fixed-route bus agencies used GPS, as did 75 percent of ferryboat agencies and 40 percent of rail systems. Railroads found GPS especially attractive. Salt Lake City's TRAX network installed a GPS tracking-and-information system for the 2002 Winter Olympics. During the next few years, British and U.S. railways tested prototype Precision Train Control (PTC) systems for track management. Not surprisingly, calls intensified for more rapid PTC implementation after a Los Angeles railway collision claimed more than two-dozen lives in September 2008.[17]

Like the railroads, ferryboat companies sought to achieve safer, more predictable operations with GPS. In 2003, NY Waterway completed installation of a US$1 million GPS-based system to instantly and continuously track the location and speed of its ferries and buses, thereby improving customer service at Manhattan's West Thirty-Eighth Street and other terminals. With more than two-dozen boats, Washington State Ferries updated the location and direction of its vessels, including those stopped or docked, every three minutes. BC Ferries, which in 2007 had forty ferryboats of various sizes plying forty routes along the coast of British Columbia, also relied on GPS for fleet tracking. Even small operators, like MetroLINK on the Mississippi River at Rock Island, Illinois, and the B Harbor Ferry at Corpus Christi, Texas—each with only two boats—used GPS-enabled AVL. The trend continued into August 2008, when Catalina–Marina Del Rey Flyer proudly announced its luxury passenger ferries all featured GPS tracking, which updated the company's computers every thirty seconds for improved passenger safety and more accurate scheduling information.[18]

AVIATION AND SPACE. The aviation industry turned to GPS for plotting more efficient air routes, managing traffic better regardless of weather, improving approaches to airports, and increasing ground safety. Ten major airports in the Democratic Republic of the Congo adopted GPS technology to erase dependency on expensive ground-based navigational aids. In 2007, Norway approved GPS-enabled approach systems at twenty-four local airports where terrain rendered traditional instrument landing systems (ILS) relatively useless. The Federal Aviation Administration announced in 2008 that GPS/WAAS approaches had surpassed ILS approaches at U.S. airports. In December 2008, New Jersey's Newark Liberty International Airport, one of the nation's worst in terms of chronic flight delays, became the first major U.S. air hub to test a GPS/Ground-Based Augmentation System to reduce congestion by allowing planes to fly closer together without compromising safety.[19]

GPS revolutionized how nations operated in space. From replacement of heavy, high-cost subsystems for control of individual satellites to precise "formation flying" of several satellites, onboard GPS units minimized intervention from ground crews. In 1992, TOPEX/Poseidon became the first satellite to demonstrate that controllers could determine a GPS-equipped spacecraft's exact orbital location. The first mission of the European Space Agency's Jules Verne unmanned transfer vehicle in 2008 used GPS to approach the International Space Station. In 2009, the French Space Agency (CNES) and several partners planned to launch a pair of small PRISMA satellites, the first European mission to validate "formation flying radio frequency" techniques using GPS.[20]

SPORTS AND RECREATION. Golf spearheaded GPS recreational use with course- and player-owned systems. Darryl Sharp, who began using GPS to map golf courses in 1997, completed fifty-five of them by May 2002. Video-game manufacturer EA SPORTS used Sharp's data to make simulated golfing on some of America's premier courses amazingly realistic. The Professional Golf Association contracted with Sharp to map courses for its ShotLink system that gave television viewers real-time information on players' shots.[21] By 2008, SkyCaddie rangefinders automatically calculated distance on more than 17,000 golf courses in more than forty-five countries, and Pro-Link Solutions boasted 10,642 GPS units distributed in 2007 for use at more than 700 courses on five continents. Player-owned systems such as Star-Caddy, MobiGolf, GoldTraxx for PocketPC and Palm PDA; iGolfScorer for cell phones; and stand-alone units such as SkyKap Advisor, SureShotGPS, and QuickRange GPS afforded the avid golfer, regardless of ability, a potentially confusing range of alternatives. Course-owned systems were no less diverse, with cart-mounted variants such as GameStar, Inforemer GPS Golf, and INOVA, or portables such as xCaddies and the XY Golf Portable GPS System.[22]

Runners and bicyclists relied on GPS. In 2004, Paula and Scott Morris used small GPS units to create a three-dimensional profile of an off-road bicycle tour covering 2,490 miles. During one stage of the 2004 Tour de France, the European navigation service experimented with GPS in vehicles driving behind the racers; during the race's fifth stage in 2005, cyclists themselves carried lightweight GPS receivers that tracked their exact positions and speed. Race organizers aimed to give spectators in future years the ability to follow competitors' positions and the race's progress live from anywhere in the world. Meanwhile, reporter Mark Schoofs used GPS devices for runners to compare distance and pace on different days over different terrains in 2007. A year later, runners with cell phones could use AllSport GPS to check time, speed, and distance traveled.[23]

GPS enhanced other outdoor experiences such as fishing, hunting, and bird-watching. Whether fishing for recreational or commercial purposes, people used GPS to plan boating routes, provide real-time navigation, record favorite spots for return trips, and advise rescuers of their position in case of trouble.[24] Hunters benefited in similar ways, using GPS to guide them to their tree stands or blinds before sunrise and to find their way back to a vehicle or camp after a long trek through the woods.[25] For bird-watchers, GPS receivers made it easy to mark sightings along migratory routes and to "geotag" digital photographs of birds for later download into a worldwide media exchange.[26]

Geocaching became the most unique GPS recreational activity. It originated in March 2000 when David Ulmer hid a bucket containing a logbook and assorted prizes in the woods near Portland, Oregon, and shared the coordinates with an Internet GPS users' group. Members had to find the cache, take a prize from it, leave something in it, and annotate the logbook. First called "GPS Stash Hunt," aversion to the possibly negative connotation of "stash" prompted Matt Stum to coin the more palatable term "geocaching" in May 2000. Over the next eight years, the popularity of geocaching skyrocketed, and specialized forms of the activity emerged. State and regional geocaching associations arose. Bookstores began stocking titles such as *The Complete Idiot's Guide to Geocaching* and *The Geocaching Handbook.*[27] Companies PlayTime, Inc. and American Outback Adventures touted geocaching for corporate team building, and the Alaska Vacation Store offered safaris to search for that state's caches.[28] In August 2008, New York's Upper Catskill Community Council of the Arts started ArtQuest, which challenged participants to find sculptures hidden at eleven sites near Cooperstown.[29] By November 2008, the number of active caches globally approached 675,000, and the number of registered geocachers exceeded 66,000.[30]

AGRICULTURE. Precision farming with GPS took many forms: nighttime cultivation; locating weed, insect, and disease infestations; fertilizer or pesticide application; mapping yield; pinpointing hail or drought damage; and tracking livestock. In 1992, Amana Farms' chief executive explained that GPS enabled determination of "crop yield by the square foot and not by the traditional bushels per acre," which dramatically changed how farmers "plant, fertilize, apply weed killers and harvest crops."[31] To further improve efficiency and increase the profitability of their operations, farmers also used GPS for detailed base-mapping of physical features such as borders, fence lines, wells, buildings, landscape particulars, irrigation canals or pipelines, and wetlands. Under a National Aeronautics and Space Administration (NASA) grant, researchers at Ohio State University worked to perfect a GPS-enabled tomato-picking robot that could significantly reduce labor costs on large corporate farms.[32]

Montana State University used the first GPS-guided fertilizer application system near Power, Montana, in August 1990. Agronomist Mitch Schefcik and engineer William Bauer devised a DGPS/GIS system in 1992 that varied herbicide application on sugar beet fields in western Nebraska, thereby increasing crop yield while satisfying government chemical-application regulations. Department of Agriculture experimentation found DGPS useful for mapping nitrogen fertilizer applications and evaluating the effect on yields in central Nebraska cornfields during 1995. Nitrogen fertilizer management also improved groundwater quality.[33]

One expert estimated in 1994 that only 5 percent of U.S. farms, mostly in the corn-producing states of the Midwest, used GPS-derived reference points for soil-specific management, but the technology was "booming." A study of soil-specific management in Stoddard County, Missouri, during that period showed reductions in fertilizer costs of US$18.70 per acre, with no loss of crop yields. University of Florida researchers touted GPS/GIS's value for locating and managing site-specific crop losses (more than US$77 billion worldwide in 1987) due to plant-parasitic nematodes. According to an Associated Press story in November 2004, up to 15 percent of U.S. farmers had GPS-controlled tractors or combines and were saving as much as 5 percent in fertilizers and pesticides by using precision guidance systems. Farmers in Kentucky, where rolling terrain led to both high and low production in the same field, realized a cost savings of US$30 per acre by using GPS-enabled yield monitors.[34]

Farmers and ranchers used GPS for greater accuracy, lower labor costs, and better documentation of aerial spraying to control insects. During 1992, the Plains Cotton Growers Diapause Program in Texas used GPS flight guidance to spray more than 450,000 acres of cotton for boll weevils. Based on that success, North Dakota's Grasshopper Integrated Pest Management Project used DGPS in 1993 when spraying 6,400 acres of rangeland. In California, reliance on DGPS in the war against the Mediterranean fruit fly brought significant savings.[35]

Unlike surveying or construction, where real-time kinematic (RTK) systems like DGPS or WAAS gained wide acceptance quite early, precision agriculture lagged in RTK usage. Equipment was not compact, and farmers had to erect a reference station near their fields; communication links were difficult to establish and sometimes required FCC approval. The cost of a complete RTK system could be as high as US$50,000, which was not cost effective for the average farmer, who seldom needed an accuracy of 2 cm day in, day out. In 2008, most farmers remained very hesitant to subscribe to an RTK service, which might cost as much as US$4,500 annually per unit. Many had ventured partway into the precision arena, however, with GPS-based solutions by equipment manufacturers who offered 1 m accuracy. Other farmers turned to companies that provided 10 cm accuracy on a regular basis and RTK solutions when needed.[36]

ENVIRONMENT. GPS/GIS applications improved environmental awareness and analysis. Nations employed GPS to regulate logging or grazing on public lands, to assess mining royalties, and to delineate borders. Meteorologists, as far back as 1992 and 1995, experimented with ground- and space-based GPS techniques to improve weather forecasts and observe tidal effects. GPS helped predict earthquakes or volcanic eruptions and enabled more efficient tracking of oil spills or other ecological catastrophes. Biologists used it to understand threats to endangered species.[37]

The Committee on the Status of Endangered Wildlife in Canada (COSEWIC) outfitted Alberta's woodland caribou with programmable collars in 1998 to understand how oil and gas development affected them. The Dian Fossey Gorilla Fund International used GPS/GIS in 1999 to track Rwanda's mountain gorillas. Save the Elephants (STE) introduced collars with embedded GPS and cell-phone technology to Kenya in February 2004. After a facial-tumor disease wiped out 90 percent of wild Tasmanian devils, zoologists equipped hand-raised orphans with GPS collars before releasing them in 2007. To learn more about how climate change affected polar bears in Southern Hudson Bay, COSEWIC scientists placed GPS collars on female bears with yearlings in 2007–2008.[38]

Indigenous people of South American and African rain forests used GPS to better manage their resources, monitor ecological change, and prevent illegal logging, mining, farming, and other encroachment. Brazilian tribes, who collected GPS data on ten million acres of their homeland, unveiled their first cultural map in 2003. Another project began during 2006 in the Democratic Republic of Congo, where loggers gave seminomadic Mbendjele pygmies GPS units to help locate and preserve critical areas.[39]

GPS equipment installed in 1999 on Hawaii's Mauna Loa volcano revealed an extension of its caldera and a rise in its summit during 2002–2005. By 2008, 14,000 GPS stations in Japan, another 465 in Southern California, and hundreds elsewhere allowed scientists to calculate tectonic plate movement with millimeter accuracy. Beyond increasing scientific understanding about seismic activity, data derived from those stations enabled civil authorities to alert residents to possible destructive events.[40]

PUBLIC HEALTH AND SAFETY. Emergency responders found GPS invaluable. In 1992, Amoco tested a GPS-aided response system in its Crossfield natural gas field north of Calgary, Alberta, and reported noteworthy cost and safety improvements due to prompt identification of an alarm site, the nearest field personnel, and the best route to the site.[41] To speed response and save lives, LaSalle Ambulance Service of Buffalo, New York, adopted a GPS tracking-and-dispatch system in 1993 for its fleet of forty-two vehicles, two aero-medical helicopters, and one fixed-wing aircraft.[42]

Stolen vehicle recoveries became easier with GPS. Businessman Jameel Yusuf began using GPS in 1998 to find stolen cars in Karachi, Pakistan, and located more than 1,000 by October 2003. When, in 2002, a thief stole a GPS-equipped taxicab in Colorado Springs, Colorado, a dispatcher telephoned the moving vehicle's exact whereabouts to police officers. Using integrated GPS/cellular tracking systems, Texas recovered more than fifty truck tractors and seventy-five trailers in 2004. GEOTrac International joined the Royal Canadian Mounted Police in 2007 to apply GPS/GIS and satellite communications technology to recover equipment stolen from oilfields.[43]

GPS helped reduce crime in other ways. North Carolina strapped GPS ankle bracelets on sex offenders in January 2008. A frequently burglarized store in Indianapolis, Indiana, placed GPS locaters inside merchandise to track the thieves. With Sprint Family Locator, beginning in 2006, parents could track children. Using GPS units and palm-size computers programmed with detailed travel instructions, blind walkers could find their way with relative safety. On college campuses (for example, University of North Carolina at Chapel Hill, State University of New York at Oswego, Colorado State University, and Montclair State University in New Jersey), the Rave Guardian personal mobile alarm system gave students, faculty, and staff members carrying cell phones a direct connection to campus police whenever and wherever needed; GPS features embedded in the caller's cell phone showed police exactly where to respond. Occasionally, GPS data could give families sufficient evidence to refute unjust speeding charges against teenage drivers. Cell phones with embedded GPS helped Washington state police locate a woman whose car had plunged down a steep ravine, and led Colorado Springs, Colorado, police to a kidnapped woman.[44]

Medical practice and public health also benefited from GPS. West Bengal, India, first used GPS/GIS in February 2003 to monitor polio eradication. Denver, Colorado, adopted similar technology to prevent untreated sewage from reaching streams. In 2005, staff members in Swedish hospitals and health care units gained the capability, through a GPS-enabled service called BodyKom, to monitor patients from a distance who did not require hospitalization. That same year, Canadian and French companies collaborated to unveil the Columba bracelet that automatically detects any departure of Alzheimer's patients from a predetermined security zone surrounding their residence or nursing home and alerts a medical assistance center. An Idaho county tested GPS/GIS in 2007 for West Nile virus abatement, and Navajo

Nation health care providers began using GPS to locate isolated diabetics. By 2008, various forms of telemedicine (principally telecardiology, teleradiology, and telepsychiatry) capitalized on GPS.[45]

DISASTER RESPONSE AND RECOVERY. In conjunction with GIS and other technologies, GPS benefited disaster response. After Hurricane Andrew devastated Florida in 1992, the Federal Emergency Management Agency (FEMA) contracted with survey crews to inventory the damage using GPS/GIS technology instead of the traditional, manual assessment that involved house-by-house interviews. Based on encouraging experimental results, in July 1993 FEMA partnered with the U.S. Army Corps of Engineers and a private contractor with GPS/GIS expertise to produce maps for disaster response, recovery efforts, and risk mitigation after Mississippi River flooding inundated more than thirteen million acres, destroyed crops worth billions of dollars, and left hundreds homeless. Following a GPS-equipped helicopter survey, a pair of two-person observer teams on the ground with GPS/GIS handheld receivers inventoried structures in communities south of Quincy, Illinois; they transferred the data to the Corps of Engineers' Rock Island, Illinois, base station, where more than 1,500 maps/data sheets were produced within a week. With traditional methods, it would have taken a team of fifty people several years to complete the same task. GPS/GIS technology enabled FEMA officials to begin meeting sooner with local leaders and, within a few months, to start rebuilding above the hundred-year flood elevation.[46]

Faced with a daunting "Ground Zero" cleanup after terrorists destroyed the World Trade Center in September 2001, workers used handheld devices with a GPS receiver, bar-code reader, and scroll-down descriptive list to catalog thousands of pieces of evidence. For removing 1.8 million tons of debris, a contractor installed off-the-shelf components that integrated GPS positioning with communications, camera monitoring, and Internet data in 235 trucks and a control center. This helped finish the project earlier and at one-tenth the cost city officials initially estimated.[47]

In August 2005, Hurricane Katrina revealed other GPS applications. Aircraft deployed GPS-enabled dropsondes to help scientists accurately predict the storm's strength, speed, and direction. GPS/GIS helped rescuers find stranded survivors. Nationwide DGPS enabled precise repositioning of 1,800 buoys, dredging of critical waterways, and navigating of silt-clouded waterways.[48]

Regional and global disaster relief organizations relied heavily on GPS and wireless technology by 2008. To increase the speed and rate of emergency-response units in 170 countries worldwide, British Telecommunications partnered with the British Red Cross in 2007, investing US$600,000 to equip relief vehicles with GPS, satellite telephones, and information technology (IT) equipment. The ReliefAsia organization, conceptualized in the wake of the 2004 Indian Ocean tsunami, delivered GPS personal tracking and SOS devices to relief workers after the Sichuan earthquake and Guangzhou floods in 2008. The equipment not only improved deployment of relief teams and delivery of humanitarian supplies, it allowed relief workers trapped by aftershocks or landslides to signal rescuers.[49]

SURVEYING AND MAPPING. GPS facilitated all types of surveying and mapping, significantly reducing the amount of equipment and labor. Between 1998 and 2002, workers realized considerable cost savings by using GPS technology for cadastral surveys in numerous countries: El Salvador, Indonesia, Morocco, Botswana, Namibia, Jordan, and several eastern European nations. Compared to "optical total station" technology, surveyors with GPS reported productivity gains of 50 percent to more than 1,000 percent. With GPS, work could continue regardless of weather or daylight, and it was not limited to line of sight. Use of RTK GPS in particular significantly improved the quality of collected data and reduced the occurrence of blunders over large distances.[50]

The creation of a globally consistent continental reference system benefited from GPS/GIS. Europe and South America first used it to establish regional reference frames. By 2008, geographers began unifying fifty-three different national coordinate reference systems into a single Africa Reference Frame.[51] Global MapAid (GMA), a project hatched by Rupert Douglas-Bate at Stanford University, supplied international aid organizations with detailed, up-to-date maps of disaster zones. Residents of a disaster area used a field kit containing handheld computers with built-in GPS, satellite phones, and innovative software to collect detailed information about "everything from the location of hospitals and washed-out bridges to the population of neighborhoods and the prevalence of contaminated drinking water." Every few hours, they transmitted the collected data to GMA field headquarters, where disaster maps were updated and reloaded onto a website for use by the aid organizations.[52]

Maps prepared through GMA's efforts benefited emergency crews and relief organizations after hurricanes, tsunamis, and earthquakes, even as GMA explored other ways to assist aid organizations. After a December 2004 earthquake and tsunami devastated Banda Aceh in northern Sumatra, GMA partnered with a task force from Banda Aceh's University of Syiah Kuala to successfully perform disaster mapping. Within a few months, GMA teams also prepared and distributed thousands of "disaster response" maps in the wake of Hurricanes Katrina and Rita on the U.S. Gulf Coast, and they created humanitarian "donor/aid worker" maps after a devastating earthquake in northern Pakistan. During August 2008, GMA volunteers undertook a pilot survey to map the daily movement and activity of a selected group of Port Elizabeth, South Africa, street children over seven days, but that was only a step toward the long-term goal of mapping the conditions of South Africa's 800,000 AIDS orphans.[53]

Archaeologists around the world took advantage of GPS for surface mapping, subsurface excavations, and underwater surveys. In 1997, preliminary fieldwork in North Kohala on the island of Hawaii took one-fourth as long as traditional techniques would have taken. At prehistoric Anasazi ruins in southwestern Colorado, GPS and total-station instrumentation combined to allow recognition of surface and subsurface structures with little or no disturbance to the sites, thereby preserving such features for later generations. Experiments carried out in Rome and its environs between December 2004 and February 2005 substantiated the feasibility of using RTK GPS techniques for fairly rapid, low-cost data collection on underground sites in an urban environment and in situations demanding high precision. The U.S. National Park Service relied on GPS/GIS during the summers of 1993 through 1995 for surveying shipwreck locations in Florida's Dry Tortugas National Park, and underwater archaeologists working on the submerged city of Alexandria in the eastern Mediterranean in 1994 combined GPS with sonar to map the contours of the seabed with 1 cm accuracy. In April 2008, researchers equipped with DGPS, sonar, and magnetometers began inventorying submerged cultural resources—for example, tanks, landing craft, barges, ships, and other material—on the World War II invasion beaches of the Northern Mariana Islands, especially Saipan.[54]

TIMING. People sometimes overlooked GPS timing and frequency applications, which entailed delivering time within 100 nanoseconds and frequency within a few picoseconds.[55] The Pacific Northwest's Bonneville Power Administration began methodically integrating GPS technology into its operations in 1988, because measuring voltage and current at selected sites and time-tagging the data with microsecond precision helped stabilize the electrical grid.[56]

Networked computers execute billions of financial transactions per second. An international investment banking firm began using Trimble Navigation's Palisade network time protocol (NTP) synchronization kit in 1999 to ensure simultaneous recording of transactions at London, New York, and Tokyo offices. By mid-2002, the New York Stock Exchange, World Bank, and other institutions used it. GPS became important enough as a time-reference standard for the world's aggregate financial network that the Heritage Foundation, a conservative U.S. public policy research organization, called it a "critical infrastructure."[57]

Conclusion

Measured statistically, GPS had a remarkable impact in a relatively short time. Its patent families (a family being a collection of related patents) grew from fewer than twenty in 1988 to nearly eighty in 1991. Procurement of GPS receivers by the U.S. military zoomed from 7,253 from 1986 through 1992, to 19,086 in 1993. Firms providing GPS-related goods or services nearly tripled in five years, from 109 in 1992 to 301 in 1997. Worldwide user equipment sales, which totaled US$3.39 billion in 1996, rose to US$6.22 billion in 1999. During the same period, industry employees increased from 16,688 to 30,622.[58]

With 50 percent of the U.S. consumer market in 2006, GPS equipment manufacturer Garmin's revenue totaled US$1.77 billion, up 73 percent from 2005; with 45 percent of the European market, Tom Tom boasted US$1.8 billion, up 89 percent from 2005. In January 2008, ABI Research estimated sales of more than 900 million nonmilitary GPS units by 2013. Another report predicted GNSS products and services would grow between 19 percent and 23 percent annually between 2008 and 2012.[59]

Within three years of its becoming fully operational, GPS had achieved recognition worldwide as an indispensable technology. Its unprecedented accuracy, in any weather, and the creativity of its users pushed the use of GPS into some surprising realms. In December 2004, a White House fact sheet described GPS-derived information and applications as "an engine for economic growth, enhancing economic development, and improving safety of life, and the system is a key component of multiple sectors of U.S. critical infrastructure."[60] Two years later, an industry research study identified "improved precision in readily available GPS and related technology as the number-one trend" and, by a substantial percentage, the "most important technological development" affecting both manufacturers and consumers in the early twenty-first century.[61] As GPS pioneer Bradford Parkinson said back in 1980, the "potential uses are limited only by our imaginations."[62]

NOTES

1 Some people identified "NAVSTAR" as an acronym derived from either "Navigation Signal Timing and Ranging" or "Navigation Satellite Timing and Ranging." Apparently, TRW Corporation once advocated a system for which NAVSTAR was an acronym (NAVigation System Timing and Ranging). Bradford W. Parkinson, however, said the Joint Program Office considered the term "simply a nice-sounding name." Bradford W. Parkinson, "GPS Eyewitness: The Early Years," *GPS World* 5(9) (September 1994): 32–45.

2 Bradford W. Parkinson, "Introduction and Heritage of NAVSTAR, the Global Positioning System," in *Global Positioning System: Theory and Applications*, Vol. 1, ed. Bradford W. Parkinson and James J. Spiker Jr. (Washington, D.C.: American Institute of Aeronautics and Astronautics, 1996), 3–28; Ivan A. Getting, *All in a Lifetime: Science in the Defense of Democracy* (New York: Vantage Press, 1989), 574–599; Richard Easton, "Who Invented the Global Positioning System?" *Space Review*,

May 22, 2006, accessed March 16, 2011, http://www.thespacereview.com/article/
626/1; Don Bedwell, "Where Am I?" *Invention & Technology* 22(4) (Spring 2007):
20–30.

3 Michael Russell Rip and James M. Hasik, *The Precision Revolution: GPS and the
Future of Aerial Warfare* (Annapolis, Md.: Naval Institute Press, 2002), 68–79.

4 Parkinson, "GPS Eyewitness," 42–44; Rip and Hasik, *Precision Revolution*, 9–10,
429–441; Anthony R. Foster, "GPS Strategic Alliances, Part I: Setting Them Up,"
GPS World 5(5) (May 1994): 34–42; National Research Council, *The Global Positioning System—A Shared National Asset: Recommendations for Technical Improvements
and Enhancements* (Washington, D.C.: National Academies Press, 1995), 1–12.

5 Robert O. DeBolt et al., *A Technical Report to the Secretary of Transportation on a
National Approach to Augmented GPS Services*, U.S. Department of Commerce,
December 1994, accessed March 16, 2011, http://www.navcen.uscg.gov/pubs/gps/
gpsaug/auggps.doc.

6 Rosalind Lewis et al., *Building a Multinational Global Navigation Satellite System: An
Initial Look* (Santa Monica, Calif.: RAND Corporation, 2005); Brian Evans, "Setting
a Course: Beyond GPS," *Aviation Today*, May 1, 2006, accessed March 16, 2001,
http://www.aviationtoday.com/av/issue/feature/Setting-a-Course-Beyond-
GPS_896.html; J. R. Wilson, "EGNOS Enhances Positioning Accuracy," *Aerospace
America* 46(1) (January 2008): 38–41; Todd Humphreys, Larry Young, and Thomas
Pany, "Considerations for Future IGS Receivers," paper presented at the IGS
Analysis Center Workshop, Miami, June 2–6, 2008.

7 John Tirpak, "The Secret Squirrels," *Air Force Magazine* 77(4) (April 1994), accessed
March 16, 2011, http://www.airforce-magazine.com/MagazineArchive/Pages/1994/
April%201994/0494squirrels.aspx; John T. Nielson, "The Untold Story of the
CALCM: The Secret GPS Weapon Used in the Gulf War," *GPS World* 6(1) (January
1995): 26–32; "GPS-Guided Artillery Used in Afghanistan," *GPS World* (March 20,
2008), accessed March 16, 2011, http://www.gpsworld.com/defense/news/gps-
guided-artillery-used-afghanistan-3283.

8 Bill Lisbon, "GPS-Guided Paradrop," *Special Operations Technology* 2(6) (2004):
29–31; Tom Vanden Brook, "'Smart' Airdrops May Save Lives of U.S. Troops," *USA
Today*, January 8, 2007; Phil Hattis and Steve Tavan, "Precision Airdrop from High
Altitude," *Aerospace America* 45(4) (April 2007): 38–42.

9 Jason I. Thompson, "A Three Dimensional Helmet Mounted Primary Flight Reference for Paratroopers" (master's thesis, Air Force Institute of Technology,
2005), 1–8; "Marine Corps Picks Rockwell Collins for Sky-Diving Nav," *GPS World*
(February 13, 2008), accessed March 16, 2011, http://www.gpsworld.com/defense/
news/marine-corps-picks-rockwell-collins-sky-diving-nav-3278.

10 "UAV Annual Report FY 1997," Defense Airborne Reconnaissance Office, November 6, 1997, accessed March 16, 2011, http://www.fas.org/irp/agency/daro/uav97/
toc.html; John Pike, "RQ-1 Predator MAE UAV," November 6, 2002, accessed
March 16, 2011, http://www.fas.org/irp/program/collect/predator.htm; Ken Jordan, "REMUS AUV Plays Key Role in Iraq War," *UnderWater Magazine* (July/August
2003), accessed December 9, 2008, http://www.underwater.com/archives/arch/
034.07.shtml.

11 James Madeiros, "New Tracking System Keeps Eye on Troops," *Air Force Link*,
October 25, 2001, accessed October 26, 2001, http://www.af.mil/news/Oct2001/
n20011025 1525.shtml; Marty Whitford, "Friend or Foe? FBCB2 Enhances Battle
Planning, Reduces 'Friendly Fire,'" *GPS World* (February 1, 2005), accessed March
16, 2011, http://www.gpsworld.com/defense/security-surveillance/friend-or-foe-
940; "Comtech Telecommunications Corp. Receives US\$3.2 Million of Movement

Tracking Systems Orders," Space Newsfeed, December 9, 2008, accessed March 16, 2011, http://www.spacenewsfeed.co.uk/2008/14_12_2008/14December2008_14 .html; "Globecomm Systems Receives Additional Contract Modification from NATO Valued at US$7.3 Million for GPS-Based Force Tracking System," Space Newsfeed, December 9, 2008, accessed March 16, 2011, http://www.spacenews feed.co.uk/2008/14_12_2008/14December2008_13.html.

12 Clement Driscoll, "What Do Consumers Really Think?" *GPS World* (July 1, 2002), accessed March 16, 2011, http://www.gpsworld.com/lbs/in-vehicle-services/what-do-consumers-really-think-755?page_id=2; Rafe Needleman, "Navigation PDAs Look for Their Market," *Business 2.0* (June 19, 2003), accessed October 24, 2003, http://www.business2.com/subscribers/articles/web/0,1653,50392,00.html?cnn= yes; "Nowhere to Hide: GPS Is Spreading," *Business 2.0* (March 2002), accessed October 24, 2003, http://www.business2.com/subscribers/articles/mag/0,1640, 37777,00.html?cnn=yes; Office for Outer Space Affairs of the Secretariat, United Nations, "Satellite Navigation and Location Systems," paper presented at the Third United Nations Conference on the Exploration and Peaceful Uses of Outer Space, Vienna, July 19–20, 1999, accessed March 16, 2011, http://web.rosa.ro/com-mon/documents/UNBP/bp4_navigation.pdf; Ed White, "USAF Space Command's GPS Operation Supports Critical Aspects of National and International Life," *GPS Daily*, July 10, 2008, accessed March 16, 2011, http://www.gpsdaily.com/reports/ USAF_Space_Command_GPS_Operation_Supports_Critical_Aspects_Of_ National_And_International_Life_999.html.

13 W. Eric Martin, "Mobilizing the Fleet," *Mobile Government* (May 2003): 6–9; Karen Van Dyke, "GPS Applications in United States Transit," Asia-Pacific Economic Cooperation Innovation Summit, May 26–27, 2008, accessed March 16, 2011, http://pnt.gov/public/2008/2008–05-APEC/vandyke-transit.pdf; Glen Gibbons et al., "Automatic Vehicle Location: GPS Meets IVHS," *GPS World* 4(4) (April 1993): 22–26; Clement Driscoll, "Finding the Fleet: Vehicle Location Systems and Tech-nologies," *GPS World* 5(4) (April 1994): 66–70.

14 "Rocky Mountain Tracking Is Saving Businesses Money," May 18, 2006, accessed March 16, 2011, http://www.rmtracking.com/press,html; Associated Press, "Track-ing Devices in Official Vehicles Help Cut Waste," *Colorado Springs (Colo.) Gazette*, November 16, 2007; Stephen Colwell, "Go Green with GPS," *GPS World* 19(10) (October 2008), 30–31; Mike Gibney, "Orkin Pest Control Cuts Crash Rate by 40% with GPS Monitoring," accessed March 16, 2011, http://www.safeteendrivingclub .org/reading_article.php?ID=22.

15 Lara Brenckle, "GPS Units Speed Emergency Response," *Patriot-News*, February 24, 2008, accessed March 16, 2011, http://www.pennlive.com/midstate/index.ssf/ 2008/02/gps_units_help_save_lives.html; Stephen C. Brown et al., "GIS and GPS Emergency Response Lessons Learned from the Space Shuttle Columbia Disas-ter," *Journal of Extension* 41(4) (August 2003), accessed March 16, 2011, http://www.joe.org/joe/2003august/iw1.shtml; Staff Writers, "North Las Vegas Fire Department Launches Enterprise GIS Platform," *GPS Daily*, October 31, 2008, accessed March 16, 2011, http://www.gpsdaily.com/reports/North_Las_Vegas_ Fire_Department_Launches_Enterprise_GIS_Platform_999.html; "HARMLESS Demonstrates GNSS Benefits to Emergency Response," *GPS World* (October 3, 2008), accessed March 16, 2011, http://www.gpsworld.com/lbs/news/harmless-demonstrates-gnss-benefits-emergency-response-2383.

16 Aberdeen Group, *The Impact of Location on Field Service* (December 2007), 15, accessed March 16, 2011, http://www.actsoft.com/news/research/Aberdeen-Report-Impact-of-Location-on-Field-Service-Report.pdf.

17 "Salt Lake City Light Rail Enhances Customer Service for Olympic Visitors: Geo-Focus LLC Installs GPS Tracking and Information System for UTA TRAX System," *Business Wire*, February 6, 2002, accessed March 16, 2011, http://www.allbusiness .com/technology/software-services-applications-information/5897047–1.html; William Vantuono, "BNSF Starts Positive Train Control Trial: North American Viewpoint," *International Railway Journal* (March 2004), accessed March 16, 2011, http://findarticles.com/p/articles/mi_moBQQ/is_/ai_114629906?tag=artBody; col1; Paul Rincon, "UK Rail Chiefs Eye Sat-Navigation," BBC News, February 11, 2005, accessed March 16, 2011, http://news.bbc.co.uk/1/hi/sci/tech/4247721.stm; Brian Caine, Lockheed Martin Corporation, "Rail Systems: Improving Efficiency and Safety in the Rail Industry," July 2008, accessed March 16, 2011, http:// www.designnews.com/file/2152-Lockheed_Martin_PTC.pdf; Grady C. Cothen Jr. et al., "Positive Train Control—Ready to Go?" *Mass Transit Magazine*, December 2007/January 2008, accessed March 16, 2011, http://www.masstransitmag.com/ publication/article.jsp?pubId=1&id=4990&pageNum=2; Alfonso A. Castillo, "Metrolink Crash Sparks Look at GPS to Prevent Accidents," *Newsday*, September 17, 2008, accessed March 16, 2011, https://www.ble-t.org/pr/news/pf_headline .asp?id=23666.

18 Washington State Department of Transportation, "San Juan Islands Vessel Watch," http://www.wsdot.wa.gov/ferries/ accessed December 22, 2008; B.C. Ferries, "Vessel Positions," accessed March 16, 2011, http://orca.bcferries.com:8080/ cc/conditions/maps.asp; "NY Waterway Installs GPS-based Tracking System," October 14, 2003, accessed March 16, 2011, http://www.marinelog.com/DOCS/ NEWSMMIII/MMIIIOct14b.html; Federal Transit Administration, "Intelligent Transportation Systems: Core Technologies for Ferry Boat Transit Agencies," September 2007, accessed March 16, 2011, http://www.pcb.its.dot.gov/factsheets/ modal/modFer.pdf; "New GPS Tracking of Passenger Ferries," Reuters, August 31, 2008, accessed March 16, 2011, http://www.reuters.com/article/prressRelease/ idUS59707+31-Aug-2008+MW20080831.

19 "Global Positioning System—Serving the World," National Space-Based PNT Office, accessed March 16, 2011, http://www.gps.gov/applications/aviation/index .html; "Norway Opts for SCAT-1 Airports Following Successful Landing," *GPS World* (November 14, 2007), accessed March 16, 2011, http://www.gpsworld.com/ transportation/news/norway-opts-scat-1-airports-following-successful-landing-6694; Northrop Grumman Corporation, "Northrop Grumman Provides Airport Ground Station for First Satellite-Based Landing System," November 22, 2007, accessed March 16, 2011, http://www.energy-daily.com/reports/Northrop_ Grumman_Provides_Airport_Ground_Station_For_First_Satellite_Based_ Landing_System_999.html; "GPS/WAAS Approaches Outnumber ILS at Airports," *GPS World* 19(12) (December 2008): 29; "Newark Airport First Hub to Test Satellite System," *USA Today*, December 19, 2008.

20 "Ocean Surface Topography from Space," Jet Propulsion Laboratory, July 30, 2008, accessed December 9, 2008, http://topex-www.jpl.nasa.gov/technology/ instrument-gps.html; "Jules Verne Using GPS to Get Around—in Space," *GPS World* (April 2, 2008), accessed March 16, 2011, http://www.gpsworld.com/ defense/news/jules-verne-using-gps-get-around-ampndash-space-3287; Thomas Grelier et al., "GNSS in Space: Formation Flying Radio Frequency Missions, Techniques, and Technology," *Inside GNSS* 3(8) (November/December 2008): 40–45.

21 Krista Stevens, "GPS Fore the Golf Course," *Point of Beginning*, April 30, 2002, accessed March 16, 2011, http://www.pobonline.com/CDA/Archives/6c07817cac0f 6010VgnVCM100000f932a8c0.

22 Doug Pike, "High Tech Takes a Big Step Forward," *Houston Chronicle*, October 18, 2005, accessed August 30, 2006, http://www.uplinkgolf.com/cms/website/medi-acoverage/HoustonChron10_18_05.pdf; Roger Graves, "From the Dailies—Golf Cars Drive Up Course Revenues with New Models, Technology," *PGA Magazine*, January 26, 2006, accessed August 30, 2006, http://www.pgamagazine.com/article.aspxUS$id=2750; Tim Schooley, "Tracking a New Market: Global Position-ing Systems Find Their Way into Golf Carts, Providing Much More Than Distance Readings," *Pittsburgh Business Times*, June 12, 2006, accessed March 16, 2011, http://pittsburgh.bizjournals.com/pittsburgh/stories/2006/06/12/focus1.html; "ProLink Sees Record Year for Golf GPS System Installation," *GPS World* (January 21, 2008), accessed March 16, 2011, http://www.gpsworld.com/consumer-oem/news/prolink-sees-record-year-golf-gps-system-installation-2780; "Golf GPS Systems for Player Enhancement and Course Management," Golf GPS Systems, accessed March 16, 2011, http://www.gps-practice-and-fun.com/golf-gps.html.

23 Dan Koeppel, "Reasons You Need a GPS," *Bicycling*, accessed March 16, 2011, http://www.bicycling.com/beginners/bikes-gear/reasons-you-need-gps; Scott Morris, "Paula and Scott's 2004 Ride along the Great Divide Mountain Bike Route," accessed March 16, 2011, http://www.topofusion.com/divide/; "Competi-tors in the Tour de France Tracked by Satellite," ESA Navigation, July 7, 2005, accessed March 16, 2011, http://www.esa.int/esaNA/SEM8176DIAE_index_2.html; Mark Schoofs, "Running with the Satellites," *Wall Street Journal*, December 1, 2007; Staff Writers, "New App for AT&T and GPS-Enabled Mobile Phones," *GPS Daily*, November 6, 2008, accessed March 16, 2011, http://www.gpsdaily.com/reports/New_App_For_AT_And_T_GPS_Enabled_Mobile_Phones_999.html.

24 Angie Carter, "GPS Technology and Fishing," accessed December 9, 2008, http://CartersGPS.com; Faye Spencer, "Using GPS for Fishing!," accessed March 16, 2011, http://searchwarp.com/swa29232.htm.

25 Rhonda Percell, "My 3 Favorite Things to Do with My Hunting GPS," accessed December 9, 2008, http://www.goarticles.com/cgi-bin/showa.cgi?C=1233395.

26 "Birding GPS and Geotagging Photos," accessed March 16, 2011, http://www.gps-central.ca/gps-birding-gps.htm.

27 Jack W. Peters, *The Complete Idiot's Guide to Geocaching* (East Rutherford, N.J.: Alpha Books, 2004); Layne Cameron, *The Geocaching Handbook* (Guilford, Conn.: Falcon, 2004).

28 "Geoteaming—High-Tech Team Building Using GPS Technology," PlayTime, Inc., accessed March 16, 2011, http://www.playtimeinc.com/products.htm; "GPS Geo-cache Challenge," American Outback Adventures, accessed March 16, 2011, http://www.americanoutback.net/team_geocaching.php; "Geocaching," Alaska Vacation Store, accessed March 16, 2011, http://www.alaskavacationstore.com/Geocaching.html.

29 Staff Writers, "A Whole New Approach to Modern Art," *GPS Daily*, August 15, 2008, accessed August 18, 2008, http://www.gpsdaily.com/reports/A_Whole_New_Approach_To_Modern_Art_999.html.

30 "Geocaching," Wikipedia, accessed March 16, 2011, http://en.wikipedia.org/wiki/Geocaching; "History of Geocaching," accessed March 16, 2011, http://www.geocacheguide.com/history_of_geocaching.html; "Geocaching—The Official Global GPS Cache Hunt Site," accessed March 16, 2011, http://www.geocaching.com/.

31 Mark W. Brown, "GPS Helps Farmers Reap Higher Crop Yields," *Space Trace* (November 1992): 4.

32 "GPS Applications," MSU Global Positioning System (GPS) Laboratory, May 20, 2003, accessed March 16, 2011, http://www.montana.edu/gps/index.html; Robert "Bobby" Grisso et al., "Precision Farming Tools: Global Positioning System (GPS)," July 2003, Virginia Cooperative Extension, accessed March 16, 2011, http://www. ext.vt.edu/pubs/bse/442-503/442-503.html.

33 Carolyn Peterson, "Precision GPS Navigation for Improving Agricultural Productivity," *GPS World* 2(1) (January 1991): 38–44; William D. Bauer and Mitch Schefcik, "Using Differential GPS to Improve Crop Yields," *GPS World* 5(2) (February 1994): 38–41; Tracy M. Blackmer and James S. Schepers, "Using DGPS to Improve Corn Production and Water Quality," *GPS World* 7(3) (March 1996): 44–52; Jim Rich et al., "Using GIS/GPS for Variable-Rate Nematicide Application in Row Crops," University of Florida IFAS Extension, accessed August 21, 2006, http:// edis.ifas.ufl.edu/IN464.

34 Hale Montgomery, "GPS Down on the Farm," *GPS World* 5(9) (September 1994): 18; Rich et al., "Using GIS/GPS for Variable-Rate Nematicide Application"; Associated Press, "GPS Goes Down on the Farm: Technologies Can Help Some Farms Cut Costs," November 29, 2004, MSNBC, accessed March 16, 2011, http://www.msnbc .msn.com/id/6608881.

35 Mike W. Sampson, "Getting the Bugs Out: GPS-Guided Aerial Spraying," *GPS World* 4(9) (September 1993): 28–34. Mike W. Sampson, "No Small Affair: DGPS Battle the California Medfly," *GPS World* 6(2) (February 1995): 32–38.

36 Eric Gakstatter, "Business Outlook—RTK Crops Up in Precision Ag," *GPS World* (May 1, 2008), accessed March 16, 2011, http://www.gpsworld.com/machine-control-ag/precision-ag/news/business-outlook-rtk-crops-up-precision-ag-3630.

37 Christian Rocken et al., "Atmospheric Water Vapor and Geoid Measurements in the Open Ocean with GPS," *Geophysical Research Letters*, January 27, 2005, accessed March 16, 2011, http://www.cosmic.ucar.edu/~rocken/GRL_2005_ explorer_color.pdf; Paul Poli et al., "Weather Report: Meteorological Applications of GNSS from Space and on the Ground," *Inside GNSS* 3(8) (November/December 2008): 1–39; E. Bonelli, "Attenuation of GPS Scintillation in Brazil Due to Magnetic Storms," *Space Weather* 5(4) (2008): 23–31.

38 Jack O'Neill et al., "Using GPS and GIS Analysis to Investigate the Effects of Industrial Development on an Endangered Species," June 2, 2000, accessed March 16, 2011, http://gis.esri.com/library/userconf/proc00/professional/papers/PAP116/ p116.htm; John Toon, "Technology Boosts Efforts to Protect Mountain Gorillas, Rebuild Rwandan Economy," *Whistle* 24(16) (April 24, 2000), accessed March 16, 2011, http://www.whistle.gatech.edu/archives/00/april/24/; Stephen Blake et al., "GPS Telemetry of Forest Elephants in Central Africa: Results of a Preliminary Study," *Journal of African Ecology* 39(2) (June 2001): 178–186; Jake Wall et al., "Elephants Avoid Costly Mountaineering," *Current Biology* 16(14) (July 25, 2006): R527–529; Katharine Houreld, "Kenya's Elephants Send Text Messages to Rangers," *Mobile Tech Today*, October 14, 2008, accessed March 16, 2011, http:// www.physorg.com/news142961636.html; "GPS: Saving the World One Elephant at a Time," *GPS World* (October 17, 2008), accessed March 16, 2011, http://sc. gpsworld.com/gpssc/Natural+Resources/GPS-Saving-the-World-One-Elephant-at-a-Time/ArticleStandard/Article/detail/560271; "GPS Satellite Tracking for First Wild Release of Orphaned Devils, 6 December 2007," Tasmanian Devil Conservation Park, accessed March 16, 2011, http://www.tasmaniandevilpark. com/news.html; Tracy Cozzens, "A Devil of a Program," *GPS World* (February 1, 2008), accessed March 16, 2011, http://www.gpsworld.com/gps/seen-heard-a-devil-a-program-1172; "Addressing Climate Change: Research on the Polar Bears of Southern Hudson Bay," Ontario Ministry of Natural Resources Website, July 19, 2007, accessed March 16, 2011, http://www.mnr.gov.on.ca/en/Newsroom/Latest News/MNR_E004159.html.

39 "New Map Helps Protect Ten Million Acres of Rainforest," Amazon Conservation Team, February 25, 2002, accessed March 16, 2011, http://www.moore.org/init-newsitem.aspx?id=522; Rhett A. Butler, "Amazon Conservation Team Puts Indians on Google Earth to Save the Amazon: Amazon Natives Use Google Earth, GPS to Protect Rainforest Home," November 14, 2006, accessed March 16, 2011, http://news.mongabay.com/2006/1114-google_earth-act.html; Tracy Cozzens, "GPS a Big Hit with Rainforest Pygmies," *GPS World* (September 1, 2007), accessed March 16, 2011, http://www.gpsworld.com/gps/seen-heard-gps-a-big-hit-with-rainforest-pygmies-1157.

40 "Summary of Monitoring Data (1970–May 2005)," Hawaiian Volcano Observatory, June 22, 2005, accessed March 16, 2011, http://hvo.wr.usgs.gov/maunaloa/current/longterm.html; Janice Partyka, "Using GPS to Track Tectonic Plates," *GPS World* (October 14, 2008), accessed March 16, 2011, http://www.gpsworld.com/consumer-oem/uampc-insights-october-2008-7220.

41 Edward J. Krakiwsky and Donald Chalmers, "Emergency Response with GPS in Oil and Gas Fields," *GPS World* 4(4) (April 1993): 48–51.

42 Margaret Ferrentino, "Code Red: GPS and Emergency Medical Response," *GPS World* 5(4) (April 1994): 28–38.

43 Reuters, "Satellites Help Slash Karachi Car Thefts, Kidnaps," CNN.com/Technology, October 23, 2003, accessed March 16, 2011, http://www.infowar-monitor.net/2003/10/satellites-help-slash-karachi-car-thefts-kidnaps/; Anslee Willett, "Global Satellite System Tracks Down Suspect in Theft of Taxicab," *Colorado Sprints (Colo.) Gazette*, January 18, 2002; Marty Whitford, "Thief Relief: GPS/Cellular Combo Acquires Abducted Assets," *GPS World* (October 1, 2004), accessed March 16, 2011, http://www.gpsworld.com/transportation/fleet-tracking/thief-relief-923?page_id=3; "GEOTrac/Iridium System Netting Canadian Oil-Field Thieves," *GPS World* (June 20, 2007), accessed March 16, 2011, http://www.gpsworld.com/transportation/news/geotraciridium-system-netting-canadian-oil-field-thieves-6639.

44 Adam Owens, "GPS Program Tracks Convicted Sex Offenders," January 30, 2008, accessed March 16, 2011, http://www.wral.com/news/local/story/2371727/?print_friendly=1; "Oft-Burglarized Store Hits Back with Hidden GPS," WRTV Indianapolis, September 24, 2007, accessed March 16, 2011, http://www.theindychannel.com/print/14195732/detail.html; Perry Swanson, "Cell Phone Aids Captive," *Colorado Springs (Colo.) Gazette*, May 11, 2008; Staff Writers, "Sprint Customers Use GPS to Locate Loved Ones," *GPS Daily*, October 31, 2008, accessed March 16, 2011, http://www.gpsdaily.com/reports/Sprint_Customers_Use_GPS_To_Locate_Loved_Ones_999.html; Staff Writers, "University Adds Personal Mobile Alarm System to Campus Safety Program," *GPS Daily*, December 10, 2008, accessed March 16, 2011, http://www.gpsdaily.com/reports/University_Adds_Personal_Mobile_Alarm_System_To_Campus_Safety_Program_999.html; "GPS vs. Radar: It's Not Over Yet," *GPS World* (July 18, 2008), accessed March 16, 2011, http://www.gpsworld.com/gnss-system/news/gps-vs-radar-it039s-not-over-yet-4511; L. A. Carter, "Teen Tries GPS Defense to Fight Speeding Ticket," *Santa Rosa (Calif.) Press Democrat*, July 12, 2008, accessed March 16, 2011, http://www1.pressdemocrat.com/article/20080712/NEWS/807120355; Carol J. Williams, "Spies Among Us: Tracking Devices Can Implicate, Acquit," *Colorado Springs (Colo.) Gazette*, November 26, 2008.

45 Tapas Kumar Ghatak, "Micro Level Utility Mapping for Health and Social Infrastructure by GPS Survey," ca. 2005, accessed March 17, 2011, http://www.gis development.net/application/health/overview/grao69pf.htm; Andrew G. Roe, "Collaboration Yields Cleaner Water," February 5, 2008, accessed March 17, 2011, http://www.cadalyst.com/gis/collaboration-yields-cleaner-water-9237; Kenneth Wong, "Fight the Bite, Track the Truck," August 21, 2007, accessed March 17, 2011, http://www.cadalyst.com/gis/fight-bite-track-truck-9200; Cyrena Respini-Irwin, "Network in the Sky," *Geospatial Solutions*, January 6, 2008, accessed January 7, 2008, http://www.geospatial-solutions.com/geospatialsolutions/content/print/ ContentPopup.jsp?id=482097, subsequently accessed March 17, 2011, http:// chromatographyonline.findanalytichem.com/lcgc/article/articleDetail.jsp?id=48 2097; "GPS Found Its Way into Telemedicine," accessed March 17, 2011, http:// www.gps-practice-and-fun.com/telemedicine.html.

46 Harry Bottorff, "Rapid Relief: GPS Helps Assess Mississippi River Flood Damage," *GPS World* 5(3) (March 1994): 22–26.

47 John Rendleman, "GPS Aids Recovery Effort," *InformationWeek*, November 12, 2001, accessed March 17, 2011, http://www.informationweek.com/showArticle .jhtml;jsessionid=YH2WRBQHCAKCCQSNDLPSKH0CJUNN2JVN?articleID= 6507967; Raymond J. Menard and Jocelyn L. Knieff, "GPS at Ground Zero: Tracking World Trade Center Recovery," *GPS World* (September 2, 2002), accessed March 17, 2011, http://www.gpsworld.com/transportation/fleet-tracking/gps-ground-zero-761.

48 Marty Whitford, "Hurricane Hunters: GPS Dropsondes Trace Katrina's Course," *GPS World* (October 1, 2005), accessed March 17, 2011, http://www.gpsworld.com/ government/emergency-response/hurricane-hunters-990; Bob Henson, ed., "Hurricane!" *SN Monthly* (October 1998), accessed March 17, 2011, http://www. ucar.edu/communications/staffnotes/9810/hurricane.html; "The Nationwide Differential Global Positioning System's Role with Hurricane Katrina's Recovery," U.S. Coast Guard Navigation Center, January 2006, accessed December 9, 2008, http://www.navcenter.org/misc/HurricaneKatrina.pdf.

49 "BT Launches Global Disaster Relief Program in Partnership with the British Red Cross," CSRwire, June 26, 2007, accessed March 17, 2011, http://www.csrwire.com/ News/9023.html; "Relief.Asia Deploys GPS Units for Relief Teams in China," Relief.Asia, June 17, 2008, accessed March 17, 2011, http://www.dotasia.org/press releases/DotAsia-PR-2008-06-17_EN.pdf.

50 Bryn Fosburgh and Joseph V. R. Paiva, "Surveying with GPS," accessed December 20, 2008, http://www.acsm.net/session01/surveygps.pdf; Matthew B. Higgins, "Guidelines for GPS Surveying in Australia," accessed March 17, 2011, https:// www.fig.net/pub/proceedings/Korea/full-papers/pdf/ws_com5_1/higgins.pdf; George OnerOgalo, "GPS in Cadastres: A Case Study of Kenya," paper presented at the FIG XXII International Congress, Washington, D.C., April 19–26, 2002; "Global Positioning System—Serving the World," National Space-Based PNT Office, accessed March 17, 2011, http://www.gps.gov.

51 Ruth Neilan, "Expert Advice: Reference Frame for Africa," *GPS World* (September 1, 2008), accessed March 17, 2011, http://www.gpsworld.com/gnss-system/expert-advice-reference-frame-africa-4250.

52 David R. Baker, "Global MapAid Seeks Clearer Disaster Maps," *San Francisco Chronicle*, January 17, 2005.

53 Global MapAid, "Live Projects," accessed March 17, 2011, http://www.globalmap aid.org/project.htm; Anton de Witt et al., "Global MapAid Street Children

Pilot-Survey, Port Elizabeth, South Africa," August 2008, accessed March 17, 2011, http://globalmapaid.org/project.htm.

54 Thegn L. Ladefoged et al., "Integration of Global Positioning Systems into Archaeological Field Research: A Case Study from North Kohala, Hawai'i Island," *SAA Bulletin* 16(1), accessed March 17, 2011, http://www.saa.org/Portals/0/SAA/publications/SAAbulletin/16–1/SAA17.html; M. Crespi et al., "The Archaeological Information System of the Underground of Rome: A Challenging Proposal for the Next Future," paper presented at the CIPA 2005 XX International Symposium, Torino, Italy, September 26–October 1, 2005; Steven M. Dinaso et al., "Total Station and GPS Methodologies Aid Surface Mapping of Archaeological Site," *Point of Beginning* (August 2004): 26–27; Larry Murphy, "Shipwrecks, Satellites, and Computers: An Underwater Inventory of Our National Parks," *Common Ground* 1(3/4) (Fall/Winter 1996), accessed March 17, 2011, http://www.nps.gov/archeology/Cg/vol1_num3-4/satellites.htm; Colin Clement, "Mapping the Treasures," NOVA Online, accessed March 17, 2011, http://www.pbs.org/wgbh/nova/sunken/mapping.html; Jason Burns and Michael Krivor, "Underwater Archaeology Invasion Beaches Survey 2008," Museum of Underwater Archaeology, April 16, 2008, accessed March 17, 2011, http://www.uri.edu/artsci/his/mua/project_journals/saipan/saipan_intro.shtml.

55 Peter H. Dana and Bruce M. Penrod, "The Role of GPS in Precise Time and Frequency Dissemination," *GPS World* 1(4) (July/August 1990): 38–43; Harold Hough, "A GPS Precise Timing Sampler," *GPS World* 2(9) (October 1991): 33–36; Edgar W. Butterline, "Reach Out and Time Someone," *GPS World* 4(1) (January 1993): 32–40; Hewlett-Packard Company, "GPS and Precision Timing Applications," 1996, accessed December 9, 2011, http://www.fgg.uni-lj.si/~/mkuhar/Zalozba/HP1272.pdf, subsequently accessed March 17, 2011, http://www.agilent.com/metrology/pdf/AN1272.pdf.

56 Kenneth E. Martin, "Powerful Connections: High-Energy Transmission with High-Precision GPS Time," *GPS World* 7(3) (March 1996): 20–36; Dennis C. Erickson and Carson Taylor, "Pacify the Power: GPS Harness for Large-Area Electrical Grid," *GPS World* (April 1, 2005), accessed March 17, 2011, http://www.gpsworld.com/government/pacify-power-954.

57 Trimble Navigation Ltd., "Trimble Launches GPS Clock for Micro-Second Synchronization of Network Computers and Internet Applications," *Directions Magazine*, July 27, 1999, accessed December 9, 2008, http://www.directionsmag.com/press.releases/index.php?duty=Show&id=742&trv=1, subsequently accessed March 17, 2011, http://www.trimble.com/news/release.aspx?id=072799a; "Trimble Offers Internet GPS Time Standard," *Space Daily*, August 3, 1999, accessed March 17, 2011, http://www.spacedaily.com/news/gps-99j.html; "Acutime 2000 GPS Smart Antenna for Precise Timing and Synchronization," Trimble Navigation Ltd., 2000, accessed March 17, 2011, http://www.trimble.com/products/pdf/acu2000b.pdf; Alan Cameron, "Billions per Second: Timing Financial Transactions," *GPS World* (July 31, 2002), accessed March 17, 2011, http://www.gpsworld.com/wireless/timing/billions-second-760; "GPS Tutorial: Putting GPS to Work—Timing," Trimble Navigation Ltd. (2006), accessed March 17, 2011, http://www.trimble.com/gps/gpswork-timing.shtml.

58 Scott Pace et al., *The Global Positioning System: Assessing National Policies* (Santa Monica, Calif.: RAND, 1995), 114–118; National Academy of Public Administration, *The Global Positioning System: Charting the Future* (Washington, D.C.: NAPA, 1995), 62–68; Scott Pace and James E. Wilson, *Global Positioning System: Market Projections and Trends in the Newest Global Information Utility*, U.S. Department of Com-

merce, September 1998, accessed March 17, 2011, http://www.space.commerce
.gov/library/reports/1998-09-gps.pdf; Frost & Sullivan, "Price Points, Product
Integration Fuel GPS Revenue Growth," *Directions Magazine*, April 26, 2000,
accessed March 17, 2011, http://www.directionsmag.com/pressreleases/price-
points-product-integration-fuel-gps-revenue-growth/97622; Futron Corporation,
Trends in Space Commerce, U.S. Department of Commerce, May 2001, accessed
March 17, 2011, http://www.space.commerce.gov/library/reports/2001–06-trends
.pdf.

59 "A White-Hot Market for Consumer GPS," *GPS World* (April 1, 2007), accessed
April 9, 2011, http://cp.gpsworld.com/gpscp/content/printContentPopup.jsp?id=
416714, subsequently accessed March 17, 2011, http://spectroscopyonline.find
analytichem.com/spectroscopy/article/articleDetail.jsp?id=416714&sk=&
date=&pageID=3; "GPS Applications to Hit 900M Units by 2013, Says ABI," *GPS
World* (January 16, 2008), accessed March 17, 2011, http://www.gpsworld.com/
consumer-oem/news/gps-applications-hit-900m-units-2013-says-abi-2774;
"North American Telematics Market to Hit US$6.47B," *GPS World* (March 14,
2008), accessed March 17, 2011, http://www.gpsworld.com/transportation/news/
north-american-telematics-market-hit-647b-6746; "Mobile Nav Subscribers to
Number 70M by 2014," *GPS World* (October 23, 2008),accessed March 17, 2011,
http://www.gpsworld.com/lbs/news/mobile-nav-subscribers-number-70m-
2014–2406; "ION: GNSS Market to Grow to US$6B to US$8B by 2012," *GPS World*
(September 19, 2008), accessed March 17, 2011, http://www.gpsworld.com/
survey/news/ion-gnss-market-grow-6b-8b-2012–3677.

60 Dana J. Johnson, "Overcoming Challenges to Transformational Space Programs:
The Global Positioning System (GPS)," October 2006, accessed March 17, 2011,
http://www.northropgrumman.com/analysis-center/paper/assets/Overcoming-
Challenges-to-Trans.pdf.

61 Ventana Research, "GPS and Related Technology: Trends and Strategies for Location Intelligence," September 2006.

62 Bradford W. Parkinson, "Overview," in *Global Position System: Papers Published in
Navigation*, Vol. 1, ed. P. M. Janiczek (Washington, D.C.: Institute of Navigation,
1980), 1–2.

6

SATELLITES, OIL, AND FOOTPRINTS

EUTELSAT, KAZSAT, AND POST-COMMUNIST TERRITORIES IN CENTRAL ASIA

LISA PARKS

In their study of energy, water, and telecommunications systems, Stephen Graham and Simon Marvin note, "infrastructure networks and the socio-technical processes that surround them are strongly involved in structuring and delineating the experiences of urban culture and what Raymond Williams termed the 'structures of feeling' of modern urban life." Furthermore, they write, "As capital that is literally 'sunk' and embedded within and between the fabrics of cities, [infrastructure networks] represent long-term accumulations of finance, technology, know-how, [and] organizational and geopolitical power."[1] Rather than examine infrastructures that are "sunk" beneath Earth's surface, in this chapter I consider the accumulations of finance, technology, know-how, and organizational and geopolitical power that orbit Earth. More specifically, I explore how satellites and their footprints have played a role in reorganizing the economic and cultural flows across the post-Communist territories of Central Asia in recent years.[2]

Central Asia underwent major political and economic transformations after the breakup of the Soviet Union in 1991. Once in the orbit of Soviet control, republics or countries from Kazakhstan to Mongolia and from Ukraine to Bulgaria experienced democratization campaigns, gradually liberalized their economies, and began to integrate within the consumer cultures of Western Europe and the United States. In the midst of these political and economic transitions, foreign investors have also scrambled to access Central

Asia's natural resource wealth. Western corporations began investing in the region in the early 1990s and have dramatically expanded operations since then in an effort to plumb the region's oil, gas, metal, and mineral deposits. Central Asia has also gained geopolitical importance in recent years as the United States has attempted to reposition itself strategically in relation to Russia and China, and has maneuvered to stake out new positions in the region in the context of the war on terror.

These historical conditions set the backdrop for satellite developments and orbital slot assignments in Central Asia since the Soviet Union's dissolution. As international communications scholar Monroe Price has argued, there is a "need for a richer understanding of the geopolitical, economic, and technical factors that determine who controls which orbital slots, what satellites gain access to those slots, and what program services are actually carried." He continues, "an understanding of infrastructure—the way satellite routes come into being and are regulated—is necessary for an assessment of the consequences of information and entertainment flows. One way to begin this process is to examine some examples—small case studies—of the differentiation of satellite patterns."[3] Following Price's suggestion, I examine the strategies of two very different satellite operators doing business across post-Communist territories of Central Asia—Eutelsat and Kazakhstan. My research on this topic began when I attended the Caspian Telecoms Conference in Istanbul in April 2007, where representatives from global telecom corporations and state officials from new nations of the Caspian region and Central Asia, including representatives of Eutelsat and Kazakhstan, met to discuss the future of the region's telecommunications infrastructure. The contrasting positions of Eutelsat (one of the world's largest and most established satellite operators, headquartered in France) and Kazakhstan (one of the smallest and newest satellite operators) were especially intriguing. They provide a unique opportunity to consider the relationship between satellites and post-Communism.

While Eutelsat has expanded into this region to cultivate new markets in what the company refers to as "extended Europe," Kazsat was launched primarily to offer satellite services to regional clients. The satellites of both of these operators have been used to circulate broadcast and telecommunication signals, facilitate flows of finance and capital, and reshape geographic imaginaries. While satellites have bolstered the broadcast and telecom sectors of the region, it is important to note that satellite operators have also provided orbital platforms for the Caspian's booming oil industry. Satellites are used to support everything from surveying oil fields to monitoring drilling operations, from constructing oil rigs to maintaining pipelines. Since the same satellites can be used to support very different industries, it is important to develop an analytical model that can account for the more "cultural" uses of satellites (such as broadcasting) *in relation to* their more "extractive" uses (such as natural resource development).

In an effort to develop such a model, I concentrate on the *footprint*—the territorial boundary in which a signal from a given satellite can be received. It is within this zone that the multifarious uses of a satellite can become intelligible and therefore can be described and analyzed. In this essay I provide a preliminary analysis of the footprints of Eutelsat's W2, W4, and Sesat 1, and Kazakhstan's Kazsat 1, all of which began operating in Central Asia after the fall of the Soviet Union. To do so, I develop an approach called "footprint analysis," which blends critical approaches from media and cultural studies, cultural geography, critical geopolitics, and science and technology studies, and attempts to establish a field of inquiry between orbit and the ground, cartography and ethnography, spectral and social space. It is a critical practice that emphasizes the material and territorializing effects of satellite technologies and involves examining footprint maps, foregrounding the eclectic ways in which satellites are developed and used, and studying the political implications of such uses. Footprint analysis assumes that the satellite is situated within a field of power relations and that, because of this, its uses have territorializing effects. As I have suggested elsewhere, satellite footprints are not inert technical boundaries; rather, they "represent the power to transform, redefine and hybridize nations, territories, and cultures in a most material way."[4] Here, I want to think about footprints specifically in relation to post-Communist conditions in Central Asia, and especially in relation to new countries with vast territories, limited terrestrial infrastructures, and natural resource wealth. Kazakhstan, for instance, has rapidly privatized its resource-rich landscapes, has launched its own satellite and Internet services, and is actively participating in regional and global economies. While major reforms have occurred in such post-Communist countries, we still know little about the role satellites have played in relation to these transitions. The chapter begins with a discussion of footprint analysis as a conceptual framework and moves on to discuss satellite footprints in Central Asia.

Footprint Analysis and Geopolitics

Footprint analysis is a critical practice that investigates the territorial, infrastructural, and geopolitical aspects of satellite use. The term "footprint" is invoked here in two ways. First, a footprint is a mapped geographic boundary that designates where on Earth a signal from a given satellite can be received. For most communication satellites, footprint maps are available that reveal the zones of coverage from that satellite's orbital position. Such maps are useful in that they illustrate the domain of potential transmission across a given territory, but they are ultimately insufficient because they do not specify the infrastructures or processes on the ground that enable signal distribution, identify the organizations that own and operate these systems, or provide information about the way various actors negotiate the effects of satellite use across different locales.

Second, a footprint is conceptualized as a critical space in which the material and sociohistorical conditions related to a satellite's operations can be described and analyzed. This second meaning is an attempt to annotate the first meaning—that is, to infuse coverage maps, which only show boundaries of potential signal circulation, with details of the diverse, complex, and contradictory sociohistorical conditions in which satellites are used. Footprint analysis moves across these two meanings, then, by first engaging with maps and then by investigating details about the satellite's development and uses, whether through on-site research or from afar. The goal is to bring the names, histories, coverage boundaries, and infrastructures of satellites into greater critical awareness and to encourage further discussion of the territorial, infrastructural, and geopolitical implications of their use.

Because footprints transect sovereign nation-state boundaries and because satellites have historically been developed and often used without the consent of nation-states, they are inherently involved in geopolitical matters. Satellite footprints are much more than static maps—they are politically charged documents that showcase previous, existing, or desired political alliances, trade relations, and/or intercultural campaigns. They are symptoms of the power of the transnational corporation in the age of globalization, in that they visualize the corporation's technological capacity to operate across nation-state boundaries, while providing little sense of the limits on this power. They are visualizations of technical processes that are imperceptible and yet increasingly set the conditions for politics, trade, and culture in our world. Footprint maps bring the satellite's geopolitics into relief, but they are only starting points that demand further critical inquiry.

Rather than assume the footprint could be described in its totality, footprint analysis engages only with portions of it in order to provide a sense of the complexity and impossibility of a complete picture. In other words, rather than set out to describe and document all parts of the system that make a footprint possible, the analysis focuses on a selection of localized sites or issues as suggestive parts of a broader system that is imperceptible in its entirety. It is a critical practice that emphasizes the discourses, manifestations, and traces of satellite technology rather than being invested in presenting the thing itself. Footprint maps imply the presence of ground stations and satellites in orbit, but do not actually represent them. They require us to infer and imagine their presence, locations, and operations.

In this way, footprint analysis is an interpretive practice informed by the culturally inflected work on geopolitics by scholars such as Gerard Toal. As Toal explains, "rather than being an objective recording of the realities of world power, geopolitics is an interpretive cultural practice."[5] Toal goes on to suggest, "Geopolitics should be studied as both structure and culture as part of the *longue durée* of the modern world and the cultural life of its most powerful institutions."[6] Following this logic, footprint analysis assumes that satellite corporations are among the world's most powerful institutions, and explores how satellites function as part of the structure and culture of the modern world. As a study of satellite geopolitics, footprint analysis is also concerned with the ways satellite uses have altered the meanings of territory and territoriality. In his discussion of geopolitics and technology, Toal asks, "How do certain technological developments in transportation, communications and military technology transform the functionality, connectivity, and meaning of territory? . . . [H]ow do certain technological paradigms and systems enable new forms of territorializations and territoriality?"[7] By transforming territories into footprints, satellite operators anticipate and track the movement of capital through the world. They are implicated in the valuation of different areas on Earth. The very activation of a satellite's footprint is a territorializing gesture that represents the power to regulate and alter the kinds of signals that move across it and practices that can occur within it. Scholars such as Harold Innis, Herbert Schiller, and Armand Mattelart have offered detailed discussions about the historical relations between communication infrastructures and the formation of national political and economic systems, but few have considered how these relations are being redefined in post-Communist conditions.[8]

It is vital that the satellite's geopolitics be reassessed in the aftermath of the Cold War and in relation to post-Communist conditions. During the past twenty years the world has shifted from a satellite economy largely dominated by the USSR and the United States to become a much more international system that includes a host of new operators. Some of these new players are operating across territories once thought of as "behind the Iron Curtain" and beyond the pale of Western enterprise. With the breakup of the Soviet Union, new satellite corporations have scrambled to enter orbital slots conducive to tapping new markets in the former Communist bloc. In the following sections, I describe events within and conditions related to the footprints of communication satellites owned by Eutelsat and the Kazakh Space Agency. The analysis concentrates on the satellites W2, W4, Sesat1, and Kazsat 1, and discusses the divergent agendas guiding their development and use.

Infrastructure Planning

For the past decade the Caspian Telecom Conference has been held each spring in Istanbul. The event is sponsored by multinational and regional telecommunications corporations such as Fintur, Eutelsat, Hughes Networks,

Gilat, Rostelecom, iDirect, iBasis, Turkcell, Satlynx, Dialogic, Turkcell, Turk Telecom, and others, and it involves participants from post-Communist countries throughout the region including Azerbaijan, Armenia, Belarus, Kyrgyzstan, Georgia, Mongolia, Turkmenistan, Kazakhstan, Uzbekistan, Moldova, Tajikistan, Ukraine, Russia, Romania, and Bulgaria. The primary goal of the gathering is to promote the liberalization of telecommunications sectors in these countries and throughout the Caspian region, and to use this liberalization process to upgrade and/or replace older fixed-line systems with wireless and satellite-based systems to make the region more standardized and integrated within global information economies. The conference is critical to infrastructural change in that it provides a regular time and space where government officials and corporate leaders articulate their visions of the region's technological future and informally negotiate deals.

Situated on the boundary between Europe and the Middle East, host country Turkey has a complex relationship with Europe. This relationship is currently being renegotiated, as the country sits on the cusp of possible integration within the European Union. Nevertheless, Turkey has thriving telecommunications and broadcast sectors that liberalized throughout the 1990s. The country offers state-of-the-art wireless and Internet services; owns four satellites, two of which are stacked at 42°E and two at 31°E; and beams Turkish television and radio channels to footprints throughout Europe, Central Asia, the Middle East, and other parts of the world. Turkey's telecommunications success has led the corporate sponsors of the Caspian Telecom Conference to describe the country as "the gateway for investment into the Caspian and Central Asian Telecommunications industry."[9]

While Turkish leaders play a major role in organizing and hosting the event, the conference brings together several other players as well. Representatives from telecommunication and information ministries of post-Communist states throughout the region attend to provide country status reports. Business officials already operating in the region discuss their success stories and goals for the future. And leaders of global telecom conglomerates headquartered in western Europe and the United States pitch their products and services and encourage adoption of entrepreneurial business models. For three days these various players deliver brief PowerPoint presentations in English, Russian, or Turkish, and multilingual translation services are available. The overall purpose of the event is to promote telecom privatization by pointing to proven success stories in the region, showcasing new services and products, and providing assurance that it is safe and secure to invest in formerly Communist countries.

Significantly, the conference also provides a forum where representatives of newly independent states can publicize their distinct national positions and priorities while expressing their willingness to work collaboratively to implement infrastructural changes that will enhance overall economic development in the region. Several speakers from post-Communist countries emphasized their use of the Internet to create e-government systems that allow citizens open access to information; they described such projects as exemplifying their commitment to democratization. Others commented on measures taken to ensure the safety and security of foreign investments, ranging from new banking procedures to data encryption systems. A Romanian representative stressed the importance of building regional autonomy and strength in the telecom sector and suggested that post-Communist countries of the region should organize infrastructure autonomously rather than rely on Western corporations for technical and financial aid. From the perspective of Western companies, Turkey serves as a model for telecom and broadcast privatization on the edge of Europe, but leaders from post-Communist countries may regard Turkey as a model for other reasons. Some of these countries may want to entertain the possibility of formally becoming part of Europe, yet may, at the same time, be concerned about maintaining their independence and distinct identities and agendas. Turkey functions as a model for such post-Communist countries in that it sustains a vibrant cultural heritage, has achieved relative economic success within a liberal capitalist framework, and has maintained a degree of autonomy and power in relation to the West. It is not surprising, then, that Turkey has become the gathering place in which leaders from post-Communist countries and Western telecom corporations convene to deliberate the region's infrastructural future.

Infrastructures do not emerge out of nowhere; as large-scale projects, they take shape through various processes over time. Conferences such as Caspian Telecom bring together the players and the possibilities for such projects and in so doing can also result in specific infrastructure developments. These changes can take different forms, from corporate contracts to interstate agreements to network construction. In recent years, for instance, Fintur—a subsidiary of the Scandinavian telecom giant TeliaSonera and jointly owned by Turkey's leading mobile operator, Turkcell—has subsidized the development of wireless networks in Azerbaijan, Georgia, Kazakhstan, and Moldova, and the company now owns part of Azercell, Geocell, Kcell, and Moldcell.[10] Mongolia recently signed a deal with Gilat to develop very small aperture terminal (VSAT) systems that will expand wireless and Internet connectivity across its vast and remote countryside. And satellite operators such as Eutelsat and Kazakhstan lease transponder space to broadcasters and businesses from Central Asia and beyond.

The conference can be understood as part of this footprint analysis, then, in that it provides a time and space for corporate leaders and state officials to convene and articulate their aspirations and plans for regional technological development and to bring particular infrastructural possibilities into being. The event also raises awareness about the communication satellites that are positioned to serve the region and pushes leaders of new states to contemplate how or whether to enter the satellite economy. The conference is designed to garner support for the satellite footprints that have already been established and to gauge the need for new satellites in the region. Two conference participants in particular have emphasized the importance of satellite technologies in Central Asia: Eutelsat and Kazakhstan.

Eutelsat's "Extended Europe"

Eutelsat is the largest satellite operator in Europe and third-largest operator in the world. The company is based in France and has ground stations, subsidiaries, and sales offices around the world. Eutelsat has been operating since 1983 and underwent major financial restructuring in 2001. As of 2008, Eutelsat operates a fleet of twenty-four satellites and claims to reach up to 90 percent of the world's population.[11] The company's satellites broadcast more than 3,000 TV channels and 1,100 radio channels in forty-five languages. In addition, Eutelsat provides international enterprise data networks to 850 business clients from around the world. During the 1990s, Eutelsat established a lucrative "video neighborhood" by installing a series of Hotbird satellites at 13°E. By stacking several satellites in the same position and adopting a "multi-satellites strategy," the company was able to use a single orbital location to relay 1,100 TV channels and service more than 120 million homes in Europe, the Middle East, and North Africa.[12]

In recent years Eutelsat has made a concerted effort to target markets in what it calls "extended Europe," which includes western, central, and eastern Europe; Central Asia, the Middle East, and North Africa.[13] When presenting at the Caspian Telecom conference, Eutelsat's regional sales director, Ali Korur, focused on Central Asia and characterized it as a vast region with relatively high populations in urban areas and largely untapped markets. Further, he suggested, since much of the region is not served by upgraded terrestrial infrastructure, satellite systems are an efficient way to provide broadcast, telephony, and Internet services to consumers scattered throughout the area. By activating footprints across Russia and Central Asia, Eutelsat sought to bring post-Communist territories into range of its satellite services with the hope that newly independent nation-states, broadcasters, telecom carriers, businesses, and consumers alike would become clients.

Indeed, not only is this "extended Europe" imagined in terms of regional geographic categories, it is also identified as a particular stretch of geosynchronous orbit that spans from 8°W to 40°E, an area running laterally from the territory of Ireland to eastern Ukraine. There are a total of twenty-one geostationary satellites within this stretch of orbital space; Eutelsat owns nine of them, giving the company a dominant position in what it calls the "extended Europe video market." Eutelsat has identified four satellites in particular as being crucial to the development of new markets in this region, explaining that the company "has partnered with broadcasters to develop other premium positions with strong regional coverage, notably for Russia, the Ukraine, and sub-Saharan Africa (W4/Sesat 1), central and Eastern Europe (W2), and Turkey (W3A)."[14]

From Eutelsat's vantage point, an "extended Europe" means having the capacity to distribute signals into areas in eastern Europe and Central Asia that have historically been off limits due to Soviet influence. It also involves the power to occupy particular orbital addresses and to lease transponder space to broadcasters and businesses that want to operate throughout the region. Eutelsat has colocated two satellites, W4 and Sesat 1, at 36°E in an effort to maximize regional opportunities from this orbital position. Launched in 2000, W4, which Eutelsat describes as "stretching from Scandinavia to the Caspian Sea," carries television and radio channels ranging from the pay-TV services of Ukraine's Poverkhnost Plus and Russia's NTV Plus, to TNV Tatarstan and TriColor TV. The same satellite also carries a large volume of channels to a spot beam in Nigeria and to parts of North Africa, giving W4 a transcontinental footprint. Eutelsat promotes the other satellite in this location, Sesat 1 (Siberian-European satellite), as "a satellite connecting East and West, from the Atlantic to the Urals." It has transponders filled with unspecified feeds, likely of businesses operating throughout the region, as well as television channels such as Armenia 1 TV; Kazakhstan's national channel, Caspionet; Al Iraqiya; Azerbaijan's Lider TV; Kurdish channels Zagros TV and Silemani TV; a Polish channel called TV Silesia; and Al-Jazeera, among others. Eutelsat's W4 and Sesat 1 footprints circulate satellite and radio channels in languages from Russian to Turkish to Arabic to English.

In addition to these satellites, Eutelsat has entered agreements with the Russian Satellite Communications Company (RSCC) that enable the company to expand its services in the region. For instance, Eutelsat has a lifetime lease of twelve Ku-band transponders on the RSCC's Express AM22 satellite at 53°E.[15] This contract expands Eutelsat's capacity in Central Asia even further, as Express AM22 provides coverage of Europe, Africa, the Middle East, and Central Asia through a highly flexible configuration of fixed and steerable beams.[16] Prior to this, Eutelsat leased five transponders on RSCC's Express A at 11°W. In addition to long-term leases of Russian transponder space, Eutelsat launched a new satellite, W7, in 2009. W7 provides wide-beam coverage over Eurasia, and is particularly well suited to TV feeds, public and corporate telephony, and data services. The satellite's Central Asian beam is targeted at the oil and gas sector for voice and data services.[17]

By installing satellites into orbital positions that reach Central Asia and leasing capacity on Russian satellites, Eutelsat is attempting to build what it calls a "video neighborhood" in an area where, historically, western signal traffic has been relatively limited. The loosening of state control over broadcasting and telecommunication sectors in post-Communist countries has enabled this western European satellite operator to establish a stronghold in emerging media markets. The large populations in the region make Eutelsat's orbital positions and footprints more valuable. As Price reminds us, "Particular orbital slots are often more important than others because of the particular terrestrial footprint a satellite can reach from those slots. A footprint that reaches a vast population or a wealthy population or a politically important one can be more valuable than one that does not."[18] In this region, the footprints of W4, Sesat 1, and W7 are significant and potentially lucrative in that they enable a European satellite operator to extend into areas that have transitional economies, large populations with a growing interest in consumerism and the West, and extensive natural resource development.

Historically, the International Telecommunications Union has assigned orbital addresses to satellite operators on a first-come, first-served basis, which has meant that the countries and corporations with the most capital have been the major players. This is not to say that communications satellites never previously operated in Communist countries or within Central Asia. During the 1960s and 1970s, the Soviet Union launched into orbit a series of satellites designed to serve and interconnect Communist countries. The INTERSPUTNIK network, for instance, delivered television and radio broadcasts throughout the Communist bloc and linked countries as far away as Cuba and Mongolia.[19] INTERSPUTNIK continues to operate satellites across the territories of the former Soviet Union, but the organization has been privatized and now competes with other satellite operators in the region, including Eutelsat.

In media and cultural studies, the use of satellites for direct broadcasting is relatively well understood. What is not as well known is that the same satellites used to downlink television and radio signals are also used for natural resource development. In Central Asia, satellite transponders not only circulate media signals, they also transmit corporate communications related to oil and gas development. Eutelsat published a special brochure, "Eutelsat Satellite Solutions for Oil and Gas Companies," indicating that whether "drilling offshore, exploring in the desert, [or] monitoring pipelines in rough environments, when the time comes for reporting to headquarters, dealing with a medical emergency, accessing the Web, or simply keeping in touch with homes, forget fixed phones, forget terrestrial infrastructures," and then suggesting several specialized satellite services for oil and gas companies operating in remote areas.[20] Some of the Western oil companies currently operating in Central Asia include Shell Group, ExxonMobil, ChevronTexaco, ConocoPhillips, Dragon, Burren, BG Group, and Unocal. These companies can attend an annual conference, cosponsored by Eutelsat, called "Oil and Gas Communications," which was held in Cairo in 2008 and in London in 2010, to promote high-speed networks and remote monitoring systems for oil and gas corporations. Satellites require so much capital that their uses must remain flexible in order to maximize the potential for return on the investments necessary to develop, launch, operate, and insure them.

Thus Eutelsat and other satellite operators sell transponder space not only to broadcasters and telecom carriers but also to Western oil corporations for the purpose of monitoring and administering remote oil fields from afar. In 2003, Eutelsat signed a contract with CapRock, a company based in Houston, Texas, that specializes in providing remote satellite services to customers in "extreme locations," including oil and gas companies in Asia. Having established itself in the Gulf of Mexico during the 1980s and 1990s, this U.S.-based firm leases transponder space from Eutelsat and creates customized satellite-based communications networks to support oil production in the region and has leased transponder space on Sesat 1 to provide services to clients within the satellite's footprint. Eutelsat's CEO Ron Samuel claimed, "Eutelsat strengthens its commitment to the oil and gas industry through its relationship with CapRock."[21] As CapRock's publicity indicates, "In a time where energy exploration is moving into newly discovered, isolated regions and the technical workforce is continuing to tighten, the need to keep remote sites connected with headquarters becomes even more essential. CapRock's satellite solutions provide tools that enable the oil and gas industry to operate more efficiently in today's environment."[22] The company uses satellites to provide "ruggedized" communications networks to monitor and manage oil infrastructures stretching across vast territories.[23] Tasks once performed by workers on the ground are now conducted remotely by sensors, computers, and satellites. CapRock is just one of many such firms offering similar satellite services in the region. Other companies, such as Hermes (United Kingdom) and Gilat (Israel), lease transponders from satellite operators and create VSAT and other communication networks for clients with geographically dispersed business operations.

This is not the first time in recent history that oil and satellite industries have converged. For example, during the mid-1990s Egypt sought to increase its satellite capacity in order to extend national television networks into remote parts of the country as well as into newly discovered oil and gas fields beyond the reach of terrestrial communication infrastructures.[24] Analogous conditions are emerging with the privatization of lands in the post-Communist countries of Central Asia, where some newly independent states are attempting to upgrade and extend their telecom systems to reach resource-rich areas.[25] In this sense, satellites can be thought of as meta-infrastructures, in that they can extend terrestrial telephone, broadcast, and Internet networks across national, transnational, and in some cases transcontinental footprints, thereby interlinking and leapfrogging some terrestrial infrastructures.

Satellites are not being used in this context to shore up the national sovereignty of post-Communist states in Central Asia; they are enabling Western corporations to redefine and annex these territories as footprints in which everything from natural resource extraction to TV audience building can take place. Eutelsat's "extended Europe" not only involves the company's capacity to distribute television and radio channels and provide business communication services across transcontinental footprints, it also supports a particular disposition toward former Communist countries as potential sites of investment, extraction, and profit. Satellite uses are implicated in the reimagining of Central Asia as the new European frontier, and the same kinds of speculation and expropriation that transformed colonial territories throughout the nineteenth century are being supplanted by the maneuvers of multinational satellite and oil giants. Yet, as we will see, the story is more complicated.

Eutelsat's vision of an "extended Europe" pushes us to recognize the geopolitics of orbits and footprints across post-Communist territories. The fall of the Soviet Union meant that Western corporations could enter the region in orbit as well as on the ground, carving out new geophysical and spectral territories in which a neoliberal hegemony could take shape. Given the dearth of existing players in the regional satellite economy, the vast size of the territory to be covered, and the paucity of state regulations on satellite operators in the region, Eutelsat has been able to enter and begin to dominate the market in Central Asia and elsewhere. At the same time, however, it is important to resist a "development" logic that posits post-Communist countries as being simply subject to or swayed by the whims and strategies of Western multinationals. The Caspian Telecom conference is significant in this regard, as it enables new post-Communist states to articulate their national telecommunication needs, priorities, and agendas publicly. And, as we shall see, in some cases these run counter to—and even challenge or compete with—the needs, priorities, and agendas of the global telecom and satellite giants.

Kazsat: Oil and Orbit

While satellite giant Eutelsat has strategized to tap and dominate markets in "extended Europe," a smaller satellite operator indigenous to the Caspian region has also emerged. On June 18, 2006, Kazakhstan, in partnership with the Russians, launched its first communications satellite (Kazsat 1) from the Cosmodrome in Baikonur, the same facility from which the Soviets launched the world's first satellite, Sputnik, in 1957. Kazsat 1 cost Kazakhstan US$65 million and was built by the Russian company Khrunichev. The satellite was designed with twelve transponders operating in the Ku band and was deployed into orbit at 103°E, where it could provide broadcasting and tele-communication services to a regional footprint that includes Kazakhstan, Uzbekistan, Tajikistan, Turkmenistan, Kyrgyzstan, parts of central Russia, and the Caucasus. By the end of 2007 Kazsat 1 was operating at 70 percent of its capacity and was carrying the signals of Kazakh broadcasters such as Astana, Kazakhstan TV, El Arna, Khabar TV, and Nau TV, as well as Russian channels such as MuzZone, Kultura +7, Rossiya +6, Rossiva +8, among others.[26]

In June 2008, two years after Kazsat's launch, the Kazakh Space Agency lost contact with its one and only satellite. A computer onboard the satellite failed, and the Kazakh TV signals it carried went off the air. A number of these stations were forced to shut down completely because of the compara-tively high cost of transponder space on other satellites. Others shifted their operations to satellite platforms such as RSCC's Express 2A, despite the higher cost.[27] The Kazakh Space Agency has since developed another satel-lite, Kazsat 2, which was launched in 2011.[28]

In the years leading up to Kazsat 1's launch, Kazakhstan had been work-ing to position itself as a player in the global space economy. In 2005, a year before Kazsat's launch, Kazakhstan's government, Aerospace Committee, and Agency of Information and Communication hosted KazSatCon, a major con-ference in the city of Almaty. They worked in conjunction with the Global VSAT Forum and the World Teleport Association, telecom trade organizations with members interested in expanding into Central Asia. Kazakhstan's prime minister, Danial Akhmetov, introduced the conference, telling the attendees, "Kazakhstan intends to spend US$345 million (280 million Euros) on its first Space Program. The Program involves launching the country's first Satellite. . . . We believe that Kazakhstan should be a space power."[29] Like the Caspian Telecom Conference, KazSatCon was attended by major corporate players such as Eutelsat, Alcatel, Nokia, Microsoft, LG Electronics, Siemens, Samsung, and Gilat, among others. Kazakhstan used the event to showcase future growth potential in the region across various sectors of the economy, and even held a special session called "Satellite Services for the Caspian and Cen-tral Asian Oil and Gas and Transport Industry." If Kazsat had remained func-tional, it, like Eutelsat's Sesat 1 and W7, also could have been used to support oil and gas connectivity.

Unlike Eutelsat's satellite fleet, which stretches across an "extended Europe," Kazakhstan has had more modest ambitions. Kazsat 1 was in-tended to support a regional (rather than a global) economy. Very few post-Communist countries in the world have the capital necessary to develop and launch a satellite, much less mount an entire space program. This possibility only emerged in Kazakhstan, however, because the country's recent oil boom increased national revenues and because Kazakhstan inherited the Baikonur space launching facilities built by the Soviets. In a matter of years, Kazakhstan went from being a place that many in the West had never heard of to a place attracting major foreign investment and attention from around the world. A country the size of western Europe, Kazakhstan has been described as having more oil wealth than the Middle East.[30] Since it gained independence, the country has privatized its natural resources, telecom, and space sectors, finding a host of foreign investors eager to buy oil fields, bandwidth, and launching services. In addition to luring companies such as Microsoft, Nokia, and Siemens to KazSatCon, oil multinationals such as Exxon, Shell, Unocal, and Agip have established operations in Kazakhstan. Major telecom corporations including Deutsche Telekom, PLD Inc., Telstra, Indosat, Alcatel, and DAEWOO Telecom also have moved into the country.

Relational economies have formed between satellite and oil industries in Kazakhstan as well. A company called Gilat, which (like CapRock) offers satellite services to clients in remote areas, leases time on a variety of satellites. One of its clients is KazTransOil, a Kazakh oil company that uses Gilat's FarAway VSAT platform to administer sixty-five sites along an oil pipeline that runs for 4,039 miles (6,500 km) through some of the country's harshest and most remote terrain. This satellite system, according to one article, allows the company to "monitor and control the remote terminals from the company's head office," which represents "a significant advantage, considering the very distant, isolated areas in which the sites are located."[31] Communication satellites, whether owned by Eutelsat or Kazsat, are not only technologies of telecommunication and broadcasting, but must be understood as technologies of natural resource development as well. When Eutelsat leases transponder space to CapRock or Gilat, companies that design and operate networks for oil companies, the satellite becomes an extension of the oil rig. One analyst goes so far as to describe satellites as "pipelines of the skies."[32] In Kazakhstan, oil revenues have not only helped to subsidize the development of the country's first satellite but also provide funding for newer satellites such as Kazsat 2. In addition to helping Kazakhstan achieve "information independence and . . . form a united information space"—two of the country's national priorities—Kazsat 2 could be used to monitor and track the country's oil production, giving Kazakhstan further economic autonomy.[33] In other words, because of its natural resource wealth, Kazakhstan has been able to develop its own satellites that eventually may support the signal traffic of regional broadcasters and telecom carriers as well as oil corporations.

That Kazakhstan has been able to enter the orbital economy is highly significant. During its short tenure, Kazsat 1 allowed the country to bolster ties with other newly independent states in the region that, in the wake of Communism, are also redefining their political, economic, and cultural agendas. Kazsat's footprint corresponds with the historically shared economic and geopolitical boundaries of certain former Soviet republics. As Martha Brill Olcott explains, "Nowhere are the states of the former Soviet Union more closely bound with one another than in Central Asia, not only because of the strong sense of common ancestry and shared cultural and religious heritage, but also simply because of the region's geography. Precise political boundaries were never drawn, and highway and railway systems pay little attention to national borders."[34] She suggests, "in addition to the road infrastructure, the Central Asian states are bound together by an integrated electrical grid and common water resources."[35] Kazsat's footprint can be understood as reinforcing existing political, economic, and technological coordination in the region, and facilitating the flow of everything from signals to crude oil. Kazsat inscribed a regionally autonomous subzone within the more expansive footprints of Eutelsat's "extended Europe." Footprint analysis involves a consideration of the comparative cartographies of satellites and their coverage, and identifies the contradictions and tensions that emerge with their activation and ongoing use.

Conclusion

Satellites are embedded within material conditions that are often difficult to perceive and challenging to describe. In media and cultural studies, we are accustomed to dealing with culture as it is readily intelligible, yet comprehending satellite technologies and their territorializing effects requires an appreciation for the obscure and the imperceptible. It is impossible to see all the satellites in orbit, all the companies that use them, all the signals they distribute, or all the locations in which those signals are received. We can use the footprint, however, to acquire a sense of the territories in which the satellites operate and to create research models for exploring the geopolitical, economic, and cultural implications of their uses.

The satellite footprints across Central Asia have been used in contradictory ways. On the one hand, they are being used to support a post-Communist version of neoliberalism, wherein state services and properties are not only being privatized but are sold off to foreign investors and managed via satellite as well. In this scenario, everything from lands to the satellites above them and the oil beneath them are owned and controlled by entities elsewhere. On the other hand, states such as Kazakhstan have used the newfound wealth generated through privatization of natural resources to develop their own satellites as part of a long-term plan to assert autonomy over the informational and geophysical spaces of the region. In this scenario, Kazakhstan maintains ownership over everything from the oil in the ground to the satellites in orbit and capitalizes on these resources to establish a regional link to the global economy.

In either case, the activities taking place in and around satellite footprints that extend to Central Asia challenge us to continue investigating three issues. First, there is a need for further research on the versions of neoliberalism that have sprouted up in central and eastern Europe, Russia, and Central Asia since the fall of Communism. Many practices that were once assumed by the Communist state not only are being privatized and taken over by corporations but are also being outsourced to satellites. Indeed, activities ranging from governance to broadcasting from health care to oil production are being automated and satellitized. The combined contexts of digitization, globalization, and post-Communism have resulted in a different formation of neoliberalism in which the unique capacities of satellites and other new communication technologies (such as the Internet) are being used to dramatically reduce the power and function of the state in some post-Communist contexts. This can be detected in the ways that privatized satellite television, mobile telephone, and Internet services have been promoted as facilitating integration with western Europe and the United States and as offering vibrant pluralist alternatives to the Communist past. Yet many people in post-Communist countries inhabit everyday orbits far removed from the Western ideals associated with technologies of distance communication, and some may prefer state-run public television to the multichannel environment, or conversation around the kitchen table to a roaming tête-à-tête via mobile phone.

Second, the footprinting of Central Asia serves as a reminder that the expansion of Europe is happening not only through the formal inclusion of post-Communist countries within political and economic communities such as the European Union and NATO but also through infrastructural campaigns designed to support interstate flows such as highways, pipelines, and footprints.[36] Of these infrastructural campaigns, satellite footprints are the most extensive and the most imperceptible, and therefore it is urgent that their activation be discussed and evaluated. The companies that control satellite footprints across post-Communist territories have the power to regulate signal traffic throughout the region, and stand to benefit financially from the carriage of such signals. The footprinting of post-Communist territories produces an "integration effect" by mapping what purports to be a pluralized mediascape in which citizen-consumers across nation-state boundaries can downlink and view television signals in multiple languages from multiple countries. Meanwhile, public broadcast services in most post-Communist countries are gravely underfunded and barely surviving. It is important to consider how satellites initiate and support new trade routes and relations, political alliances, and cultural practices. In some cases, satellite footprints have been used to help unify and standardize technological, economic, and cultural systems in preparation for post-Communist states' integration into the European Union. In other cases, satellites have enabled post-Communist states to "beam" their new national identities both regionally and abroad.[37] At the bare minimum, satellite footprints in post-Communist territories should cause us to question the very terms of "transition" and "integration," and critically evaluate how and why such processes are orchestrated.

Finally, the footprinting of Central Asia encourages us to consider how satellites function as part of both cultural *and* natural resource economies. Television and oil industries rely on the same satellite technologies to become part of the global economy, even though their commodity forms and destinies differ dramatically. Since signals related to broadcasting and oil production pass through transponders on the same satellite, it is important to create research models in media and cultural studies that enable us to explore how and why the same apparatus may be used in very different ways and how this flexibility is organized to extend and multiply the flow of capital in the world. This involves fostering relational thinking across the divides of mass culture and natural resource development, and moving toward an environmentally and territorially inflected field of media and cultural studies. It also involves continuing to investigate the disparate ways capitalism works in different parts of the world, across different economic sectors, and in relation to different technologies. Such a project can only occur within a field of cultural studies that is attentive to transformations at both global and local scales, that values interdisciplinary and experimental modes of analysis, and that puts a high priority on critical thinking. It is my hope that the footprint will become a space in which cultural studies research continues.

NOTES

Another version of this chapter appeared in the *European Journal of Cultural Studies* 12(2) (2009): 137–156.

1 Stephen Graham and Simon Marvin, *Splintering Urbanism: Networked Infrastructures, Technological Mobilities and the Urban Condition* (London: Routledge, 2001), 12.

2 Research for this chapter was conducted while I was a fellow at the Wissenschaftskolleg (Institute for Advanced Study) of Berlin from 2006 to 2007. I am grateful for the support of the institute's staff and fellows. I am also grateful to Ursula Biemann and Angela Melitopoulos for their collaboration on the Transcultural Geographies/B-Zone project, where ideas for this work began.

3 Monroe Price, "Satellite Broadcasting as Trade Routes in the Sky," in *In Search of Boundaries: Communication, Nation-States and Cultural Identities*, ed. Joseph Chan and Bryce McIntyre (Westport, Conn.: Ablex, 2002), 149.

4 Lisa Parks, *Cultures in Orbit: Satellites and the Televisual* (Durham, N.C.: Duke University Press, 2005), 70.

5 Gerard Toal, "Geopolitical Structures and Cultures: Toward Conceptual Clarity in the Critical Study of Geopolitics," in *Geopolitics: Global Problems and Regional Concerns*, ed. Lasha Tchantouridze (Winnepeg, Ont.: Centre for Defence and Security Studies, 2004), 75.

6 Ibid., 99.

7 Ibid., 80.

8 Harold Innis, *The Bias of Communication*, 2nd ed. (Toronto: Toronto University Press, 2008); Herbert Schiller, *Mass Communications and American Empire*, 2nd ed. (Boulder, Colo.: Westview Press, 1992); Armand Mattelart, *Networking the World: 1794–2000* (Minneapolis: University of Minnesota Press, 2000).

9 Caspian Telecom Conference, conference program, Istanbul, April 12–13, 2007.

10 "Investing in Communications for Continuous Development," Fintur Holdings B. V. corporate brochure, n.d.

11 Eutelsat Communications, corporate brochure (October 2006): 5.

12 "Eutelsat Commissions New W3B Satellite from Thales Alenia Space," *Red Orbit*, February 26, 2008, accessed March 24, 2011, http://www.redorbit.com/news/business/1269461/eutelsat_commissions_new_w3b_satellite_from_thales_alenia_space/index.html.

13 Ali Korur (Eutelsat Regional Sales Director, Middle East and Central Asia), "Eutelsat Presentation," Caspian Telecom Conference, Istanbul, April 12–13, 2007, 10.

14 Ibid., 7.

15 The Express AM22 was launched on December 29, 2003, and went into operation on March 9, 2004, at 53°E in geostationary orbit.

16 "Eutelsat Buys Lifetime Lease for 12 Transponders on Russia's AM22 Bird," *Space Daily*, March 18, 2004, accessed March 24, 2011, http://www.spacedaily.com/news/satellite-biz-04zo.html.

17 "Upcoming Launches," Eutelsat, 2008, accessed March 24, 2011, http://www.eutelsat.com/satellites/upcoming-launches.html.

18 Price, "Satellite Broadcasting," 148.

19 John Downing, "The Intersputnik System and Soviet Television," *Soviet Studies* 37(4) (1985): 465–483.

20 "Eutelsat Satellite Solutions for Oil and Gas Companies," Eutelsat corporate brochure, n.d.

21 "CapRock and Eutelsat Team Up to Deliver Telecommunications Solutions to Customers," *Bnet Business Network*, November 2003, accessed March 24, 2011, http://findarticles.com/p/articles/mi_qa5367/is_200311/ai_n21340416.

22 "Solutions for Oil and Gas," Caprock website, accessed March 24, 2011, http://www.caprock.com/markets/oil_gas_satellite.htm.

23 "Company," Caprock website, accessed March 24, 2011, http://www.caprock.com/company/company.htm.

24 Naomi Sakr, *Satellite Realms: Transnational Television, Globalization and the Middle East* (London: I. B. Tauris, 2001).

25 Historically, the massive size of the Soviet Union made comprehensive national telecom and broadcast coverage a great challenge. As John Downing explains, one of the reasons the Soviets developed satellite programs such as INTERSPUTNIK during the 1960s and 1970s was to access and integrate remote, resource-rich regions within the national communications system; see Downing, "Intersputnik System," 465–483.

26 Kazakh Space Agency, "KazSat-1 Satellite Is Most Likely Lost," Interfax-Kazakh-stan, Kazakhstan News Agency, June 16, 2008, accessed March 25, 2011, http://www.interfax.kz/?lang=eng&int_id=10&function=view&news_id=1900.

27 Lewis Page, "Kazakh TV Off-Air after Satellite Breathing Troubles,'" *Register*, June 18, 2008, accessed March 25, 2011, http://www.theregister.co.uk/2008/06/18/kazakh_satellite_troubles/.

28 "Kazsat," Russian Space Web, accessed March 25, 2011, http://www.russian spaceweb.com/kazsat.html.

29 KazSatCon 2005 Conference and Exhibition Website, website dated February 23, 2005, accessed March 25, 2011, http://www.satsig.net/satellite-kazakhstan.htm.

30 Christopher Robbins, *Apples Are from Kazakhstan: The Land That Disappeared* (New York: Atlas, 2008).

31 "Voice and Data Networks for Oil and Gas Companies," *Online Journal of Space Communication* 7 (Fall 2004), accessed July 17, 2008, http://satjournal.tcom.ohiou.edu:16080/issue7/cur_gilatoil.html; subsequently accessed March 25, 2011, http://spacejournal.ohio.edu/issue7/cur_oilGas.html.

32 Errol Olivier, "Pipelines of the Skies," *Online Journal of Space Communication* 7 (Fall 2004), accessed July 17, 2008, http://satjournal.tcom.ohiou.edu:16080/issue7/ov_skypipes.html; subsequently accessed March 25, 2011, http://spacejournal.ohio.edu/issue7/ov_skypipes.html.

33 "Kazakhstan to Change Equipment of Its Second Satellite," Organization of Asia-Pacific News Agencies, June 17, 2008, accessed March 25, 2011, http://www.oananews.org/view.php?id=2176&ch=AST.

34 Martha Brill Olcott, "Confronting Independence: Political Overview," in *Energy in the Caspian Region*, ed. Yalena Kalyuzhnova et al. (New York: Palgrave, 2002), 43; see also Martha Brill Olcott, "Regional Cooperation in Central Asia and the South Caucasus," in *Energy and Conflict in Central Asia and the Caucasus*, ed. Robert Ebel and Rajan Menon (Lanham, Md.: Roman and Littlefield, 2000), 123–144.

35 Olcott, "Confronting Independence," 44.

36 For a collective visual research project that explicitly makes this point in an examination of oil pipelines, highways, and footprints on the outskirts of Europe, see Anselm Franke, ed., *B-Zone: Becoming Europe and Beyond* (Barcelona: Actar Press, 2005).

37 Lisa Parks, "Postwar Footprints: Satellite and Wireless Stories in Slovenia and Croatia," in Franke, *B-Zone*, 306–347.

SATELLITE MEDIASCAPES

7

From Satellite to Screen

How Arab TV Is Shaped in Space

NAOMI SAKR

One of the most intriguing questions about Arab satellite television has always been how far the technical possibilities of transnational broadcasting could offer a real, practical escape from national regulation, since national regulation in most Arab countries is designed to prop up incumbent regimes. It was apparent by the end of the 1990s that a majority of Arab governments were complicit in an emerging pan-Arab system of governance over transnational broadcasts. Under this system, mutually agreed controls operated at the national level to limit the ability of any dissenting broadcaster (in particular, Al-Jazeera) to report or uplink within the region at will.[1] It also emerged that national regulation and intergovernmental agreements outside the Arab world could equally obstruct satellite broadcasting to Arab countries. This was amply demonstrated to Kurdish residents of Iraq and Syria when Turkish government pressure on regulators and service providers in France, Portugal, Spain, Germany, Poland, Slovakia, and the United Kingdom led to withdrawal of permissions and services from the Kurdish satellite station, Med-TV.[2] Examples of non-Arab intervention multiplied in the 2000s, with the introduction of special legislation in France to deny the station Al-Manar, associated with the Lebanese movement Hizbollah, access to the satellites of Paris-based Eutelsat. Consultations among European national audiovisual regulators then led to Al-Manar being dropped from other satellites, both U.S. and European owned.[3] These moves seemed to suggest that use of satellite technology is subject to political decisions by governments that act together, both inside the Arab region and the Global North.

Ultimately, however, the question about potential dislocation between national regulation and transnational communication remains unanswered, because regulatory action against Med-TV, Al-Manar, Al-Jazeera, and others could not block them entirely. Med-TV was reincarnated as Medya TV, based in Belgium with a satellite uplink from France, and when France cut that link in 2004, Roj TV took over as its successor from Denmark. It continued broadcasting for more than three years amid tense exchanges on the subject between the Turkish and Danish governments.[4] Al-Manar meanwhile continued to broadcast from Nilesat and Arabsat, which are both under the jurisdiction of Arab governments. By April 2008, Al-Manar had also begun broadcasts via Indonesia Telkom's Palapa C2 satellite, having been removed in January that year from the Thai satellite Thaicom after a three-day test run.[5] An administrative order by the German interior ministry in November 2008 to "ban" Al-Manar in Germany received considerable publicity, since it applied to advertising, fund-raising, and broadcasting in hotels. Yet it stopped short of banning reception in private homes with dishes large enough to receive transmissions from Nilesat and Arabsat.[6] Americans wanting to watch Al-Jazeera English, launched in late 2006, could likewise do so using a satellite dish from GlobeCast, a division of France Telecom, even though political resistance to distribution of the channel in the United States limited its availability on cable systems in the thirty-two months it took for a breakthrough to be achieved in mid-2009.[7]

One way to frame the question raised by these examples is to ask, as Lisa Parks has done, whether satellite television practices have "cut and divided the planet in ways that support the cultural and economic hegemony of the (post)industrial West," or whether there persists a more differentiated struggle over meanings and uses, in which practices are mobilized in different times and places for or against militarization, corporatization, and so on.[8] In much the same way, Colin Sparks has noted that "the continued centrality of imperialism in explanations of the contemporary world" does not mean we should go along with theories of cultural imperialism.[9] The task instead is to expose the multiple realities of power relations, whereby—for example—people who run media operations in the so-called developing world often feel more threatened by the poor and disenfranchised in their own countries than by governing elites in politically and economically powerful states.[10] The challenge, therefore, in reviewing struggles for control over satellite broadcasts is to know whether players act alone or in concert, and whether joint action reflects networks that may give the lie to what Lena Jayyusi has called the "counterfeit geographies" of the Middle East.[11] These are geographies linked to one-sided or distorted perceptions and discourses derived from a "military/security logic," which have been naturalized by the global projections of contemporary media forms.[12]

A further challenge, arising from the notion that "counterfeit geographies" marginalize some struggles while strategically pushing others to the foreground, is to detect struggles that powerful groups seek to keep off the agenda altogether.[13] Struggle may be obscured by an appearance of consensus. Monroe Price builds a distinction between the consensual and the unilateral into his "taxonomy of analytic approaches" to national media policies. He distinguishes between the activity of shielding domestic markets, on the one hand, and altering external markets, on the other, noting that each activity may be conducted unilaterally or consensually.[14] His hypothesis—that there is a general tendency to move from the unilateral to the consensual, negotiated, and multilateral—seems to be echoed in the context of European governance of communications, in which we find the increased use of what Maria Michalis terms "soft" policy instruments (recommendations, benchmarks, sharing of experience and models) and "soft" institutions (such as transnational groupings of regulators).[15] Yet the presence of consensus and multilateral negotiation should not be assumed to indicate an absence of struggle. In what follows, the use of satellite capacity to beam television programming across Arab borders is considered in terms of whether it is unilateral or consensual, and in terms of whose consent is obtained. The account begins by examining how Arab satellite operators have expanded their capacity. It goes on to consider censorship of satellite broadcasts. It ends by examining what the evidence reveals about the geopolitics of Arab satellite television.

Multilateral Relationships Forged through Expansion of Satellite Capacity

The phenomenal recent proliferation of Arab satellite television channels, which followed the September 11, 2001 attacks and the 2003 invasion of Iraq, draws attention to the satellite capacity that lies behind the phenomenon. Channels multiplied more than fivefold in the five years from 2003 to mid-2008, at which point an Amman-based consultancy, Arab Advisors Group, estimated the total number of channels at around 517, including 377 free-to-air channels and 140 subscription channels. Reflecting this rapid expansion, the Middle East and North Africa registered the highest revenue growth in the entire global commercial satellite industry in 2005, according to *Middle East Broadcasters Journal*, and this region was described in 2007 as "one of the most vibrant satellite markets on the planet."[16] The expansion worked both ways: as digital compression increased the number of channels that could be accommodated on each satellite transponder, the cost of leasing channels fell far enough to bring satellite transmission within the reach of smaller broadcasters. Such fluctuations in turn affected the gate-keeping power of satellite owners.

Competition among satellite owners became apparent with the launch of Nilesat-101 in 1998. In commissioning the manufacture and launch of Nilesat, Egypt became the first individual Arab country to acquire its own satellite system and the first to have access, through the state's majority shareholding, to a satellite with digital technology. Other Arab states had shares in the pan-Arab venture Arabsat, headquartered in the Saudi Arabian capital (Riyadh) and 36.7 percent owned by the Saudi Arabian government, with shares totaling another 40.3 percent held by Kuwait, Libya, Qatar, and the United Arab Emirates (UAE). After the Arab boycott of Egypt in the 1980s, Egypt's share in Arabsat was a mere 1.59 percent. Arabsat did not launch its first digital satellite, Arabsat-3A (later renamed Badr-3), until 1999. Making use of this time lag to boost profits, Nilesat's chairman was ready to accommodate customers rejected by Arabsat. He made overtures to Canal France International when it lost its slot on Arabsat in 1997, after Arabsat complained about sexually explicit material that had appeared on the channel for a few minutes through a technical mix-up. Similarly, Iraq was offered transponder space on Nilesat in 1998, despite having been suspended from Arabsat after the Iraqi invasion of Kuwait in 1990.[17]

In the decade following the launch of Nilesat-101 and Arabsat-3A at the end of the 1990s, Nilesat and Arabsat separately launched more satellites, but their rates of expansion could not keep up with burgeoning demand from broadcasters. Both were obliged at times to meet this demand for channel capacity by leasing transponders on satellites owned by other players.[18] Nilesat-102 went aloft in 2000, joining Nilesat-101 at 7°W. By 2005, after two years of surging demand, the two satellites were together broadcasting some 280 television channels and ninety-five radio stations. As they were nearly fully stretched, Nilesat turned to the giant operator Eutelsat for a quick fix. The resulting Nilesat-Eutelsat deal went into effect in July 2006, when a new satellite launched into orbit at Eutelsat's HotBird position at 13°E released one of the older HotBird craft to be shifted from 13°E to 7.2°W, right next to the two Nilesat craft at 7°W and alongside Eutelsat's existing AtlanticBird-2 at 8°W. There it was renamed AtlanticBird-4, with capacity leased to Nilesat under the name Nilesat-103. In 2007, after a 9 percent annual increase in overall demand for transponders serving the Middle East and North Africa, with rising expectations that high definition television (HDTV) would boost the need for bandwidth, and with the fifteen-year lifespan of Nilesat-101 coming ever closer to expiry, Nilesat had to make new provisions. It placed an order for the manufacture of Nilesat-201, equipped with twenty-eight transponders, which was launched in 2010.

According to a study by Euroconsult in 2008, Middle Eastern transponder demand grew even more steeply in 2007 than in 2006, reaching a rate of 12 percent. The same study calculated that revenues from leasing satellites serving the Middle East had risen by an average of 17 percent per year since 2003, amounting to US$752 million in 2007. This surge helps to explain why Arabsat was suddenly galvanized into stepping up its own expansion between

2003 and 2007. Despite its history as the oldest Arab-owned satellite venture, with satellites first launched in 1985, Arabsat's expansion program had a late and unlucky start. In December 2001, less than three years into its supposed thirteen-year lifespan, Arabsat-3A suffered a damaging power failure affecting eight of its twenty transponders. This led Arabsat to pay for capacity on PanAmSat's PAS-5, which was placed side by side with Arabsat-3A at 26°E. In 2003, Eutelsat's HotBird-5 (later renamed Eurobird-2) was also hired by Arabsat and slotted close by at 25.8°. That was the point at which Arabsat finally started ordering more satellites of its own, placing orders for two craft to be launched in 2006. But the first of the 2006 launches failed through a rocket booster malfunction, resulting in the total loss of Arabsat-4A. This crisis was temporarily resolved through further reliance on Eutelsat. Its aging HotBird-1 satellite was moved from 13°E to 25.5°E to back up Eurobird-2, and Arabsat's lease on Eurobird-2 was extended to March 2007, pending the November 2006 launch and subsequent entry into service of Arabsat-4B. Arabsat worked much harder for self-reliance from then on. After two successful launches, of Arabsat-4B in 2006 and Badr-6 in July 2008, the company had four satellites in action at two orbital positions, 26°E and 30.5°E. It also had an additional four satellites lined up for launching at a rate of one per year from 2009 to 2012, including one at 20°E.

Eutelsat maneuvers on behalf of Arabsat and Nilesat in 2005 through 2007 reflected a multilateral interweaving of the interests of global business and sovereign governments. Formerly an intergovernmental venture, Paris-based Eutelsat was privatized in 2001, after which its shares changed hands between private equity firms, European telecoms companies, banks, and small investors. It was thought in 2004 that Eutelsat, as one of the world's three leading satellite operators, might buy all or part of PanAmSat Corporation, the world's first privately owned global satellite service provider. It was put up for sale by News Corporation of Australia, which had acquired PanAmSat's dominant shareholder, Hughes Communications, in 2003. Instead, PanAmSat was eventually sold to Intelsat in 2006. Intelsat, headquartered in Bermuda, is—like Eutelsat—a former intergovernmental operation that was privatized in 2001.[19] In 2005, Intelsat ownership moved to a consortium of private equity funds led by Apax Partners, an international company based in the United Kingdom. Given Arabsat's existing use of PAS-5, Intelsat was ready to bolster the leasing arrangement when Arabsat-4A was lost in 2006. However, Eutelsat reportedly objected on the grounds that PAS-5 signals risked interfering with Eutelsat craft near 26°E.

Meanwhile, Eutelsat had to juggle its commitments to Arabsat and Nilesat with a promise it had already made to Noorsat ("Light" Sat), a privately owned newcomer to the satellite market with headquarters in Bahrain, marketing operations in Jordan, Saudi ownership, and a plan to lease capacity from major operators for subletting to television and telecommunications companies.[20] Omar Shoter, a former Arabsat director general who formed Noorsat in 2004, announced the planned leasing of capacity on a new Intelsat craft located at 1°W and later gained access to Eutelsat's Eurobird-2 and Hot Bird-4. Mawared Group, Saudi owners of the pay-TV network Orbit, took a 30 percent share in Noorsat, possibly influenced by their experience in 2001, when three of the eight transponders hit by power failure on Arabsat-3A were in use by Orbit at the time. But Noorsat's strategy was also to bring content to Arabic-speaking viewers in Europe, using Eutelsat facilities like Atlantic Bird-2 at 8°W and Eurobird-9A at 9°E, which would allow this content to be received with very small dishes.

Thus the activities of Arabsat, Nilesat, and Noorsat created a web of interlocking interests between private and public businesses inside and outside the Arab world. In December 2006, despite being potential competitors, Arabsat and Nilesat signed a strategic partnership agreement enabling Arabsat to provide broadcasters based in Egypt with direct uplinking and other services, similar to those already provided via Arabsat gateways in Germany, Spain, and six Arab states. The chairmen of Arabsat and Nilesat said on signing that there would be more cooperation in the future. In March 2009, a new agreement between Nilesat and Eutelsat sealed their close relationship for the next ten years. With Nilesat's orbital position at 7°W having become what Eutelsat press releases described as a shared "neighborhood," the new pact envisaged further "neighborhood" development through the launch of a tenth HotBird satellite, to operate at 7°W under the name AtlanticBird-4A, with the further addition of AtlanticBird-4R at the same location planned for 2011. An official statement said AtlanticBird-4A would enable Nilesat to continue increasing the number of transponders it leased from Eutelsat until 2019.

Following the arrival of Yahsat in 2007 and SmartSat in 2009—both based in the UAE, one with government backing and the other making much of its private ownership—the network of cross-border interests grew ever more complex. Yahsat's sole owner is Mubadala, a strategic investment vehicle established in 2002 by the government of the UAE's richest emirate, Abu Dhabi, and chaired by Abu Dhabi's crown prince. With interests in energy, heavy industry, and infrastructure, as well as telecommunications and aerospace, Mubadala describes itself in promotional material as a "patient investor" with a strong "commitment to the United States." This commitment is reflected in a US$8 billion joint venture with GE, partnerships with the Cleveland Clinic and New York University, and shareholdings in the private equity fund Carlyle Group and computer-chip maker Advanced Micro Devices. Yahsat ordered two satellites for launching in 2010 and 2011, to offer a range of interactive services besides broadcasting. Meanwhile SmartSat, owned by Smartlink (a Jordanian broadband services provider) and Al Jawahara Holding (a Kuwaiti investment company), announced its planned investment of US$500 million in the first privately owned Arab satellite, to be launched in 2011. With headquarters in Dubai, SmartSat said it intended to provide capacity for media, telecom, and military users in the region.

Opaque Forms of Collusion in Channel Blocking Practices

Any comparison of Arabsat, Nilesat, Noorsat, Yahsat, or SmartSat transponder resources (actual or planned) with those of giants like Eutelsat and Intelsat highlights a major imbalance in their respective abilities to control the content beamed by satellite to Arab viewers. Acknowledgment of the imbalance was implicit in the Arab Charter for Satellite TV, a regulatory document proposed to all Arab states by the governments of Egypt and Saudi Arabia at a meeting in Cairo in February 2008. The pan-Arab charter aimed to control satellite broadcasts from the ground by subjecting them to the same vaguely worded restrictions contained in most Arab press laws, which prohibit material deemed "offensive" to political or religious leaders or "damaging" to "social harmony" or "national unity." Under the charter, any channel flouting these or other rules would have its license and work permits withdrawn. Nilesat's chief technical officer, Salah Hamza, explained the charter as an attempt by Egypt to transfer the role of gatekeeper from satellite operators to national licensing bodies in individual Arab states. He told Chris Forrester of *Rapid TV News* that it was unreasonable to expect Nilesat to act as gatekeeper for all 335 free-to-air channels it was transmitting at that time. "We at Nilesat, and our fellow satellite operators, are really squeezed from viewers and authorities who might have justifiable complaints," Hamza said. "In effect we are judge, jury, and the only licensing body, and it is not a responsibility we want to have."[21]

In fact, the much-publicized Arab Charter for Satellite TV was not endorsed by the governments of Qatar, Lebanon, or the UAE. Undaunted, Egypt's own information minister, Anas al-Fiqi, later declared that Egypt would implement the charter anyway. But the Egyptian authorities had already started to remove selected channels from Nilesat during the early months of 2008 without recourse to any explicit content regulation. When enforcing removals they either declined to give reasons or cited bureaucratic technicalities such as incomplete or inconsistent paperwork. In the case of Al-Hiwar (Dialogue), a London-based satellite channel that had broadcast criticism of human rights abuse in Egypt, the channel was blocked from Nilesat, with no explanation, a year before its contract with the satellite operator was due to expire.[22] The Committee to Protect Journalists wrote to Nilesat's chairman on April 8, 2008, describing the closure of Al-Hiwar as "secretive" and calling for public clarification. Since Al-Hiwar managed to continue reaching the same viewership, via Eutelsat's AtlanticBird-4 right next to Nilesat's orbital slot, its suspension from Nilesat seemed little more than symbolic. It did, however, imply that Nilesat might have reasons to communicate with Eutelsat over matters of content. Al-Hiwar employees say their transmissions from AtlanticBird-4 were stopped for a few hours in January 2009 after callers to the channel criticized the Egyptian president's policy of closing the country's border with Gaza.[23] In May 2009, Al-Hiwar stated publicly that it had suffered deliberate and repeated jamming disruption to its transmissions via satellites owned by both Arabsat and Eutelsat.[24]

As the Al-Hiwar example illustrates, the production, uplinking, and transmission of programs by satellite typically involve businesses and regulatory authorities in multiple countries with different strategic interests. It seems that these differences necessitate informal bilateral or multilateral negotiations and that the techniques for deciding outcomes are not always clear-cut. For instance, the Egyptian authorities initially declared themselves unconcerned about footage of the shooting and bombing of U.S. occupation forces in Iraq being shown on a channel called Al-Zawraa, which had a contract with Nilesat. Yet Al-Zawraa, apparently funded by Iraqi politician Mishaan al-Jabburi, who was accused of embezzling reconstruction funds, was targeted for closure by the United States, both at its Iraqi offices and at its satellite positions in space.[25] Egypt's Anas al-Fiqi told an American journalist in early January 2007 that the U.S. ambassador in Cairo had approached him about the channel "in a friendly way" but lodged no formal request to remove it. Fiqi said Nilesat's contract with Al-Zawraa was "pure business" and that Egypt could only act to remove a channel in response to an Arab consensus, not U.S. pressure.[26] Within two months, however, Al-Zawraa was gone from Nilesat-101. Egyptian press reports of its removal blamed jamming as the cause. They said Nilesat official Salah Hamza had tried informally to persuade Jabburi to tone down the content, but that, in the end, the channel had been taken off the air because, according to Nilesat chairman Amin Basyouni, its broadcasts were being electronically jammed from an unknown source, which was causing interference with other channels.[27]

Lack of transparency about Nilesat's reasons for removing Al-Zawraa was compounded when the Arab Charter for Satellite TV was proposed in 2008. Hussein Amin, an American University in Cairo (AUC) professor who advises the policy committee of Egypt's ruling National Democratic Party and who drafted the Arab Charter, cited the example of Al-Zawraa when he told a journalist in April 2008 that the charter was meant to protect Arab youth from violence and "hate campaigns" run by terrorist groups.[28] This logic, together with Saudi Arabia's support for the Arab Charter, was at odds with Al-Zawraa's more prolonged presence on two Arabsat craft, Badr-3 and Arabsat-2B, and its appearance on a Eutelsat AtlanticBird-4 transponder leased by

Saudi-owned Noorsat, all observed by BBC Monitoring. In January 2007, the channel broadcast for just a day on Eutelsat's Eurobird-2 before being removed. Three months later, Eutelsat transmissions of Al-Zawraa were formally stopped under French law.[29] But it was still showing on Badr-3 in May 2007, despite U.S. State Department protests to the Saudi government.[30] According to the newsletter *Intelligence Online*, jamming eventually disrupted Al-Zawraa's transmissions to the point where it had to cease broadcasting in July 2007.[31] Arabsat's approach to Al-Zawraa can be contrasted with its approach to channels featuring self-appointed healers promising to cure illnesses and solve personal problems via TV. In June 2007, Arabsat's chief executive officer, Khaled Belkhyour, said satellite channels using Arabsat faced suspension for propagating "myths, falsehood, and charlatanism."[32] His pronouncement followed representations by a group of religious scholars who said such practices were against Islam.

The Dubai government's decision in November 2007 to block the signals of a private Pakistani network, Geo TV, operating from studios in Dubai Media City via several satellites, was attributed at the time to maneuvers between political leaders behind the scenes. The network was ordered off the air in Pakistan, along with a raft of domestic and international channels, when it refused to abide by curbs introduced under a state of emergency declared by Pakistan's then president, General Pervez Musharraf. Two weeks later it transpired that Geo TV and ARY One World had also been ordered to cease operations from Dubai in the UAE. The channels alleged that Musharraf had personally pressured his Dubai counterpart to take this step. ARY One World managed to get reinstated quickly after signing a letter promising to abide by regulations set by TECOM, the body that runs Dubai's free zones, which Dubai officials said were consistent with the UAE's policy of neutrality in foreign affairs.[33] Geo TV's reinstatement took longer and was more troubled. A Reuters report in June 2008 said the channel had been warned it could lose its license again unless it dropped two talk shows that had given wide coverage to judges dismissed by Musharraf. In the judgment of the International Federation of Journalists (IFJ), the Dubai authorities' treatment of Geo TV was not transparent. The IFJ called on the UAE government to "explain why, and on whose authority" it had asked the broadcaster to cancel the programs.[34]

Likewise, Eutelsat stood accused in 2008 of being less than clear in its treatment of a Saudi opposition station, Islah TV. Islah TV was created in 2003 (with no budget for video programming) by the Movement for Islamic Reform in Arabia, using a satellite uplink in Croatia and HotBird bandwidth leased by Deutsche Telekom. Islah's backers considered this the best way to reach households in Saudi Arabia after Islah Radio's shortwave broadcasts were jammed. But when Islah TV broadcast a call for demonstrations in Saudi

Arabia in October 2003, mysterious jamming was directed at Deutsche Telekom facilities near Frankfurt. According to David Crawford of the *Wall Street Journal*, the jamming affected five television programs broadcast on the same transponder, prompting Deutsche Telekom, which had reportedly also received anonymous telephone warnings, to cancel its contract with Islah TV.[35] Islah TV's head, Saad al-Faqih, was later added to lists of terrorists compiled by the United States and the United Nations. Noting this sequence of events, Monroe Price has observed that a "variety of informal arrangements" were implicated in what happened to Islah TV in 2003 and 2004.[36] Five years later Reporters Sans Frontières (RSF) criticized Eutelsat for informality in its treatment of channels regarded as troublesome by the Saudi Arabian and Chinese governments. Eutelsat suspended New Tang Dynasty Television (NTDTV) in June 2008, citing technical problems. But RSF recorded a conversation in which a Eutelsat representative revealed that China had put pressure on Eutelsat to drop NTDTV. The pressure had come at a critical moment for satellite broadcasting, just before the Olympic Games in Beijing, when Eutelsat was looking to increase its business ties with China.[37] The following month Eutelsat stopped carrying Islah TV after one of its HotBird transponders was subjected to interference. RSF deemed the case similar to that of NTDTV because it alleged that Eutelsat had terminated its contract with the backers of Islah TV "without trying to establish the origin of the interference."[38]

Interlocking multilateral interests were also at play in dealings with Al-Aqsa TV's satellite channel, launched by the Palestinian group Hamas after it won 74 out of 132 parliamentary seats in the 2006 elections in the West Bank and Gaza Strip. Ensuing armed conflict between Palestinian factions was reflected in struggles over broadcasting, with the Fatah-led Palestinian Authority reportedly urging the Egyptian government to ban Al-Aqsa TV from Nilesat. Nilesat's response was that it had no direct dealing with the channel because its capacity was leased to Al-Aqsa TV via an intermediary.[39] In January 2009, the channel went on air on three Eutelsat satellites (Eurobird-2, Eurobird-9, and AtlanticBird-4), but was taken off almost immediately when the French regulatory body, the Conseil Supérieur de l'Audiovisuel (CSA), intervened. According to an Agence France-Presse (AFP) report of January 9, 2009, Eutelsat then explained that the space for Al-Aqsa TV had been leased to Noorsat and that, on receiving the CSA's warning, it had asked Noorsat to respect international and national laws on channel content. AFP reported that the Simon Wiesenthal Centre, based in New York, had written to the CSA voicing concerns about Al-Aqsa programs.[40]

Conclusion

Western writing about militant channels such as Al-Zawraa, Al-Manar, and Al-Aqsa often gives the impression of a struggle in which Arab and non-Arab interest groups are on opposite sides, with Arab state-owned satellite opera-tors defending their sovereignty and control over satellite transmissions.[41] Indeed, a draft resolution (number 1308) presented to the U.S. Congress in June 2008 even sought to have the two main Arab satellite platforms classi-fied as terrorist organizations.[42] Yet the evidence presented here suggests that any such supposed struggle may belong with other "counterfeit geogra-phies" of the Middle East, since the deployment of satellite capacity and the gate-keeping process have involved multilateral interactions that are some-times quite opaque in terms of where decision-making power resides and how decisions are enforced.[43] For Arab satellite operators, the logistics of launching satellites into space and maximizing capacity for transmission require a willingness to deal with non-Arab operators as well as with each other. For non-Arab operators, cooperation with entities like Nilesat and Arabsat has enhanced their access to a larger number of orbital positions. Such deals create or strengthen networks of shared interests, which under-mine the notion of a divide reflecting divergent security concerns.

The evidence also showed that during the period of Arab satellite chan-nel expansion that followed 9/11 and the 2003 invasion of Iraq, "Arab" cen-sorship of channels was mainly conducted through informal methods. Several cases mentioned here involved unexplained jamming or private negotiations, and were denounced by media freedom advocates as nontrans-parent. The outcomes in these cases may indicate a trend away from unilat-eral assertion of sovereignty to negotiated multilateral consensus, as discussed in the introduction. Yet the means by which they were achieved were coercive, not "soft" like the "soft" consensual informal institutions and policy instruments discerned in governance of regional communications elsewhere. In the very diverse cases of Al-Hiwar, Geo TV, Islah TV, and even Al-Zawraa, informal coercion voided each power struggle of questions about legal rights and responsibilities and shielded those aspects of the struggle from public scrutiny.

NOTES

1 Naomi Sakr, *Satellite Realms: Transnational Television, Globalization and the Middle East* (London: I. B. Tauris, 2001), 163.

2 Amir Hassanpour, "Diaspora, Homeland and Communication Technologies," in *The Media of Diaspora*, ed. Karim H. Karim (London: Routledge, 2003), 84–85.

3 Foreign policy dimensions of the French decision are highlighted in Olfa Lam-loum, "De 'la Nocivité' des Chaînes Arabes," in *Médias, Migrations et Cultures Transnationales*, ed. Tristan Mattelart (Paris: de Boeck, 2007), 127. For other aspects of the decision, see Naomi Sakr, "Diversity and Diaspora: Arab Commu-nities and Satellite Communication in Europe," *Global Media and Communication* 4(3) (December 2008): 283–287.

4 Yigal Schleifer, "Denmark, Again? Now It's Under Fire for Hosting Kurdish TV Station," *Christian Science Monitor*, April 21, 2006. Tension surfaced strongly in

March 2009 when Turkey reportedly threatened to veto the Danish prime minister's appointment as NATO secretary general.

5 Komsan Tortermvasana, "Thaicom Cuts Transmission of 'Terrorist' TV Channel," *Bangkok Post* website in English, January 17, 2000, accessed March 18, 2011, http://blogs.rnw.nl/medianetwork/lebanese-al-manar-tv-launches-on-indonesian-palapa-c2-satellite; Andy Sennitt, "Lebanese Al-Manar TV Launches on Indonesian Palapa 2 Satellite," Radio Netherlands Worldwide Media Network website, April 8, 2008, accessed April 12, 2011, http://blogs.rnw.nl/medianetwork/lebanese-al-manar-tv-launches-on-indonesian-palapa-c2-satellite. See also press reports from Jakarta at www.adnkronos.com, April 9, 2008, accessed April 12, 2011; and from Bangkok in the *Nation*, January 16, 2008.

6 Benjamin Weinthal, "Germany Bans Hizbullah Television," *Jerusalem Post*, November 20, 2008.

7 In June 2009, Al-Jazeera English signed its first major U.S. distribution deal with a cable TV company in the Washington, D.C., area.

8 Lisa Parks, *Cultures in Orbit: Satellites and the Televisual* (Durham, N.C.: Duke University Press, 2005), 2, 174, 182.

9 Colin Sparks, *Globalization, Development and the Mass Media* (London: Sage, 2007), 215.

10 Ibid., 224.

11 Lena Jayyusi, "Internationalizing Media Studies: A View from the Arab World," *Global Media and Communication* 3(3) (December 2007): 253.

12 Ibid., 251–253.

13 Stephen Lukes, *Power: A Radical View* (Basingstoke, U.K.: Macmillan Press, 1974).

14 Monroe E. Price, *Media and Sovereignty: The Global Information Revolution and Its Challenge to State Power* (Cambridge, Mass.: MIT Press, 2002), 230.

15 Maria Michalis, *Governing European Communications: From Unification to Coordination* (Lanham, Md.: Rowman and Littlefield: 2007), 4.

16 Chris Forrester, "Contradictions and Challenges in the Middle East," *Cable & Satellite International* (May–June 2007): 34.

17 Sakr, *Satellite Realms*, 161.

18 Unless otherwise stated, information for this section was obtained from company press releases and the trade press, notably *Rapid TV News*, *Space News*, and *Cable & Satellite International*.

19 For more on the privatization, see Daya Kishan Thussu, "Privatizing Intelsat: Implications for the Global South," in *Global Media Policy in the New Millennium*, ed. Marc Raboy (Luton, U.K.: University of Luton Press, 2002), 39–53.

20 Chris Forrester, "NourSat, the New Satellite in the East," *Transnational Broadcasting Studies* 13 (Fall–Winter 2004), accessed March 18, 2011, http://www.tbsjournal.com/Archives/Fall04/forrester.html.

21 Chris Forrester, "Mideast TV 'Must Be Licensed,'" *Rapid TV News*, June 3, 2008.

22 Saad Guerraoui, "Al-Hiwar Another Victim of Egypt-Media War," *Middle East Online*, April 4, 2008, accessed March 18, 2011, www.middle-east-online.com/english/?id=25225=25225&format=0.

23 Anonymous Al-Hiwar employees, personal communication, London, April 12, 2009.

24 *Satnews Daily*, May 19, 2009, accessed March 18, 2011, http://www.satnews.com/cgi-bin/story.cgi?number=1583368063.

25 Marc Santora and Damien Cave, "On the Air, the Voice of Sunni Rebels in Iraq," *New York Times*, January 21, 2007.

26 Lawrence Pintak, "War of Ideas: Score Another One for the Bad Guys," January 3, 2007, accessed March 18, 2011, http://uscpublicdiplomacy.com/index.php/news room/pdblog_detail/war_of_ideas_score_another_one_for_the_bad_guys/.

27 See reports by *Al-Gumhuriya* and *Al-Masry al-Yaum* between February 26 and March 5, 2007.

28 Liam Stack, "Arab TV Feels the Pinch of New Broadcast Limits," *Christian Science Monitor*, May 2, 2008.

29 Monroe E. Price, "Orbiting Hate? Satellite Transponders and Free Expression," paper for "'Hate Speech' and Incitement to Violence Workshop Series," Columbia University, Spring 2009, 14, accessed March 18, 2011, http://www.law. columbia.edu/null/Orbiting+Hate_Monroe+Price?exclusive=filemgr.download &file_id=151394&showthumb=0.

30 Alistair Coleman, "Iraq 'Jihad TV' Mocks Coalition," BBC Monitoring, BBC Online, accessed March 18, 2011, http://news.bbc.co.uk/1/hi/world/middle_east/6644103 .stm.

31 Price, "Orbiting Hate?" 15. Price is quoting the article "U.S. Pulls Plug on Insurgent TV," *Intelligence Online*, January 17, 2008.

32 Ibrahim al-Thagafi, "Saudi Arabia: Channels Under Fire for Promoting Black Magic," *Asharq al-Awsat* English website, June 21, 2007, accessed March 18, 2011, http://www.aawsat.com/english/news.asp?section=7&id=9336.

33 Lynne Roberts, "Pakistan TV to Resume Dubai Broadcasts," *Arabian Business*, November 19, 2007; Ashfaq Ahmed, "Geo TV Also Plans to Move Out of Dubai," *Gulf News*, November 25, 2007.

34 International Federation of Journalists, "IFJ Demands UAE Overturn Ban on Pakistan TV Programs," June 12, 2008, accessed March 18, 2011, http://www.canada freepress.com/index.php/article/3473.

35 David Crawford, "A Battle for Ears and Minds," *Wall Street Journal*, February 4, 2004, accessed March 18, 2011, http://ics.leeds.ac.uk/papers/vp01.cfm?outfit= pmt&folder=1259&paper=1323.

36 Price, "Orbiting Hate?" 11.

37 Reporters Sans Frontières alert, July 10, 2008, accessed April 2, 2011, http://en.rsf .org/china-european-satellite-operator-10-07-2008,27818.html.

38 RSF alert, September 2, 2008, accessed March 18, 2011, http://en.rsf.org/saudi-arabia-european-satellite-operator-02-09-2008,28388.html.

39 Amr Al-Masry, reported in *Al-Mesryoon*, July 11, 2007.

40 In Chris Forrester, "Hamas Loses European TV Coverage," Rapid TV News, January 11, 2009, accessed March 29, 2011, http://www.rapidtvnews.com/index.php/ 20090111 2888/hamas-loses-european-tv-coverage.html.

41 For example, some of the articles referred to in these notes.

42 Ali Jaafar, "U.S. Bill Targets Arab TV Channels," July 7, 2008, accessed March 18, 2011, http://variety.com/article/VR1117988568.html?categoryid=1445&cs=1.

43 Jayyusi, "Internationalizing Media," 253.

8

Beyond the Terrestrial?

Networked Distribution, Multimodal Media, and the Place of the Local in Satellite Radio

Alexander Russo and Bill Kirkpatrick

For most people in the United States, "satellite radio," means direct broadcast satellite radio—Sirius and XM, which merged in 2008. These are relatively new players in the broadcasting world, beginning to beam their programming only at the start of the twenty-first century. Surrounding this form of satellite radio are discourses of newness and difference from "terrestrial radio"—new technologies, new choices, new possibilities for niche programming, and new business models. "Radio has been stuck in an engineering time warp for two generations," wrote Mike Langberg in the *Philadelphia Inquirer*, "[and] not much has changed since the introduction of FM about forty years ago." But now satellites are ushering in a "space-age radio revolution."[1]

Such rhetoric is misleading. Although satellite broadcasting may be colloquially thought of as a new technology, a perception encouraged by the "satcasters" themselves, such a notion elides a much longer relationship between U.S. radio and satellite technologies. Indeed, the rhetoric of the revolutionary "newness" of satellite radio, which conceives of satellite radio in opposition to long-established practices of terrestrial radio, silences the many ways in which satellite has a long and important role in American broadcasting, and masks a number of contradictions within radio practice today. This chapter seeks to reclaim satellite radio's history to attend to the ways it acts in conjunction with terrestrial radio as a multimodal distribution technology profoundly affecting what listeners have been hearing for decades. This focus on the relationship between distribution and content follows Lisa Parks's recent challenge to the fields of television and media studies to examine the impact of distribution technologies. She suggests, "if television technology is a historically shifting form and set of practices, then it is necessary to consider more carefully how the medium's content and form change with different distribution systems."[2] The same is true of post-1950s radio, the histories of which have largely conceived of the medium in terms of the station and its local audience, despite the fact that this narrative omits how and why certain kinds of program content reached the station. Satellite radio is one such system, consisting of a hybrid of distribution technologies: traditional terrestrial station-to-receiver broadcasts as well as the systems required for stations to obtain the programming and commercials they then rebroadcast. In its initial iterations, satellite radio was conceived as a means for two-way program exchanges as well as unidirectional program distribution. A policy context of deregulation and the acceleration of radio formats contributed to the dominance of the latter model in the 1980s and into the 1990s, when satellites were used to syndicate programs to revived radio networks. More recently, personal satellite receivers and program providers like XM and Sirius have emerged, operating in ways that are both in opposition to and in accordance with long-standing norms of terrestrial radio.

A historically informed look at the role of satellites in U.S. radio reveals that satellite technology has been central to "terrestrial" radio for decades, participating in major industrial shifts since the 1970s. These shifts include the proliferation of national networks, the increasing centralization and automation of programming, the intensification of audience segmentation, and an ongoing crisis in the supposed ontological qualities of "good" radio, namely "liveness" and localism. In this sense, the satellite radio of Sirius XM does not represent a strong break with terrestrial broadcasting—a "revolution" as popular imagination and marketing discourses would have it—but rather a continuation of long-standing trends and tensions in radio practice. This perspective also reveals how personal digital satellite radio's replication of the dynamics of terrestrial radio has made it particularly vulnerable to competition from the technologies of media convergence and helps account for its tenuous future.

Satellites and Radio Broadcasting: Multimodal Media

Although the first commercial telecommunications satellites went into orbit in 1962, satellites were not regularly used for program distribution in the United States until the 1970s. Western Union's launch of Westar I in 1974 provided new possibilities for distribution but was at first largely limited to television networks. Less discussed was interest in satellite distribution by radio news networks, which at the time delivered their national programming primarily over AT&T's landlines. Although some of these companies bore the names of iconic radio networks of the 1930s and 1940s, they were radically diminished entities in the age of television. In the late 1970s, commercial radio networks provided relatively small amounts of programming, usually just five minutes of news per hour and assorted features. In addition, while stations as a whole were enjoying large profits, networks were increasingly losing money. In 1976, a year in which radio stations enjoyed profits of US$172 million, a 70 percent increase over the prior year, the major networks (CBS, Mutual, NBC, and ABC's four networks) lost US$5 million, doubling their losses of the previous year. For networks that owned affiliates, station revenues offset the losses.[3] The lone exception, Mutual, did not own its affiliates and, not surprisingly, was the first to turn to the new technology of satellites to distribute its programs.

The networks saw four principal advantages to satellite radio: cost savings over landlines, multiplexing (delivering multiple content streams), improved sound quality, and (at least initially) the possibility of two-way content streams. All of these appealed to National Public Radio (NPR), one of the earliest adopters of satellite distribution. When the Corporation for Public Broadcasting (CPB) inaugurated plans to distribute PBS television programs via Westar I, NPR sought to "piggyback" its signal on the TV network's transponders.[4] As Jack Mitchell has chronicled, in 1978 NPR was able to appropriate 25 percent of the CPB budget to fund its satellite build-out. For news networks with numerous affiliates and a need for live national distribution, satellites offered substantial cost savings; in NPR's case, this reduced distribution costs for some programs from US$1,500 to US$50.[5] For nonlive programming, producer stations often relied on "bicycling," or physically sending tapes to individual affiliates, and NPR was understandably eager to cease being "a radio network that depends on the goddamn postal service," as its president, Frank Mankiewicz, colorfully noted in 1978.

Mutual, one of the other early advocates of satellite distribution, shared with NPR a larger and wealthier parent as well as a nontraditional affiliate structure. The network moved toward satellites shortly after its 1977 purchase by Amway. The large direct-merchandising company provided the network with an infusion of cash to finance its build-out via Westar IV.[6] Unlike NPR, Mutual was interested in a downlink-only system with small-size dishes. In 1979, Mutual was joined by another news network that did not own its affiliates, the Associated Press (AP). The AP's initial plans involved a 660-station network with thirty-seven uplink stations. It claimed that satellite transmission would save US$760,000 per year over phone lines.[7] Most other commercial broadcasters were more ambivalent about satellite program distribution.[8] NBC, ABC, and CBS embraced satellite distribution between their major production studios by 1979, but did not equip their affiliates with satellite dishes to receive the programming directly.[9] These networks had long-term contracts with AT&T and continued to use wire lines to send programming to affiliates until the mid-1980s.[10]

Nonetheless, the rising importance of FM and format consultants (and the subsequent institutionalization and standardization of much of the FM dial) would decisively affect how satellites were used. In the late 1960s, FM was, as Susan Douglas describes it, "not just a technical reaction against AM; it was a cultural and political reaction as well."[11] According to Douglas, the "free-form" or "underground" stations rejected the high-tempo, strong-sell antics of 1960s Top 40 in favor of a laid-back approach that emphasized long cuts of counterculturally oriented "progressive rock" music. Commercial station owners were initially reluctant to embrace such formats and deemed them unprofitable, but by the mid-1970s new audience research focusing on narrowly defined demographic groups and the reimposition of tightly formatted playlists homogenized and rationalized FM broadcasting. In 1973, FM had a 28 percent share of the radio audience. Five years later it was up to 49 percent, and in 1979 it would overtake AM.[12] These changes, often driven by a reliance on professional consultants such as Lee Abrams and California's Drake-Chenault Enterprises, dramatically increased the profitability of FM stations and established the template for program philosophies that would become increasingly important in the satellite era.

At the same time, consultants and owners were eagerly anticipating the refinement of technologies to automate and computerize stations. Rudimentary automation systems like the Gates "Autostation" had been introduced as early as the 1950s and were becoming more practical by the 1970s.[13] One-seventh of all stations were automated by the mid-1970s, although these were largely "beautiful music" stations that featured few, if any, interruptions by disc jockeys.[14] One limiting factor for automation was program distribution. FM was predicated on high-quality stereo sound that AT&T landlines were incapable of carrying. Program syndicators had to record music and vocal breaks on audio tape and then use bicycle distribution to deliver them to stations. This cost much less than wire line charges, but it limited programmers' flexibility while also violating the principle of liveness that had been a key marker of "quality" radio since the 1920s. The resolution of such issues became an important part of how satellite technology found a home in terrestrial broadcasting.

Programming syndicators were initially wary of satellite distribution because of the cost involved, but this quickly changed. In a panel discussion titled "The Syndicated Program Revolution" at the 1979 National Association of Broadcasters Annual Radio Programming Conference, one syndicator dismissed satellite distribution as "untried, untrue, and very expensive."[15] Tom Rounds, president of Watermark, the syndication firm that produced Casey

Kasem's *American Top 40*, did not think satellite distribution would be cost effective until "a majority of stations have dishes in their parking lots."[16] This skepticism was misplaced, however. One month later, the Federal Communications Commission (FCC) relaxed its requirement that Earth stations must be capable of transmitting as well as receiving, making it much less time-consuming and expensive to equip network affiliates with receive-only dishes.[17] Within two years, so many new satellite-delivered program services had entered the arena that syndicators began talking of their industry as a "Darwinian world" of increased competition for affiliates and listeners.[18] This dramatic shift of perspective among syndicators was replicated across the broadcasting industry as U.S. radio found both new uses as well as older adaptations for satellite distribution.

The Integration of Satellites into Radio Broadcasting

By 1981, it was clear that satellites would be an integral component of much radio broadcasting, and the first half of the 1980s became an era of resolving the regulatory issues involved, consolidating new business models and industrial relationships to take advantage of satellite distribution, and squaring the philosophical contradictions that satellite-delivered programming represented to an industry that still privileged liveness and localism in its notions of quality, however half-heartedly or even hypocritically. It was not that every station owned a satellite dish by the mid-1980s, or even necessarily wanted one, but the political, economic, and cultural structures of the satellite's role in U.S. radio were largely in place by that time. This set the stage for the consolidation and nationalization of programming trends that characterized the second half of the 1980s and the 1990s.

Although the rapid adoption of satellite distribution was not inevitable, several aspects of the era facilitated the emergence of the satellite boom of the early 1980s. First, the broader recession of the late 1970s had resulted in an ongoing economic crisis in the radio industry, but in a way that fell differently on different sectors. Although there was a downturn in the industry as a whole, in 1980 network advertising sales increased 33 percent and national spot sales rose 45 percent.[19] National advertisers turned to radio as a cheaper alternative to television, thus making satellite networks an ever more viable alternative to the national spot advertising market, a shift encouraged by the increasing demographic segmentation that radio provided. Satellite networks and syndicators therefore stood at the intersection of two key commercial imperatives for large sponsors: efficient national saturation and the ability to effectively target niche audiences. In this, satellite radio offered the same advantages as cable television networks, which were also gaining popularity around this time, but at lower cost and with hundreds of millions of radio receivers already in place.

A second important contextual element in the growth of satellites was the ongoing deregulation of the media industries that began under President Jimmy Carter and accelerated during the Reagan era. Federal regulatory support for localism was weakened during this period, easing the adoption of nonlocal satellite programming by local stations. Dating to the 1920s, localism describes a regulatory philosophy that required stations to justify how their programming served community interests of the area in which they were located, policies that broadcasters had long viewed as a thorn in their sides.[20] In 1981, the FCC reduced station obligations to meet with community leaders and air public affairs programs that addressed local issues. This allowed broadcasts to slash news staffs or even eliminate locally originated news programming altogether. The cumulative effect of these changes was a regulatory climate in which stations faced substantially less political pressure to produce local content or justify the public-interest merits of their programming choices. This, in turn, allowed them to feel increasingly confident in using more nonlocal content. With satellite technology maturing and lower costs as a key selling point, satellite distributors were perfectly poised to take advantage of this new regulatory attitude; as one executive of the Satellite Music Network put it in 1981, "We didn't have deregulation in mind when we started the network, but I think it will help us by making it easier for hopeful customers to change their formats and switch to our service."[21]

Against this backdrop, several radio networks followed NPR's and Mutual's lead by moving heavily into satellite distribution in 1981 and 1982, most notably RCA. Although some networks, like RKO, provided satellite dishes to their affiliates in exchange for long-term contracts, the price of dishes was falling precipitously enough that many stations began purchasing them on their own. So many distributors wanted to get into the game that there was a temporary shortage of transponder space, a condition that was made more acute when RCA's Satcom III was lost in space in 1979. By 1983, there was again enough capacity for a multitude of companies to affordably turn to satellites, including ABC, NBC, and CBS all using Satcom I-R.[22] Satellite distribution offered not merely cost savings and higher quality but also solved the problems of timeliness and flexibility faced by syndicators in bicycling distribution. Importantly for the medium's development, downlinks were cheaper and more plentiful than uplinks, privileging a model of a few centralized content providers feeding programming to individual stations rather than an open market of hundreds of individual stations potentially providing satellite-delivered productions to hundreds of other stations.

The technological possibilities of national satellite distribution dovetailed with the industry's already strong and growing reliance on market research and programming consultants; indeed, many of these emerging satellite networks were connected with the same radio consultants who had pioneered formats in the 1970s. Satellite networks claimed to provide one-stop shopping for on-air talent, expert research, and playlist consultation, at a lower cost than individual stations could reproduce on their own. An important part of this shift was the move from a few minutes of satellite-delivered news to longer-form programming; increasingly, radio networks offered play-by-play sports coverage, concerts, and other special features.[23] At the extreme end of this trend, 1981 and 1982 saw the launch of fully automated twenty-four-hour "turnkey" services, such as Chicago's Satellite Music Network (SMN) on Satcom III-R and ABC's New York–based Superadio on Westar III; these services potentially eliminated the need for a local programming staff altogether. One SMN executive boasted, "[a]ll you need is your sales department and a production guy," since his company offered complete programming in the adult contemporary, country, or MOR (middle-of-the-road) format for less than US$1,000 a month.[24] In practice, most affiliates continued to program the lucrative drive time on their own, as well as local inserts during the daytime hours. Turnkey operations first took off in the overnight shift in smaller markets, where it rarely paid to have live local talent running the station.[25]

Satellite technology was thus developing hand in hand with research-fueled niche marketing and computerized automation, each driving the other through technological development, shifting programming philosophies, reduced regulatory enforcement of localism, and above all the economic advantages of these new ways of doing radio. However, such changes challenged long-held ideas about "quality" radio as being local and live—ideas that might have seemed antiquated or obsolete, but that held powerful sway over broadcasters, regulators, and audiences. As automation and turnkey services increased, so did the need to reconcile them to the industry's ideas about itself. The satellite boom of the early 1980s thus also included the uneasy justification and legitimation of these practices on the part of several different formations within the industry and the public at large.

In particular, the diminishment of the value of local content during this period required a fair amount of ideological negotiation. The postwar era had allowed the appearance of a comfortable and easy understanding of the social role of media forms: TV was supposedly primarily (although never exclusively) "national," while radio was primarily "local." In that context, broadcasters had been told for decades that local service was not just their duty as federal licensees but also good for the bottom line—yet these new distribution possibilities challenged broadcasters to rethink their operations as smart businessmen, responsible trustees of the public interest, and "showmen" offering quality radio.[26] "More and more stations," said one consultant in 1981, "are having to decide whether to go with nationally distributed program formats or increase their emphasis on community needs."[27] Although that either-or formulation was overstated, the early 1980s were rife with discussions about how to reconcile such tensions.

On the policy side, even as the FCC voted six to one to relax local content requirements in 1981, regulators claimed that they were not abandoning local public-service programming. As one commission staffer claimed, "[T]he issues have to be local [but that] is not to say that the only way to serve a local outlet is through local programming."[28] Likewise, FCC commissioner Joseph Fogarty stated that he would not want "the religious principle of localism to stand in the way" of satellite build-out.[29] Such pronouncements worried some in the industry as stations embraced satellite syndication. In 1982, a National Association of Broadcasting (NAB) panel of radio network executives promised that their programs "are merely tools and not necessarily the end of local programming."[30] Two years later, in the context of moves to relax ownership caps on radio stations that were de-localizing content and ownership, Bernie Mann, head of the National Radio Broadcasters Association, tried to reassure broadcasters that "radio is a very local business. . . . Let Sears try to run radio stations like they run department stores. It just is not the same business."[31]

Similar ambivalences were found in industrial discussions of the value of localness as a register of quality—a definition that existed uneasily with emerging practices of national satellite distribution. Satellite networks and syndicators themselves bent over backward to reassure potential affiliates that they could air satellite-delivered content and still be—or at least *sound*—local. For example, a 1982 ad for SMN read: "Local identification is another area of great concern to station operators. . . . Local I.D.s, local news, traffic reports and even special locally-produced shows can be easily accommodated. Most listeners . . . aren't even aware that they're listening to a radio network."[32] Many stations did indeed create exactly such hybrids, repackaging satellite-delivered programs with local deejays and opportunities for local listeners to call into the station. KTRH in Houston, Texas, for example, would have its local on-air staff announce a nationally distributed sports program, play the program and a network spot commercial, return to local staff commenting on the program, and then open the phone lines to local listeners.[33] The head of ABC Radio, Ben Hoberman, touted this kind of synergistic relationship between his networks and their affiliates as a selling point for satellite services: "Stations may now take satellite feeds from national program sources and seamlessly cut in and out, adding the key ingredient—local flavor and identification."[34] The early 1980s were full of such pronouncements that local identity still mattered and that satellite distribution was compatible with a station's commitment to localism. Although these struggles over radio continued and played out differently on a market-to-market basis, they did not markedly slow the incorporation of satellite technology into terrestrial broadcasting.

One reason for the lack of consensus on the value of "the local" was the decline of mass-oriented broadcasting and its replacement by niche-oriented narrowcasting. The multiplexing capabilities of satellite distribution challenged the popular equation of "network" with "mass." ABC alone had six different demographically distinct networks in 1982 spread over 1,800 affiliates.[35] Possibilities of multiple content streams provided a powerful additional rationale for early adopters of satellite radio programming. Ostensibly, they saw it as a way to combine national content with local station choice, but their plans suggest ways in which industrial and regulatory definitions of localism fit uneasily with the developments of formats and national consultants. For example, Mutual planned to offer three or four programs per market to 500 of the stations within its network of 700 affiliates. The company's president, Gary Worth, described these plans as "extending to a national level the previously local idea of multiple formats." In a similar fashion, NPR noted that satellites would enable the network to offer its affiliates breaking news or feature-length cultural programming for the same time-slot.[36] Segmenting audiences by age, religion, political orientation, or race and ethnicity, multiplexing recast localism as the satisfaction of local preferences from among nationally produced offerings—a state of affairs resonant of network-affiliate relationships in the 1930s and 1940s.

From the mid-1980s through the 1990s, the industry pursued strategies of consolidation facilitated by a deregulatory environment that reduced restrictions on the number of stations a single entity could own. In 1984, San Antonio–based Clear Channel Communications, up until then a fairly small ownership group concentrated in the Midwest and South, went public and purchased Broad Street Communications, giving it twelve radio stations.[37] The following year, industrial conventional wisdom was upended when Capital Cities, a company largely focused on small-market stations, purchased ABC, including its radio networks. This purchase shocked the industry, which considered Capital Cities a small-market group, hardly the equal of the legacy network. However, as Cap Cities, Clear Channel, and, later, Infinity would also show, smaller-market stations were inexpensive and, collectively, could produce significant profits. These station groups further developed centralized modes of radio production that made use of satellite-distributed programming.

The increasing size of ownership groups oriented toward second-tier markets dovetailed with the increased use of satellites for distribution of national music and talk formats. For the declining AM band, hampered in the competition for music listeners by its poorer sound quality, satellite distribution made feasible nationally syndicated live talk shows, particularly sports and right-wing political programs (especially following the elimination of the Fairness Doctrine in 1987).[38] At the same time, the use of satellite networks to distribute music intensified and branched out from the fairly anonymous adult-contemporary and beautiful music formats to pop, rock, country, and metal. Critically for these small-market stations, the ability to obtain inexpensively what they perceived as higher-quality programming drove their decisions.[39] Moreover, by this time the cost of a receive-only satellite dish was less than US$5,000, making the move into satellite-distributed programming affordable for nearly any station.[40]

These debates over the nature and quality of terrestrial radio and trends toward national distribution prefigured many of the struggles that U.S. radio would face again in the 1990s and 2000s with the emergence of XM and Sirius. The local continued to exist as an abstract, albeit contested, value; although no longer enforced in practice, localism remained important within industrial and popular rhetoric as a register of the connection between a station and its audience. Still, discourses of taste, cultures, and audience preferences began to supplant ideals of a public interest at the same moment that centralized national satellite distribution began to offer the demographically distinctive programming that could sound local. The tensions between technological possibilities, industrial practice, and promotional discourse would play out in the development of direct broadcasting services in the 1990s and 2000s.

Satellite Digital Audio Radio Service (SDARS)

In 1981, an article in *Broadcasting* peered into the future to predict the state of broadcasting in 2001. "The cornerstone of the broadcasting medium in the future will not be the local station alone, however. It will be joined . . . by high-powered communications satellites that, hurtling through space at ten times the speed of sound, will beam multiple channels of programming to virtually every home in the country."[41] This prediction was uncannily accurate: in 2001 XM Satellite Radio began sending signals from its Washington, D.C., studios to its XM-1 and XM-2 satellites. However, the radio industry resisted rather than embraced the path to individualized satellite radio reception, and the eventual emergence of satellite-based digital audio radio service, or SDARS, was fraught with conflict and hesitation. Ultimately, this tension had less to do with conventional understandings of radio as an aural medium than in the investments of a wide variety of actors who conceived of the use of satellite technology in widely divergent ways.

Neither of the two major organizations that were ultimately responsible for developing SDARS had primary interests in broadcasting. Both Satellite CD Radio (which soon dropped "Satellite" and, later, became Sirius) and

American Mobile Radio (AMR) (later to become XM) had leadership with backgrounds in cellular telephony and other nonbroadcast use of satellites. Satellite CD Radio was half-owned by Martin Rothblatt, who had run GeoStar, a vehicle tracking system; its CEO, Dave Margolese, had a background in cellular telephony. AMR was a subsidiary of American Mobile Satellite, which had interests in cellular telephony and other mobile data-tracking systems and was itself partially owned by General Motors subsidiary Hughes Aeronautics; AMR's CEO, Hugh Panero, came out of the cable television industry.[42] These industrial connections suggest ways in which distribution concerns were the primary focus for these companies during the initial years of the development of SDARS, with content a distant second.

As it had with direct-to-home satellite television service, the traditional broadcasting industry viewed SDARS as a threat and opposed it from its inception in the late 1980s and early 1990s.[43] Despite industrial trends toward national programming distributed to stations via satellites, the NAB immediately and consistently invoked the threat to localism as its principal objection to SDARS and predicted "possible dire consequences" should satellite digital radio be developed.[44] One particularly apocalyptic station owner was Saul Levine of KKGO in Los Angeles, who called SDARS "diabolical" and predicted that his station would be "destroyed" by it.[45] All the while Satellite CD Radio and AMR denied that they planned to offer local news, weather, and traffic reports.[46]

Although these objections did not prevent CD Radio and AMR from receiving authorization from the FCC, the obstruction significantly delayed their plans, both directly and indirectly.[47] Initially, CD Radio had hoped to begin operations in 1995 but had to convince the FCC that SDARS would not harm the public interest.[48] Additionally, the objections by terrestrial broadcasters highlighted the limited market for SDARS radio, making it more difficult for CD Radio and AMR to secure the capital necessary to develop SDARS receivers and launch satellites.[49] Indeed, much of the decade's press and trade coverage displayed a marked skepticism toward the success of SDARS: would listeners be willing to pay for radio—that most humble of media they were so used to receiving for free?

In response to industrial resistance and investor skepticism, the SDARS industry articulated an evolving series of rationales for its product. These included historical analogies to the success of cable: "cable for the car," as *Automotive News* put it.[50] The possibility of beaming a cablelike variety of content into cars helped the industry secure investment and partnerships (General Motors and Honda would each invest US$50 million in XM in 1999 and 2000). Other rationales positioned satellite radio as the answer to what the SDARS firms perceived as the source of the public's dissatisfaction with terrestrial radio: too few genres, too many commercials, and too narrow playlists. They also crafted appeals to ill-served rural audiences and the passionate fandom of niche audiences whose favored genres were getting no airplay: "There are around forty-five million people in this country who live in markets that have five or fewer radio stations," pointed out a Sirius marketing executive. "In Detroit, there is no classical radio station. In New York City there is no reggae station."[51] Some analysts suggested that the reduced advertising alone would be enough to get listeners to pay: "[T]hey are paying for [terrestrial radio] now—they're paying for it with their time."[52]

Ultimately, it took satellite DARS until 1997 to secure authorization from the FCC. Soon thereafter "satcasters" turned to programming. AMR hired famed programming consultant Lee Abrams, in many ways an ideal choice. Abrams had pioneered formatting and market research in the 1970s and satellite network distribution in the 1980s, making him familiar with both the technological and content aspects of satellite broadcasting. AMR thought that Abrams's reputation as a format guru would enable XM to develop the large number of program formats that would comprise its service. Abrams's history developing satellite networks also placed him squarely against localism, which he noted "doesn't mean anything anymore," since "with a few exceptions, local radio died years and years ago."[53]

Abrams applied many of the same procedures used in FM broadcasting to satellite radio, with the crucial difference of bandwidth. He replicated existing logics of segmentation and psychographics in hopes that enough niches could equal a mass. There is some evidence that, at least initially, Abrams de-emphasized his research techniques (retail callback cards, focus groups, and polling at live concerts) in favor of more free-form programming, although XM's financial woes later caused it to emphasize its research-driven "hits" channels that more closely resembled terrestrial stations. In another irony, Abrams used disc jockey talent that had been displaced by the voice-tracked programming strategies that came out of satellite network distribution.[54] Abrams also made programming pacts with national content providers like *USA Today*, Bloomberg News Radio, and C-SPAN, but focused primarily on individual music channels.[55] This strategy of programming brands was exceeded by XM's New York City–based rival Sirius that, most famously, hired shock-jock Howard Stern for a five-year, US$500 million contract, touching off a bidding war for high-profile talent that added to both networks' heavy debt load. Despite such offerings, by the time XM launched its satellites in 2000, people had numerous other means to listen to a wide variety of programming, including car CD players, Internet radio stations, and MP3 players.[56] Even before launch, some skeptics worried that satellite radio was "behind the times" and "too late" for its business model.[57] Still, there was enough confidence in the service among potential investors that, in addition to General Motors' US$50 million, Clear

Channel invested US$75 million in XM, and DirecTV (linked with Hughes) invested US$50 million in 1999 as well.[58]

By the time XM and Sirius were ready to begin service in 2001 and 2002, respectively, public discontent with terrestrial radio was widespread, with listeners and media reformers alike lambasting the standardization and automation of contemporary radio. Such complaints were as old as broadcasting itself, of course, but they had been intensifying throughout the 1980s and 1990s in rough correlation to the consolidation of the radio industry. Even before the 1996 Telecommunications Act, companies such as Clear Channel and Infinity were butting up against the FCC's relaxed ownership limits, which grew from seven in 1981 to forty in 1996. Following the 1996 act and the resulting waves of conglomeration and centralization in radio, these complaints took on a new political urgency as influential media critics, most prominently Robert McChesney, drew political-economic connections between media consolidation and the power of conservative politics in the United States.[59] By the time of the emergence of XM and Sirius, terrestrial radio had, for many, come to represent not just standardization within the corporate music industry, or cultural standardization in American life more broadly, but also something of a standardization of political discourse dominated by the Right.

Against this backdrop, the revolutionary rhetoric that accompanied the rollout of the XM and Sirius SDARS systems seized on the idea of "satellite radio" as a potential (if necessarily only partial) solution to the terrestrial radio of the day. Lee Abrams, for example, promised that XM would break with past broadcasting models to allow each channel to offer a unique "point of view, without compromise."[60] Yet even at this moment of literal and figurative launch, there were already indications that satellite radio marketers were fighting the previous (and thus the wrong) war. More precisely, satellite radio and terrestrial radio were fighting each other over shares of a music industry that was undergoing transformations that stood to leave both of them behind. As Jody Berland has pointed out, the programming philosophy perfected by Abrams and others was predicated on a conception of audiences as more or less preformed demographic typologies that could be attracted by playing the music they liked—the listeners as the "target" of targeted programming: "Format music programming styles thus appear to spring from and articulate a neutral marriage of musics (country and western, Top 40, etc.) and demographics, and to correspond opportunistically to already established listener tastes, whose profiles are discovered through the neutral science of market research."[61]

The problem with this conception, argues Berland, is that listener tastes "are an effect, as much as cause, of this specialization process."[62] In other words, satellite radio might have offered more channels, but did nothing to alter this industrial conception of audiences as taste communities at whom radio pushed content. At the same time, however, new media technologies were allowing listeners to discover for themselves, through a wide range of "pull technologies" (in which the request for data, such as a specific web page, originates with the client) and socially networked modes of music discovery, the extent to which formatted radio had theretofore constrained their musical experiences.

Thus radio broadcasting's eighty-year history as a one-to-many, time-based medium, and its thirty-year history as a push-based provider of narrowly formatted programming (pushed at the client by the industry), cast a powerful shadow on both traditional and satellite broadcasters' ability to recognize the threat from new media technologies. Believing they understood the medium and the audience, the question broadcasters asked themselves was whether enough listeners would pay for radio to make direct satellite radio profitable, when the longer-term challenge was whether these time-honored models were still viable at all. The recording industry, for its part, was in no mood to support satellite radio, fearing that the increased sound quality and lossless reproducibility of digital signals would lead to illicit recording and distribution of music (fears that appeared to be realized when XM began incorporating recording capabilities into its receivers, leading the Recording Industry of America [RIAA] to file suit in 2006).[63] It is telling that XM and Sirius executives repeatedly described their competition as terrestrial radio, CDs, and cassettes, all but ignoring the shift to hyperpersonalized, portable musical choice represented by digital media players, most formidably the Apple iPod (released just a month after XM began its service in 2001). Although both companies offered a substantial amount of content unavailable on either terrestrial radio or portable digital media players, such as live sports or popular programming like Stern's show, their primary model of pushing narrowly segmented content streams remained vulnerable to the new possibilities of MP3 music and podcasts: privatized, mobile, and increasingly embedded in social relations that bypassed the mechanisms of taste production perfected by what Yochai Benkler calls the "industrial information economy."[64]

One more legacy of terrestrial radio returned to help shape SDARS: the turn to localism. Initially, as discussed earlier in this chapter, it was the terrestrial broadcasters who made an issue of satellite radio's provision of local content. Noting XM's and Sirius's use of terrestrial repeaters to augment the satellite signal and maximize coverage, especially in urban areas, the NAB darkly warned of a satellite conspiracy to take over local as well as national radio. As NAB president Eddie Fritts put it in 2001, "The time for subterfuge is over. These companies must come clean with regulators and the American people on their true intentions. . . . If XM and Sirius want to provide a traditional over-the-air radio service, they should apply for over-the-air licenses like everyone else."[65] While the NAB's rhetoric may have been over-blown, some of these repeaters were indeed in violation of FCC regulations, with at least a third of XM's 800 terrestrial antennas either placed in un-approved locations or operating above their approved power.[66]

What the NAB really objected to was not the technical violations, nor even the use of local repeaters for the delivery of national content (perhaps suggesting how little respect terrestrial broadcasters had for their own research-driven, consultant-polished national programming); rather, it was XM's and Sirius's provision of local content, especially traffic and weather reports, in major markets. Claiming that the 1997 FCC authorization of SDARS had prohibited satcasters from offering such local content, the NAB repeat-edly (and unsuccessfully) petitioned the FCC to put a stop to it. The potential hypocrisy inherent in this complaint—after two decades of incorporating satellite-delivered national programming into local radio, terrestrial broad-casters were now complaining about XM and Sirius incorporating terrestrial-delivered local programming into national radio—did not appear to faze broadcasters. The satcasters responded that they were allowed to offer such content as long as they did so nationally: "There is a difference between locally generated broadcasts and local information that happens to be broad-cast nationwide," a spokesman for Sirius claimed, meaning that they could offer New York City traffic reports as long as listeners anywhere in the nation were receiving those reports.[67] Some thirty years after the advent of the satel-lite radio, the technology still had the power to disrupt notions of the local and the national, even as it continued old patterns of radio programming into a far less certain digital age.

Conclusion

As of this writing (mid-2009), the long-term future of direct satellite radio broadcasting is in doubt. Burdened by tremendous debt, buffeted by com-petition from newer technologies, and following extensive lobbying of the FCC and the Department of Justice (for monopoly exemptions), XM and Sir-ius merged in 2008.[68] In order to achieve economies of scale, Sirius XM pur-sued a "merged monopoly" strategy under Sirius CEO (and former Infinity Broadcasting chief) Mel Karmazin, slashing its workforce by 22 percent and combining the two services' programming offerings. In the aftermath of this consolidation, many of the formerly innovative programming strategies that gave deejays creative autonomy were replaced by an emphasis on program "brands" and other hits-based channel formats that featured shallower playlists and smaller music rotations. The high-cost, star-driven channels remained, but the consolidation engendered "mass cancellations" by sub-scribers who were upset over the loss of their favorite channels.[69] With a hostile takeover looming, in early 2009 Sirius XM received a cash infusion from Liberty Media, the owner of the DirecTV satellite television service, in exchange for a 40 percent stake. By the end of the first quarter, it reported revenue growth over 2008 and began moving toward more diversified distribution—for example, by offering subscribers streamed content over their cell phones for a small premium.[70] However, the company still faced a debt load of US$2.3 billion and had lost some 400,000 subscribers, even as plunging car sales weakened one of its most important sources of new customers and increasing music royalty rates promised further hikes in subscription fees in the midst of a severe recession.[71] At the same time, the growing "smart phone" market enabled services like Slacker.com and Pandora to offer not just more preformatted channels than Sirius XM but also user-customized channels, streamed almost anywhere over wi-fi or cell phone networks, at a fraction of the cost of a Sirius XM subscription or even free with advertising.[72]

What these struggles illustrate is that, more than at any time since the 1920s, the very question of what constitutes radio is at stake, and it remains to be seen whether Sirius XM's answer to that question will be compelling enough to preserve a place for satellites in twenty-first-century "radio." Throughout its four decades, satellite offered new possibilities for program distribution, but none were so radical as to displace the existing radio models of research-driven formats and nationalized program syndication. Heavily formatted radio stations embraced the economies of scale allowed by satellite distribution in the 1980s, but a decade later switched to high-speed ISDN lines that permitted real-time voice tracking. This allowed one deejay to broadcast to many areas while sounding "local" to each, which further eroded the relationship between radio stations and the geographic area they served. Developments like this have led some observers to suggest that the future of radio might lie in a return to greater localism as a way to offer listeners original content that they cannot find elsewhere.[73] To the extent that radio is integral to the health of local music scenes or might still function to help construct local identities and invigorate local public spheres, a return to radio localism could represent a rebuke to the national formatting strategies in U.S. radio to which satellite distribution was so central. At the same time, the history of satellite radio distribution suggests the ways in which geographically oriented programming can, but does not always, easily mesh with taste-based sound cultures.

In summary, then, while direct satellite broadcasting technology ultimately was able to deliver on its initial promises, the long and costly development cycle prevented satellite radio from establishing itself fully within the media landscape before competing services emerged to threaten it with obsolescence. Moreover, Sirius's and XM's reactions to competition involved recourse to "proven" programming strategies and business models of terrestrial broadcasters (even as those techniques were somewhat beholden to an earlier notion of satellite distribution), making them especially vulnerable to the possibilities of customization and personalization afforded by new technologies of convergence. These programming choices also belied the popular discourses of newness and difference that sought to distinguish satellite from terrestrial radio. Instead of rejecting the claims of these discourses outright, however, it is more productive to view them as rationally foreseeable outcomes of thirty years of satellite radio. Indeed, as the Internet and terrestrial wireless distribution become increasingly central to both audio and video content distribution, it may soon prove that SDARS was in fact the last throes of a much older era of satellite radio, rather than the beginning of a new one.

NOTES

The authors would like to thank Jack Mitchell, John Farrell, Brendan Smith, and Teresa Young for their assistance with this article.

1 Mike Langberg, "Driving with Pay Radio: Clear Sound, Big Choices," *Philadelphia Inquirer*, September 27, 2001, in *Lexis-Nexis*; Cara Beardi, "Radio's Big Bounce," *Advertising Age*, August 27, 2001, S2.

2 Lisa Parks, "Where the Cable Ends," in *Cable Visions: Television Beyond Broadcasting*, ed. Sarah Banet-Weiser et al. (New York: New York University Press, 2007), 114.

3 The seven national radio networks lost nearly US$5 million on revenues of US$64.3 million in 1976 even as individual station profits were up 70.2 percent over 1975. "It Was an Incredible Year," *Broadcasting*, December 12, 1977, 28.

4 "CPB Says It's Going to Cost a Bundle to Put NPR on Satellite," *Broadcasting*, April 26, 1976, 38.

5 "Satellites: Tomorrow Is Here Today," *Broadcasting*, March 27, 1978, 57.

6 "Amway Agrees to Buy Mutual Broadcasting," *New York Times*, August 18, 1977, in ProQuest Historical Newspapers; Ernest Holsendoph, "Mutual Radio Applies to FCC to Be First All-Satellite Network," *New York Times*, November 22, 1977, in ProQuest Historical Newspapers.

7 "AP Radio Gets Down to Satellite Business," *Broadcasting*, April 9, 1979, 64–65; "Up, Up and Away for Radio Networking," *Broadcasting*, March 17, 1980, 37.

8 "Satellites: Tomorrow Is Here Today," 57.

9 "ABC Radio Goes Satellite," *Broadcasting*, March 26, 1979, 106.

10 "ABC Radio Networking: Five Years Down, 20 Years Upward," *Broadcasting*, February 13, 1978, 74–76; see also Andrew Inglis, *Behind the Tube: A History of Broadcasting Technology and Business* (Boston: Focal Press, 1990), 436–437.

11 Susan Douglas, *Listening In: Radio and the American Imagination* (New York: Times Books, 1999), 258.

12 "FM: The Great Leap Forward," *Broadcasting*, January 22, 1979, 32.

13 "New Gates Tape-Disc System Promises Entirely Automatic Radio Operation," *Broadcasting/Telecasting*, July 23, 1956, 71–72.

14 Douglas, *Listening In*, 276–280; "It Was an Incredible Year," 28–29.

15 Jim Kefford, vice president and general manager of syndicator Drake-Chenault, quoted in "Programming: Covering the Radio Waterfront at the Riverfront," *Broadcasting*, September 17, 1979, 53.

16 Quoted in ibid.

17 "FCC Lifts Rules on Antennas in Bid to Spur Satellite Service," *New York Times*, October 19, 1979, in ProQuest Historical Newspapers.

18 "RPC IV: Satellites and Syndicators," *Broadcasting*, August 24, 1981, 34–35.

19 "National Advertisers Are Beginning to Discover Cinderella," *Broadcasting*, June 30, 1980, 56–57; see also "Overview: National Radio Sales Are Booming," *Broadcasting*, August 25, 1980, 43ff.

20 Sam Cook Diggs, quoted in "Gathering Time for CBS Radio," *Broadcasting*, October 20, 1980, 24.

21 Roy Bliss, quoted in "RKO, Burkhart Spawn New Satellite Radio Programming Services," *Broadcasting*, April 13, 1981, 118.

22 "After 10 Years of Satellites, the Sky's No Limit," *Broadcasting*, April 9, 1984, 43–68.

23 See, for example, "Overview: Jockeying for Position in the Marketplace," *Broadcasting*, August 15, 1980, 50.

24 Ivan Braiker, quoted in James A. Smith, "Satellites & Syndication: Will It Be Boom or Bust?" *Radio Only*, November 1982, 42.

25 "Speaking of and for Radio," *Broadcasting*, April 27, 1981, 62.

26 As a McCann Erickson executive rhetorically asked in 1981, "With satellite programs beamed to two hundred or more radio stations, won't the need for local stations lessen? The station's role may be reduced to a carrier of programs." Gene DeWitt, quoted in "2001: Advertising," *Broadcasting*, October 12, 1981, 245. For more on the discourse of radio showmanship going back to the 1930s, see Jennifer Hyland Wang, "Convenient Fictions: The Construction of the Daytime Broadcast Audience, 1927–1960" (Ph.D. diss., University of Wisconsin, 2006).

27 Ed Shane, quoted in "The Fickle Business of Formats," *Broadcasting*, August 17, 1981, 60.

28 Richard Shiben, quoted in "Freer at Last," *Broadcasting*, January 19, 1981, 33.

29 Quoted in "The FCC on the Firing Line in Las Vegas," *Broadcasting*, April 21, 1980, 42.

30 "Satnets," *Broadcasting*, April 12, 1982, 73.

31 Quoted in "Bullish on Radio and NRBA," *Broadcasting*, April 2, 1984, 46.

32 "Satellite? 'Let's Wait and See!'" *Broadcasting*, February 22, 1982, 53–55. Some broadcasters disputed the sonic transparency of satellite distribution. Recalled one deejay, "You could always tell a satellite station by the 'deadness' of the sound; it didn't sound very dynamic because all those triggering tones had to have time to work getting all those elements on the air." John Farrell, e-mail to Bill Kirkpatrick, December 6, 2008.

33 Ed Shane, "How to Make Non-Local Programming Sound Local," *Radio Only*, January 1983, 46.

34 "Hoberman Forecasts Radio's Changes," *Broadcasting*, January 18, 1982, 74.

35 "ABC, the Uncommon Common Denominator," *Broadcasting*, February 1, 1982, 8–9; Smith, "Satellites & Syndication," 39.

36 "Satellites: Tomorrow Is Here Today," 58–60; see also "MBS Plans Satellite Service to Its Affiliates," *Broadcasting*, November 13, 1977, 41; "NBC Raises Objections to Mutual's Filing for Earth Stations," *Broadcasting*, May 1, 1978, 64.

37 Alec Foege, *Right of the Dial: The Rise of Clear Channel and the Fall of Commercial Radio* (New York: Faber & Faber, 2008), 67–68.

38 For the emergence of the talk radio format, see Jane H. Bick, "The Development of Two-Way Talk Radio" (Ph.D. diss., University of Massachusetts, 1987); Douglas, *Listening In*, 84–327; and Wayne Munson, *All Talk: The Talk Show in Media Culture* (Philadelphia: Temple University Press, 1993), 19–62. So apparent were the advantages of satellite distribution for many stations that even AT&T (which had a long-standing policy of investing in potential rivals) partnered with a producer to launch a twenty-four-hour satellite radio service. "AT&T Files for Satellite Radio Distribution Service," *Broadcasting*, February 1, 1982, 64.

39 Nancy Bishop, "Abrams Brings Special Touch to SMN Hard Rock Format," *Adweek*, December 12, 1988, Southwest Edition, in *Lexis-Nexis*; Dean Foust, "Lee Abrams: Out to Fill the Air with Heavy-Metal Fare," *BusinessWeek*, June 5, 1989, 107; Craig Rosen, "Lee Abrams' Classic Rock Rolling into Smaller Markets via Satellite Music Net," *Billboard*, January 19, 1991, 25.

40 Inglis, *Behind the Tube*, 150.

41 "2001: Technology," *Broadcasting*, October 12, 1981, 249.

42 Stan Hinden, "CD Radio Plans Satellite-to-Car Broadcast System," *Washington Post*, March 7, 1994, in *Lexis-Nexis*; Peter Passell, "Coast-to-Coast Radio without a Squawk," *New York Times*, November 27, 1994, in Proquest Historical Newspapers, in *Lexis-Nexis*; Snigdha Prakash, "American Mobile Satellite Set to Market Produce Monitor," *Washington Post*, December 30, 1991, in *Lexis-Nexis*.

43 Sydney Head et al., *Broadcasting in America* (New York: Houghton Mifflin, 1994), 86.

44 "A Dispute Over Radio Technology," *New York Times*, August 23, 1990, in ProQuest Historical Newspapers; Jay Mallin, "Digital Audio Promises Coast-to-Coast Radio," *Washington Times*, September 11, 1990, in *Lexis-Nexis*. For other public pronouncements of the threat of SDARS to localism through the 1990s, see, for example, Doug Abrahms, "Radio Stations Face Challenge from Space," *Washington Times*, January 5, 1995, in *Lexis-Nexis*; Edmund Andrews, "FCC Plan for Radio by Satellite," *New York Times*, October 8, 1992, in ProQuest Historical Newspapers; Edmund Andrews, "FCC Backs Digital Radio," *New York Times*, January 13, 1995, in *Lexis-Nexis*; Doug Abrahms, "Radio Signals a Revolution; Satellites Will Allow National Stations," *Washington Times*, December 10, 1998, in *Lexis-Nexis*; and Michael Rozansky, "Satellite Radio: Problem or Solution?" *Philadelphia Inquirer*, April 23, 1998, in *Lexis-Nexis*.

45 "FCC Considering Rules for Radio Satellite Services," *Satellite Week*, April 29, 1996, in *Lexis-Nexis*; Rozansky, "Satellite Radio."

46 See, for example, the comments of CD Radio president David Margolese in 1997 in "American Mobile Radio and Satellite CD Radio Win Licenses at DARS Auction," *Satellite Week*, April 7, 1997, in *Lexis-Nexis*; and those of XM representative Vicki Stearn in 1998 in Abrahms, "Radio Signals."

47 Edmund Andrews, "Has the FCC Become Obsolete? Technology Moves Fast; Determining the Public Interest Takes Time," *New York Times*, June 12, 1995, in ProQuest Historical Newspapers.

48 "FCC Begins Considering Rules for Satellite Radio Services," *Satellite Week*, April 29, 1996, in *Lexis-Nexis*; James Kim, "Satellite Radio Up for Approval," *USA Today*, March 3, 1997, in *Lexis-Nexis*; Christopher Stern, "FCC OKs 4 Bidders for Digital Radio Rights," *Daily Variety*, March 4, 1997, 8, in *Lexis-Nexis*.

49 Bill Holland, "Digital B'Casting Applicant Space Systems Drops Out," *Billboard*, April 3, 1993, 87, in *Lexis-Nexis*; Hinden, "CD Radio Plans Satellite-to-Car Broadcast System"; Passell, "Coast-to-Coast"; "Despite Setbacks and Risks, Satellite Industry Is Called Good Investment," *Satellite Week*, April 1, 1996, in *Lexis-Nexis*; "AMSC Investors Begin 11th-Hour Debate Over Emergency Financial Aid," *Mobile Communications Report*, April 8, 1996, in *Lexis-Nexis*.

50 "XM Ready to Introduce 'Cable for the Car,'" *Automotive News*, March 5, 2001, 34i, in *Lexis-Nexis*.

51 Doug Wilsterman, quoted in "Sirius Offers Buffet for Music-Hungry Drivers," *Automotive News*, March 5, 2001, 36i, in *Lexis-Nexis*. There are many examples of this rhetoric; see, for example, Stern, "FCC OKs 4 Bidders"; and Rozansky, "Satellite Radio."

52 Robert Berzins, analyst for Lehman Brothers, quoted in Carla Beardi, "Radio's Big Bounce: Satellite Services Set Wide Target, But Will Consumers Pay for What's Always Been Free?" *Advertising Age*, August 27, 2001, S2, in *Lexis-Nexis*.

53 Andrea Adelson, "Jacor a Rising Star among Radio Networks," *New York Times*, June 23, 1997, in ProQuest Historical Newspapers; Bruce Haring, "Tuning into the Next Radio Wave," *USA Today*, September 30, 1998, in *Lexis-Nexis*.

54 Marc Fisher, *Something in the Air: Rock, Radio, and the Revolution That Shaped a Generation* (New York: Random House, 2007), 296–301; Richard Martin, "Would You Buy the Future of Radio from This Man?" *Wired*, October 2004, 132–136; Carrie Borzillo, "Hi-Tech Tools Changing Face of Promo Biz," *Billboard*, October 1, 1994, 7, in *Lexis-Nexis*.

55 Cynthia Littleton, "Radio Satcaster Pacts for Programs, Changes Name," *Variety*, October 19, 1998, 58, in *Lexis-Nexis*; David Hinckley, "Dialing for Dollars Opera to Sports: Satellite Radio Will Create a Niche for Everyone," *New York Daily News*, August 31, 1999, in *Lexis-Nexis*.

56 Frank Ahrens, "The Radio Waves of the Future: Internet Stations Give Listeners a New Way to Tune In," *Washington Post*, January 21, 1999, in *Lexis-Nexis*.

57 "Programming Newsline," *Billboard*, May 29, 1999, in *Lexis-Nexis*.

58 "Worldspace Sells Stakes in XM Radio for AMSC Shares and Cast," *Satellite Week*, June 14, 1999, in *Lexis-Nexis*; "XM Radio Files for IPO; CD Radio Signs on Panasonic," *AudioWeek*, August, 2, 1999, in *Lexis-Nexis*.

59 See, for example, Robert W. McChesney, *Rich Media, Poor Democracy: Communication Politics in Dubious Times* (New York: New Press, 2000).

60 "Satellite Radio to Roll Out Nationally in Nov.," *AudioWeek*, July 30, 2001, in *Lexis-Nexis*.

61 Jody Berland, "Radio Space and Industrial Time: The Case of Music Formats," in *Critical Cultural Policy Studies: A Reader*, ed. Justin Lewis and Toby Miller (Malden, Mass.: Blackwell, 2003), 232.

62 Ibid.

63 "RIAA Requests DAB C'right Safeguards; Seeks FCC 'Surrogate' to Hoped-For Amendment," *Billboard*, November 28, 1992, 5, in *Lexis-Nexis*; Stephen H. Wildstrom, "Copyrights and Wrongs," *BusinessWeek*, July 3, 2006, 24, in *Lexis-Nexis*. Sirius avoided a similar lawsuit by agreeing to higher royalties for the record companies; XM and the major labels eventually settled the suit in 2007–2008.

64 Yochai Benkler, *The Wealth of Networks* (New Haven, Conn.: Yale University Press, 2006).

65 Quoted in "NAB Calls for FCC Action Against Satellite Radio," *Billboard*, September 8, 2001, 83, in *Lexis-Nexis*.

66 Kim Hart, "XM Radio Hits Some More Interference," *Washington Post*, October 5, 2006, in *Lexis-Nexis*. As part of the approval of the merger between XM and

Sirius in 2008, the satcasters agreed to pay fines of US$19.7 million for these infractions and take down 100 antennae that were still in violation. See Jim Puzzanghera, "Sirius, XM Settle with FCC Over Violations," *Los Angeles Times*, July 25, 2008, in *Lexis-Nexis*.

67 James Collins, quoted in Anitha Reddy, "XM to Air Traffic, Weather for Region; Channels Mark Debut of Local Content," *Washington Post*, February 28, 2004, in *Lexis-Nexis*.

68 This merger occurred despite the strenuous objection of terrestrial broadcasters; NAB president Fritts joked in 2008 that the organization should be renamed "the National Association of Satellite Radio Killers." See Ron Orol, "Satellite Killers," *Deal*, April 25, 2008, accessed December 28, 2008, http://www.thedeal.com/dealscape/2008/04/satellite_killers.php; subsequently accessed March 20, 2011, http://www.freepress.net/node/38973.

69 Tim Arango, "Satellite Radio Still Reaches for the Payday," *New York Times*, December 28, 2008; Olga Kharif, "Sirius XM's Dual Concerns: Debt, Delisting," *BusinessWeek*, December 12, 2008, accessed March 20, 2011, http://www.businessweek.com/technology/content/dec2008/tc20081212_917411.htm; Olga Kharif, "Sirius-XM: A Long, Challenging Road Ahead," *BusinessWeek*, November 11, 2008, accessed March 20, 2011, http://www.businessweek.com/technology/content/nov2008/tc20081110_729795.htm; "Karmazin Still Sees Growth in 2009," *Radio Business Review*, December 18, 2008, accessed March 20, 2011, http://www.rbr.com/radio/11927.html; Mike Snider, "As Sirius, XM Signals Merge, Customers Are Confused," *USA Today*, November 17, 2008, accessed March 20, 2011, http://www.usatoday.com/life/lifestyle/2008-11-17-sirius-xm_N.htm. For ongoing discussions of Sirius-XM, see www.orbitcast.com, www.siriusbuzz.com, www.savesirius.org, and www.xmfan.com.

70 Sirius XM, Inc., "SIRIUS XM Radio Reports First Quarter 2009 Results" (press release), May 7, 2009, accessed March 20, 2011, http://investor.sirius.com/releasedetail.cfm?ReleaseID=382353.

71 Carolyn Okomo, "Sirius Faces Its Next Challenge in May," *Deal*, February 18, 2009, accessed June 29, 2009, http://www.thedeal.com/dealscape/2009/02/sirius_faces_its_next_challeng.php, subsequently accessed March 20, 2011, http://markets.financialcontent.com/mi.charlotte/news/read?GUID=8033101; Steve Guttenberg, "Sirius XM Sticks It to Subscribers," *CNET*, May 9, 2009, accessed March 20, 2011, http://news.cnet.com/sirius-xm-sticks-it-to-subscribers; Greg Sandoval, "Sirius XM Must Raise Prices to Pay Music Royalties," *CNET*, June 25, 2009, accessed March 20, 2011, http://news.cnet.com/8301-1023_3-10273078-93.html.

72 Sirius XM's foray into iPhone apps was quickly downloaded more than one million times, but the rollout was marred by the news that Howard Stern and other flagship programs, such as Major League Baseball, would be unavailable on the application due to licensing issues. Don Reisinger, "Sirius XM's Latest Blunder: Its iPhone App," *CNET*, June 18, 2009, accessed March 20, 2011, http://news.cnet.com/8301-13506_3-10267549-17.html.

73 See, for example, Robert L. Hilliard and Michael C. Keith, *The Quieted Voice: The Rise and Demise of Localism in American Radio* (Carbondale: Southern Illinois University Press, 2005), especially 208ff.

9

Crossing Borders

The Introduction and Legislation of Satellite Radio in Canada

Brian O'Neill and Michael Murphy

In a relatively short period, digital satellite radio has emerged as an important mold-breaker of conventional analog radio. The delivery via satellite of radio programming and audio services to fixed and portable receivers now challenges the hegemony of locally defined broadcast radio and is leading to new configurations of audience reception and of audio program services. As such, its challenge to conventional radio has often been met with strong resistance, particularly where, by virtue of its transnational nature, satellite radio has been seen to circumvent the normal regulatory or business regime applicable to local or national radio.

The introduction of satellite radio to Canada provides a useful illustration of this disruptive character. In 2005, the Canadian media regulator, the Canadian Radio-television and Telecommunications Commission (CRTC), offered licenses to Canadian operators in partnership with the U.S.-based XM and Sirius satellite platforms. Because of their extended footprint over the North American continent, reception of XM and Sirius radio signals within Canada was always going to be a thorny issue for the Canadian regulator. Fearing a chaotic situation of unregulated and illegal access to the hundreds of new channels available, just as had happened with Direct Broadcast Satellite (DBS) television services some years previously, the CRTC's strategy was to contain their impact insofar as was feasible within existing regulatory conditions. The introduction of satellite radio to Canada sparked a major public debate on cultural sovereignty, content regulation, and the impact of foreign-produced programming on Canadian culture. It also had the effect of derailing Canada's own digital radio policy and raised wider questions about Canadian strategy on the regulation of new media and communications technologies.

Canada has a long history of dealing with transnational issues related to broadcasting.[1] The United States has a population and economy an order of magnitude larger than Canada's, and with the majority of Canadians living within reception range of U.S. border stations, government policy has had to balance issues of cultural sovereignty with commercial and trade interests. In 1928, a government Royal Commission developed a scheme to counter the powerful and popular conventional radio signals emanating from the United States, but this was abandoned due to pressures of the Great Depression. However, patriotic concern that Canada's airwaves would be dominated by American commercial interests resulted in the formation of the Canadian Radio League (CRL) in 1930 by two young idealists (Graham Spry and Alan Plaunt). CRL lobbying eventually lead to the formation of the Canadian Radio Broadcasting Commission (CRBC), which was transformed into the Canadian Broadcasting Corporation (CBC) in 1936. This ensured that Canadians had access to made-in-Canada, government-funded programming as a balance to both American stations and private Canadian radio stations rebroadcasting popular American programming.

Decades later, in 1972, Canada took the controversial step of regulating that 30 percent (later increased to 35 percent) of all music played on all Canadian radio stations had to be of Canadian origin. In an era of free trade agreements and pressure to increasingly open borders to commercial interests, the Canadian approach to protectionism and cultural sovereignty continues to be actively debated. In this chapter, we develop this theme by exploring the role of satellite radio as a catalyst for change in the world of radio broadcasting, as illustrated by Canada's experience. Looking first at the phenomenon of satellite radio in a North American context, we examine the controversy leading up to and following the decision to license it in Canada. We then offer some observations on the future impact that satellite radio as a technology and a programming platform is likely to have.

Satellite Radio Services in the North American Context

Satellite radio services were proposed in the United States as early as 1990, but suffered a protracted gestation period, as the Federal Communications Commission (FCC) did not begin to process license applications until 1997. The first system to launch was XM Satellite Radio Inc. on September 25, 2001, followed by the launch of its competitor Sirius Satellite Radio Inc. on February 14, 2002. At the time of launch, both services offered 100 channels of digital audio programming for a monthly subscription cost of US$10 to US$13.

Both XM and Sirius differentiated their product from conventional terrestrial radio. In order to command a subscription fee from an audience not used to paying for radio, both companies stressed the variety of their mostly commercial-free music offerings. The audio channels were programmed and targeted to various demographic groups to offer a much wider choice of sub-genres than was available with conventional radio. For example, XM provided seven different channels dedicated to country music ("Classic Country," "Americana," "Willie Nelson's Traditional Country," "Bluegrass," "Folk," "New Country Hits" and "'80s & '90s Country"). Sirius offered ten different channels dedicated to rock music, organized by subtype ("Early Classic Rock," "Later Classic Rock," "Pure Hard Rock," and so on). In addition to music, both services also offered a wide range of talk channels including news, sports, comedy, lifestyle, and entertainment. A final benefit was continental reception. Under ideal conditions, subscribers of XM or Sirius were able to receive all of their music and talk channels anywhere in the continental United States. If the subscribers had a car or portable radio, they had the ability to travel across the continent and enjoy digital-quality audio without dropout, static, or retuning to a new station location.

The capital and investment business model for satellite radio closely resembles that for DBS television services in that both require significant capital outlay for technology at the front end, but require little capital investment to add an additional subscriber to the service. Satellite-based services are heavily front-end loaded with respect to capital investment. Sirius and XM have together invested over US$2.6 billion for "space and ground assets."[2] This satellite infrastructure has a design life of fifteen years. The marginal capital cost of adding an additional subscriber to this infrastructure is low, as no additional capacity is required at the satellite, ground station, or broadcast operation. A significant sales channel for both providers is the automotive original equipment manufacturer (OEM).[3] XM and Sirius often subsidize the cost of a radio when it is factory-installed by their partners. The OEM receives a share of subscription revenue for all buyers who activate a subscription after a free trial period. XM and Sirius also subsidize their radio manufacturers on a per-radio basis, to lower the wholesale cost to the retail industry.

Satellite radio achieved one of the fastest adoption rates of comparable mass market technology products, taking only 3.6 years to achieve market penetration of 5 million units.[4] In the United States at the end of 2006, XM had more than 7.5 million subscribers, and Sirius had over 6 million subscribers.[5] However, even with growing subscriptions, profitability was difficult to achieve. Both companies began to compete aggressively by hiring big-name entertainers, driving up programming costs beyond their original business plan. For example, Sirius signed a five-year, US$500 million deal with shock-jock Howard Stern to move his radio show exclusively to Sirius in January 2006. Other high-profile deals for Sirius included Martha Stewart, Jimmy Buffett, and Eminem, while XM retaliated by recruiting Oprah Winfrey, Bob Dylan, and Ellen DeGeneres, among others. Operating cost increases began to outstrip healthy subscription growth. As an example, between 2005 and 2006, while Sirius was able to increase its total revenue from US$242 million to US$637 million (an increase of 163 percent), its programming and content costs ballooned from US$118 million to US$552 million (an increase of 368 percent).[6]

In February 2007, XM and Sirius announced their intention to merge, creating a single satellite radio network. Because the FCC had originally approved two satellite radio services to ensure competition, the proposal was controversial, but was eventually approved, and the merger was completed in July 2008. At the present time, both XM and Sirius continue to operate in the United States as separate brands, but they now offer programming subscriptions that can include channels from both services.

XM and Sirius Satellite Radio Technology

By the time both XM and Sirius received FCC spectrum approval in 1997, digital broadcasting using satellites was already a proven technology. However, existing technology used a relatively large fixed dish attached to a building to receive a signal, and existing systems were not intended for mobility. Car and portable radio receivers need to receive the signal while moving. These same receivers also require a much smaller antenna to receive the signal. XM's design approach is to use two high-powered geostationary satellites with a large number of terrestrial repeaters to provide reliable mobile coverage. Sirius uses three lower-powered satellites in an elliptical figure-eight pattern, with a smaller number of terrestrial repeaters, to provide mobile coverage.

XM and Sirius deliver their audio channels using digital compression to save on required bandwidth capacity, and this allows them to squeeze many more audio channels into their allocated spectrum. For its music channels, XM uses Advanced Audio Codec Plus (aacPlus) from Coding Technologies.[7] For its information channels (primarily voice), XM uses Advanced Multi-Band Excitation (AMBE) developed by Digital Voice Systems, which provides acceptable speech quality at very low bit rates (as low as 2 kilobits per second).[8] XM has the ability to set compression rates to meet the demands of each of its audio channels. For example, a classical music channel (requiring very high quality by its market demographic) can be set to a lower compression (higher quality) rate than a pop music channel, and a talk radio channel (primarily voice) can be set to a very high compression (lower quality) rate. By managing these levels of compression, at present XM is able to deliver more than 170 channels of audio content using its allocated bandwidth. Sirius uses a proprietary system known as Perceptual Audio Coder (PAC), developed by Lucent Technologies, for both its music and information channels.[9] Like XM, Sirius is able to manage levels of compression to deliver, at present, more than 130 channels. Technically, both XM and Sirius platforms can also be used to deliver other forms of digital data including text, image, and video. XM and Sirius already offer weather and traffic services in the United States, and Sirius recently introduced a mobile children's video service called Backseat TV.

Both services maintain centralized production and distribution centers. XM originates programming at its broadcast studio complex in Washington, D.C.[10] This complex contains eighty-two radio studios and is one of the largest radio production centers in the world. In addition, XM operates three more studios in New York City; another studio in Nashville, Tennessee; and another in Chicago.

XM combines (multiplexes) more than 170 channels of audio programming into a signal that is encrypted and uplinked to two very high-powered satellites in geostationary orbit above the equator. One of the XM satellites (XM Rhythm) is located at 115°W, and the second XM satellite (XM Blues) is at 85°W.[11] The XM satellites broadcast on the 2.3-gigahertz S band.[12] Either satellite, individually, can provide coverage to the continental United States and southern Canada. The two satellites are used to provide directional "space diversity" for a radio receiver on Earth; if the radio receiver cannot receive a signal from one of the satellites due to line of sight being blocked, it can attempt to receive from the second satellite, which is positioned in a different direction. Space diversity enables the receivers to be mobile, a feature not typically available with other satellite broadcast services.[13]

The Sirius infrastructure resembles that of XM. Sirius originates programming at its broadcast studios located in New York City. In addition, Sirius operates studios in Los Angeles and Memphis, Tennessee. Currently, Sirius offers its customers more than 130 channels of audio, and these channels are combined into a signal that is uplinked to three satellites in geosynchronous elliptical orbit above North and South America. Unlike XM satellites, which are in a fixed position above Earth at all times, the Sirius satellites are constantly moving in an elliptical pattern.[14] This system ensures that each of the three satellites is above the equator broadcasting to North America for sixteen hours each day and is below the equator "at rest" for eight hours each day.

This satellite configuration provides coverage to the continental United States and Canada. Like XM, Sirius uses multiple satellites to provide directional "space diversity" for a radio receiver on Earth. If the radio cannot receive a signal from one of the satellites due to line of sight being blocked, it can attempt to receive from the second satellite, which is positioned in a different direction. The Sirius system is designed to ensure that there is at least one satellite at high elevation over North America at any given time. The logic is that if a satellite is at high elevation, there is less chance for blockage of the signal by buildings and other obstructions. Sirius's satellites also broadcast using the 2.3-gigahertz S band, but are nearer to Earth than XM's satellites, allowing operation at lower power.[15] Because both XM and Sirius systems were designed for mobile reception in automobiles in large urban centers, there will be times when the radio will not be able to receive any of the satellite signals—due to blockage by tall buildings or travel in tunnels, for example. For this reason, terrestrial repeaters are deployed in urban centers. The radio receiver therefore has multiple chances of being able to receive a signal. In nonurban areas, it will be able to receive the signal from one of the satellites. In urban centers, it has the additional possibility of receiving the signal from one or more terrestrial repeaters.

Finally, both XM and Sirius radio receivers also deploy "time diversity" by storing four seconds of all signals received in digital memory. This provides a four-second safety copy to prevent dropouts. If a signal is lost from all sources for up to four seconds (for example, while driving under a bridge or through a tunnel), the receiver will continue to play the audio content from the digital copy contained in its memory. It is the combination of space diversity (multiple satellites and terrestrial repeaters) and time diversity (using digital memory) that allows a satellite radio to function effectively as a mobile device.

Proposing Satellite Radio for Canada

The entry of satellite broadcasting into the radio market was greeted by the industry as a disruptive and unwelcome competitive threat. The National Association of Broadcasters protested to the FCC, "A competing satellite service presents a potential danger to the United States' universal, free, local radio service and, thus, to the public interest it serves."[16] Appearing to usher in a whole new era in which radio's bonds with its local roots would be inevitably loosened, satellite radio seemed to promise another stage in the development of the medium, offering new formal, economic, and transnational possibilities.[17] From the outset, it was clear that U.S. satellite radio was always going to pose significant dilemmas for media regulation and cultural policy in Canada. As the satellite footprints of both XM and Sirius covered most of the continent, an extension to their service for the Canadian market was an obvious source of additional revenue for the companies with little extra investment. Under Canada's Broadcasting Act, however, its media system must be Canadian-owned and -managed, and must carry substantial amounts of Canadian content. Partnerships with Canadian firms and additional Canadian-produced content would be needed to make satellite radio eligible for licensing in Canada. The threat of a large "gray market" developing was comparable to the situation in the mid-1990s when large numbers of Canadian viewers circumvented licensing arrangements by subscribing to U.S. satellite television services using a U.S. postal address.

Following its launch in the United States, a number of expressions of interest were lodged with Canada's broadcasting and telecommunications regulator, the CRTC, to bring satellite radio to the Canadian market. The commission formally called for applications in December 2003 to provide multichannel subscription radio services with the requirement that applicants consider the needs of Canadian broadcasting policy in supporting diversity and Canadian content (CanCon).[18] Applications were submitted by two Canadian consortia, partnering respectively with the two main U.S. satellite platforms. Canadian Satellite Radio, owned by Toronto businessman John Bitove, offered a joint venture with XM Satellite Radio in the United States to bring its service, plus an additional four Canadian channels, to the market. The second applicant was a consortium made up of the more unlikely alliance of a public broadcaster—the CBC (40 percent)—and the privately owned Standard Radio, Inc. (40 percent) in partnership with Sirius Satellite Radio, Inc. (20 percent). This alliance also would offer Sirius programming plus additional Canadian content, including two existing CBC stations, two new CBC channels, and one channel produced by Standard.

Both satellite proposals offered an appealing prospect of more than 100 channels of largely commercial-free, CD-quality music across musical genres, as well as news, sports, talk, weather reports, and information for a monthly subscription fee of approximately C$12.99. Using a special satellite receiver, listeners across Canada and all of North America could enjoy precisely the same service and be able to access the full programming service regardless of location. Without licensing, however, Canadians were ineligible to subscribe but could become gray-market subscribers by using a U.S. billing address. Furthermore, satellite services would not expect to be subject to the same content quota regulations as domestic Canadian services because the satellite signal did not use the same broadcast spectrum. In both cases, though, promises were given to invest in Canadian content and to offer dedicated Canadian channels, although at a much lower proportion than the 35 percent Canadian content required of Canadian broadcasters.

Following the expressions of interest by the two satellite service providers, a third application from Canadian radio and television broadcaster CHUM, Ltd., and partner Astral Media, Inc., was also lodged, this time for a pay radio service using terrestrial digital radio technology. Initially thought to have sought a partnership with one of the satellite holdings, CHUM's decision to propose a Canadian digital radio service was based on an optimistic reading of the potential for subscription radio as a new form of radio service.[19] Without access to satellite distribution, however, it was restricted to land-based transmitters. Subscribers would get access to about fifty channels, all produced in Canada, for C$9.95 a month, although initially the service would be available only in the larger cities. One significant barrier to legalizing U.S. satellite radio in Canada was the existing satellite-use policy established in 1995 in the wake of developing DBS television services. The policy issued at that time by the Canadian government established that broadcasting undertakings should make use of Canadian satellite facilities to carry Canadian programming services and that "under no circumstances should an undertaking use exclusively foreign satellites for the distribution of its services to Canadians."[20] Seeking a clarification on the policy's implications for satellite radio, the CRTC received government acknowledgment that the policy had not envisaged a situation where specialized satellite facilities might not actually exist in Canada. The relevant government department confirmed that Canada had no satellite facility capable of distributing digital satellite radio broadcasting and was unlikely to have such a facility in the future; neither had Canada secured with the International Telecommunication Union (ITU) the required spectrum resources at the S-band to develop its own specialized satellites. Accordingly, the satellite-use policy was amended to permit the use of foreign satellite facilities to distribute Canadian programming services for home and vehicular reception.

Not surprisingly, a central issue at the November 2004 CRTC hearings was the amount of foreign content compared to the relatively small number of Canadian channels for the proposed services. Satellite radio technology offered a major challenge not only to established Canadian radio interests but also to the Canadian regulatory system of protecting Canadian culture by requiring broadcasters to offer a minimum amount of content produced by Canadians on radio and television. The CRTC's 1998 Commercial Radio Policy had in fact increased the content requirements for AM and FM radio broadcasting from 30 to 35 percent.[21] The CHUM/Astral proposal promised that Canadian content on every one of its channels would be above the level required for conventional radio—about one-third of the playlist. All channels would be produced in Canada, and a number of them would feature content in French and other languages.[22]

In contrast, Canadian Satellite Radio's proposed commercial-free offering, accessible anywhere in the country, potentially would offer a vast array of choices unencumbered by strict quota regulation. In support of its case, Canadian Satellite Radio argued that an all-Canadian satellite system was impractical given that there was no such system in place and that it would be prohibitively expensive for the size of the market. Yet, it argued, linking up with the proven U.S. satellite platforms could offer national coverage at little or no additional cost and eliminate the growing unregulated gray market for U.S. services, estimated to be about 30,000 subscribers at the time.[23] Furthermore, satellite radio offered the opportunity to bring Canadian content to the entire continent of North America. CSR was quick to remind the commission that this was the first opportunity that Canada had had to export channels into the U.S. market and to provide much-needed exposure for new Canadian talent on conventional radio channels.[24] It was also suggested that the wide-ranging music content on satellite radio would recapture younger listeners to radio, repatriating an audience that was increasingly being lost to Internet downloading and MP3 music listening.

The satellite radio proposals received support from a variety of quarters, including the auto industry, with whom the satellite companies had signed exclusive deals—XM with General Motors, Honda, and Toyota; Sirius with Ford, BMW, DaimlerChrysler, and Audi. Independent recorded music producers, through representative body Indie Pool, were also strong supporters, arguing that their clients were getting insufficient coverage on conventional radio and welcoming the opportunity offered by new music channels (such as the Canadian-talent channel CBC Radio 3) on the new satellite platform.[25]

Significant opposition to the proposed licensing of satellite radio was voiced by those who believed it posed an unacceptable threat to Canadian interests and undermined the Canadian regulatory system. The lobby group Friends of Canadian Broadcasting argued that the threat of the gray market had been overstated by the satellite companies and that a dangerous precedent would be set by proceeding to license the services, effectively under coercion. Just because it was impossible to prevent signals from entering Canadian space did not mean that regulation according to Canadian priorities could not be achieved, they argued. But importing nearly 200 foreign channels with just a handful of Canadian channels could irreparably harm the Canadian broadcasting and music industries and undermine the long-standing commitment to nurturing Canadian talent. In short, Friends of Canadian Broadcasting proposed, the satellite companies should not be let off lightly, and if they were to be given permission to enter the Canadian market, onerous conditions should be placed to minimize the damage.[26]

The Decision and Appeal

In June 2005, the CRTC announced its decision to approve all three applications and formally establish a licensing framework for satellite subscription radio services in Canada.[27] The commission argued that the decision balanced the interests of Canadian consumers, the radio industry, and the music industry, and that building on the U.S. satellite platforms was the optimal solution for Canada at this time. The licenses, it was argued, would harness new technologies for Canadians, give Canadian talent exposure to listeners across Canada and throughout North America, and provide more choice and diversity for consumers, particularly those in rural and remote areas, where broadcasting choices are limited. Conditions were attached, particularly in respect of the concerns raised throughout the hearings about the low number of Canadian channels proposed. As a result, licensees were required to offer at least eight original channels produced in Canada with a maximum of nine foreign channels to be offered for each Canadian channel. In addition, 25 percent of the Canadian channels had to be in French, 25 percent of the music on the Canadian channels had to be new Canadian musical selections, with a further 25 percent by emerging Canadian artists. Furthermore, licensees also had to contribute at least 5 percent of their gross annual revenues to initiatives for the development of Canadian talent, both English-language and French-language talent.

The conditions were stricter than anticipated, but yielded to the inevitability of providing a licensing framework for satellite radio in Canada. For its part, the CRTC had been placed in an unprecedented position in seeking to regulate an existing North American service under Canadian conditions. Applicants expressed disappointment, claiming that the threshold for Canadian content had been raised to a level that undermined the commercial viability of their proposals. In the case of CHUM/Astral, there was bitter disappointment at what was perceived as inequitable treatment; CHUM/Astral believed that the all-Canadian digital radio proposal would be subject to normal terrestrial regulation, whereas the satellite operations would have to meet less onerous conditions.[28]

Almost immediately, appeals were considered against the decision by those who thought the CRTC had "sold out" to the United States with very little in return. An alliance of arts and media organizations led by Friends of Canadian Broadcasting petitioned the government to set aside the June 16 decision and refer the matter back to the CRTC for reconsideration, as provided for under the Broadcasting Act.[29] A coalition of French-language cultural groups also appealed, decrying the small number of French channels proposed. Critics held the public broadcaster CBC out for special criticism over its involvement in the Sirius Radio Canada proposal, accusing it of abdicating its responsibility to promote Canadian programming, while praising the CHUM/Astral proposal for its positive approach and its potential to be a rival Canadian satellite service at some point in the future.[30] The Canadian Association of Broadcasters, while not overly active in the debate, queried the fairness of the decision, and expressed concern about the impact newcomers would have on the conventional radio market.[31] It was also widely implied that the decision would hasten the unraveling of the Canadian content quota system, with the likelihood that Canadian broadcasters would now look for a relaxation of the rules to enable them to deal with the new competition.[32]

In July 2005, the CHUM/Astral consortium, along with a number of smaller radio companies, also decided to appeal the decision, claiming that the low threshold of Canadian content requirements imposed on the U.S.-supported satellite licensees was a "dramatic departure from historical broadcasting precedent."[33] Larger companies such as Corus and Rogers, which are among the top radio conglomerates in Canada, remained relatively silent, indicating their confidence in the ability of local radio to compete against the new satellite radio services while also preparing to cooperate with them through content-provision arrangements.

The CBC, in the meantime, remained impervious to criticism of its involvement. Arguing that its role was to serve Canadians on all available platforms and to support new technology developments, the proposition gave the CBC an opportunity to ensure its coverage was extended across the country on satellite without any major infrastructural investment of its own, while also bringing its flagship Canadian radio services onto an international platform. The alternative, from the CBC's point of view, was that not to participate would only further marginalize Canadian radio from new platforms and new ways of accessing content.[34] Participating in the Sirius Satellite Radio bid might also be financially lucrative and allow for reinvestment in Canadian programming.[35]

Cast into uncertainty with three appeals against the CRTC's decision, the debate about satellite radio moved firmly into the political arena, from where Canadian Heritage Minister Liza Frulla was called upon to decide the matter. With forty-five days for a determination, the cabinet had until September 2005 to decide whether to rescind the decision or refer it back to the CRTC. The ongoing delay, the applicants warned, could prompt the satellite companies to reconsider the value of entering the Canadian market at all, leading to a chaotic situation for broadcasters and the public. Intense lobbying followed throughout the summer with Liberal MPs, particularly those from Québec, calling on Frulla to seek a full CRTC review. Formally delivering its petition, the cultural alliance formed by Friends of Canadian Broadcasting accused the commission of sanctioning "wholesale dumping of foreign cultural products and services."[36] The debate assumed a geographical dimension with Atlantic Canada and Québec seen as largely favoring sending the ruling back to the regulator, while southern Ontario (with significant auto industry interests) and the West (representing large rural spaces) largely supported the CRTC's decision to grant licenses.

As the issue developed into a full-scale public relations battle involving public campaigns and demonstrations, extensive newspaper advertising, and political lobbying, claims and counterclaims were offered about the public response. Sirius Canada released research that found that only 20 percent of Québec's population wanted the CRTC decision overturned, while 22 percent of Canadians said they would be interested in subscribing to satellite radio.[37] On the other hand, research for the Canadian Recording Industry Association suggested few Canadians even knew of satellite radio, with only 15 percent saying they were aware of new radio services.[38] Another study, this one for

Friends of Canadian Broadcasting, claimed that two out of three (64 percent) Canadian adults wanted the federal government to intervene in the CRTC decision.[39]

The government was reportedly split on the matter but sensitive to the content and cultural issues involved, particularly in Québec. As a final gesture to win government support for the satellite licenses, both the XM and Sirius alliances offered to provide an additional French language channel on each service, ultimately swaying sufficient government support to confirm the CRTC decision in September 2005. Both satellite operations expressed approval for the terms and committed to launching a full, legalized Canadian service in time for the Christmas market, while the CHUM/Astral consortium conceded it would almost definitely not launch under the licensing terms.

Open Borders: Satellite Radio in Canada

As a result of the partnership between Canadian Satellite Radio (CSR) and XM, XM was made available to Canadians on November 22, 2005. A basic Canadian subscription fee of C$14.99 per month allows the customer to receive 130 channels of digital audio. Of these channels, fifteen are produced in Canada, out of which five are in French. The remaining channels are produced in the United States. Through Sirius Canada, Sirius was made available to Canadians on December 1, 2005. The Canadian basic subscription fee of C$14.99 per month allows the customer to receive 120 channels of digital audio programming. Of these, twelve are produced in Canada, out of which five are in French for some or all of the day. The remaining channels are produced in the United States.

Mirroring its fortunes south of the border, satellite radio has struggled to overcome heavy start-up losses and meet its targets for new subscribers. One year after its launch in late 2005, CSR, the operator of the XM subscription radio service in Canada, had achieved a subscriber base of just 120,000 listeners and reported losses of C$102.7 million in its first fiscal year due to discounts and marketing costs incurred to attract customers.[40] By April 2007, XM subscribers had increased to 219,000, compared to a reported 300,000 at rival Sirius Canada; clearly, these numbers were well below expectations of reaching a million subscribers by 2010.[41] Confirmation that satellite radio remained a niche market player with very little encroachment on conventional radio was given in industry research for the year 2006, which showed continued revenue growth for the sector. Canadian private radio performed better than the U.S. radio sector, with a 5.3 percent growth in revenue compared to a decrease in American radio advertising revenue of 0.1 percent.[42] With 10 percent of the total advertising market, radio's profits remained healthy, and its position in the major metropolitan areas robust. For their part, the satellite operators hoped for better long-term prospects, seeking to build their base in less populated areas where their impact was most keenly felt and welcomed.[43]

In 2007, after five years of operation without a profit, Sirius Satellite Radio and XM Satellite Radio Holdings announced a proposed merger in the United States to deal with their accumulated losses of US$7 billion. Despite protests by organizations such as the National Association of Broadcasters about the dangerous monopoly that would be created, the deal received approval the following year from the U.S. Department of Justice and the FCC. However, while the U.S. companies agreed to an equal division of profits within the newly merged Sirius XM Radio Inc., the implications for the Canadian operation were less clear. The newly merged company has large minority ownership of both Canadian partners, and future programming arrangements in the United States will have significant implications for operations north of the border. At the present time, neither Canadian company, both of which have majority Canadian ownership, has expressed intentions to collaborate or merge; they continue to operate independently and in competition with one another. Canadian programming remains exclusively with either the XM or Sirius service, both in Canada and the United States. However, a merger will become inevitable once both operations are consolidated. XM and Sirius receivers are currently incompatible, but combo-radios are becoming available that can access both services. This approach could allow the offering of programming packages spanning both services, as well as the elimination of overlapping music formats.

The main barrier to a merger in Canada will be a question of an equitable division of shares. Unlike in the United States, in Canada Sirius is the more popular satellite platform, with roughly three times as many subscribers as XM can claim. Accordingly, Sirius Canada, which is privately owned by the CBC, Standard Broadcasting, and its U.S. parent, regards itself as the more valuable part of a merged entity.[44] Against this, XM Canada has argued that its content deals—in particular, its eight-year, US$69 million contract with the National Hockey League—need to be taken into account. Once shareholdings are agreed on, however, a Canadian merger will almost certainly result, with little that the regulator can do, as one company will most likely relinquish its license and a combined holding will proceed to operate as XM Sirius Canada Inc.[45] As a result, the twenty-seven Canadian channels currently on offer across the two satellite platforms could be reduced by half, with the precise lineup being subject to further difficult negotiation.[46]

A decisive feature in the FCC decision to approve the XM-Sirius merger was the question of whether the market was defined as one of satellite radio, in which a monopoly would now operate, or whether the market was more broadly defined to include Internet radio, iPods, terrestrial radio, and other competing audio media. The FCC endorsed the latter position, arguing that the marketplace had substantially changed since the two companies were formed, and that now new technologies compete with traditional media.[47] While the services are known and marketed as satellite radio, the technology really is capable of delivering any form of digital media to mobile devices via satellite. As such, it is conceivable that in the future new broadcast media services outside of radio (including increasing video and other interactive content) could be the focus of expansion, and take advantage of redundant capacity that may exist as a result of merging the technological infrastructure. Against this background, media regulation based on a nationally defined and medium-specific logic will be both less effective and less relevant.

NOTES

1 Kenn Scott and John Pellatt, *On Our Wavelength: Broadcasting History from a Canadian Perspective* (Toronto: Oblique House, 2008).

2 U.S. Copyright Royalty Board, J. Armand Musey Testimony, Docket 2006-1 CRB DSTRA, accessed December 19, 2008, http://www.loc.gov/crb/proceedings/2006-1/index.html, subsequently accessed March 21, 2011, http://www.loc.gov/crb/proceedings/2006-1/sirius-musey.pdf.

3 U.S. Copyright Royalty Board, Expert Report of Dr. John R. Woodbury, 2006 Docket 2006-1 CRB DSTRA, accessed December 19, 2008, http://www.loc.gov/crb/proceedings/2006-1/index.html, subsequently accessed March 21, 2011, http://www.loc.gov/crb/proceedings/2006-1/sirius-woodbury.pdf.

4 CSR Satellite Radio Holdings, Inc., *2005 Prospectus*, accessed December 19, 2008, http://cdnsatrad.com/downloads/1206%20csr%20final%20prospectus.pdf.

5 Satellite Standard Group, "Subscriber Breakdowns," January 5, 2007, accessed March 21, 2011, http://satellitestandard.blogspot.com/2007/01/subscriber-beakdowns.html.

6 Sirius Satellite Radio, Inc., *2006 Annual Report*, accessed March 21, 2011, http://investor.sirius.com/annuals.cfm.

7 Christof Faller et al., "Technical Advances in Digital Audio Radio Broadcasting," *Proceedings of the IEEE* 90(8) (2002): 1303–1333.

8 AMBE Product website, accessed March 21, 2011, http://www.dvsinc.com/products/products.htm.

9 Louis E. Frenzel, "Satellite Radio Gets Serious," *Electronic Design*, August 18, 2003, accessed March 21, 2011, http://electronicdesign.com/article/digital/satellite-radio-gets-serious5603.aspx.

10 Corley Dennison, "Digital Satellite Radio," in *Museum of Broadcast Communications Encyclopedia of Radio*, ed. C. Sterling and M. C. Keith (London: Routledge, 2004).

11 XM has four geostationary satellites in orbit. XM Rhythm and XM Blues are active satellites that were launched to replace XM's original two satellites, XM Rock and XM Roll, which developed a technical fault. The original satellites are not active and act as two in-orbit spares.

12 Each of the two XM satellites transmits the same signal but uses a different pair of 1.85 megahertz frequency bands (for a total allocation of 3.7 megahertz per

satellite). The terrestrial repeaters use a third pair of 2.5 megahertz frequency bands (for a total allocation of 5 megahertz). The total S band frequency allocation for the XM system is between 2332.5 megahertz and 2345.0 megahertz. These numbers are approximate.

13 D. H. Layer, "Digital Radio Takes to the Road," *Spectrum, IEEE* 38(7) (2001): 40–46.

14 F. Davarian, "Sirius Satellite Radio: Radio Entertainment in the Sky," paper presented at the Aerospace Conference Proceedings, New York, March 9–16, 2002.

15 Each of the two Sirius satellites active over North America transmits the same content signal, but each uses a different 4 megahertz frequency band. One of the satellite signals is delayed by 4 seconds (providing temporal diversity). The terrestrial repeaters use a third, 4 megahertz frequency band. The total S-band frequency allocation for the Sirius system is between 2320.0 megahertz and 2332.5 megahertz. These numbers are approximate.

16 Dennison, "Digital Satellite," 467.

17 Michele Hilmes, "Foreword: Transnational Radio in the Global Age," *Journal of Radio Studies* 11(1) (2004): 1–6.

18 CRTC, "Broadcasting Public Notice CRTC 2003–68: Call for Applications for a Broadcasting License to Carry On a Multi-Channel Subscription Radio Programming Undertaking across Canada," (2003), accessed March 21, 2011, http://www. crtc.gc.ca/eng/archive/2003/pb2003-68.htm.

19 Michael Posner, "Third Contender Enters Radio Race," *Globe & Mail*, March 27, 2004, accessed March 21, 2011, http://www.friends.ca/News/Friends_News/ archives/articles0327040I.asp.

20 Industry Canada, "Notice No. DGTP-007-04: Proposed Clarification to the Government Satellite-use Policy for the Delivery of Broadcasting Services," (2004), accessed March 21, 2011, http://www.ic.gc.ca/eic/site/smt-gst.nsf/eng/ sf08253.html.

21 Canadian Radio-television and Telecommunications Commission (CRTC), "Commercial Radio Policy 1998" (Public Notice CRTC 1998–41), accessed March 21, 2011, http://www.crtc.gc.ca/eng/archive/1998/PB98–41.HTM.

22 CRTC, "Broadcasting Public Notice CRTC 2005–61: Introduction to Broadcasting Decisions CRTC 2005–246 to 2005–248: Licensing of New Satellite and Terrestrial Subscription Radio Undertakings," (2005), accessed March 21, 2011, http://www. crtc.gc.ca/eng/archive/2005/pb2005-61.htm.

23 Paul Brent, "Brave New Radio Waves," *National Post*, September 25, 2004, accessed March 21, 2011, http://www.friends.ca/News/Friends_News/archives/articles 0925040I.asp.

24 Richard Blackwell, "Radio Hearings Hinge on Content," *Globe & Mail*, November 1, 2004, accessed March 21, 2011, http://www.friends.ca/news-item/6034.

25 Paul Vieira, "CRTC Sings the CanCon Song," *National Post*, November 6, 2004, accessed December 19, 2008, http://216.187.70.81/news-item/6117, subsequently accessed March 21, 2011, http://www.friends.ca/news-item/6117.

26 Friends of Canadian Broadcasting, "Presentation to CRTC re Pay Radio (PN CRTC 2004–6)," November 3, 2004, accessed March 21, 2011, http://www.friends.ca/ brief/444.

27 CRTC, "The CRTC Authorizes Canada's First Three Subscription Radio Services," June 16, 2005, accessed March 21, 2011, http://www.crtc.gc.ca/eng/com100/2005/ r050616.htm.

28 Graham Fraser, "Radio Ruling Draws Static from CHUM-Astral Team," *Toronto Star*, June 17, 2005, accessed March 21, 2011, http://www.friends.ca/news-item/3600.

29 Other members of the coalition included the Alliance of Canadian Cinema, Television and Radio Artists (ACTRA); the Society of Composers, Authors and Music Publishers of Canada (SOCAN); Canadian Independent Record Production Association; Communications, Energy and Paperworkers Union of Canada; Directors Guild of Canada; Songwriters Association of Canada; Writers Guild of Canada; and the National Campus and Community Radio Association.

30 Friends of Canadian Broadcasting, "CRTC and CBC Attempt to Subvert Parliament," press release, June 16, 2005, accessed March 21, 2011, http://www.friends.ca/News/news06160501.asp.

31 Fraser, "Radio Ruling."

32 Greg O'Brien, "Satellite Radio Decision Could Demolish Cancon Industry, Says Coalition," *Toronto Star*, June 28, 2005, accessed March 21, 2011, http://www.friends.ca/News/Friends_News/archives/articles06280517.asp.

33 Greg O'Brien, "Mass Appeal of Sat Rad Ruling," *CARTT*, July 12, 2005, accessed March 21, 2011, http://www.friends.ca/News/Friends_News/archives/articles07120506.asp.

34 Brian O'Neill, "CBC.ca: Broadcast Sovereignty in a Digital Environment," *Convergence* 12(2) (2006): 179–197.

35 Eric Reguly, "Get Sirius: CBC Deal Should Prove a Wise Move," *Globe & Mail*, June 25, 2005, accessed December 19, 2008, http://friendscb.org/News/Friends_News/archives/articles06250501.asp, subsequently accessed March 21, 2011, http://www.friends.ca/news-item/3828.

36 Friends of Canadian Broadcasting, "Petition to the Governor in Council to Set Aside Broadcasting Decisions CRTC 2005–246 and CRTC 20005–247," accessed March 21, 2011, http://www.friends.ca/brief/418.

37 Simon Tuck, "Industry Blasts New Review of Digital Radio," *Globe & Mail*, August 29, 2005, accessed March 21, 2011, http://www.friends.ca/News/Friends_News/archives/articles08290509.asp.

38 Simon Doyle, "Poll Suggests Canadians Lukewarm to Satellite Radio," *Ottawa Citizen*, September 2, 2005, accessed December 19, 2008, http://friendscb.org/News/Friends_News/archives/articles09020505.asp, subsequently accessed March 21, 2011, http://www.friends.ca/news-item/5058.

39 Friends of Canadian Broadcasting, "Two in Three Canadians Want the Federal Government to Intervene in Recent CRTC Approval of Satellite Radio," poll results released September 5, 2005, accessed March 21, 2011, http://www.friends.ca/poll/6937.

40 Grant Robertson, "Satellite Radio Looks to Long Haul," *Globe & Mail*, November 16, 2006, accessed March 21, 2011, https://www.friends.ca/news-item/6278.

41 Barbara Shecter, "XM Canada Ponders Subscriber Shortfall; Merger Disadvantage," *National Post*, April 17, 2007, accessed December 29, 2008, http://friendscb.org/News/Friends_News/archives/articles04170701.asp, subsequently accessed March 21, 2011, http://www.friends.ca/news-item/2410.

42 Statistics Canada, "Radio and Television Are Alive and Well," 2008 report, accessed December 29, 2008, http://www41.statcan.ca/2006/2256/ceb2256_004-eng.htm; see also Marke Andrews, "Traditional Radio Remains Alive and Well in

Canada," *Vancouver Sun*, August 10, 2007, accessed March 21, 2011, https://www. friends.ca/news-item/4645.

43 Shawn Ohler, "Radio's Space Invasion Not So Out of This World," *Edmonton Journal*, December 28, 2006, accessed December 31, 2008, http://friendscb.org/News/ Friends_News/archives/articles12280602.asp, subsequently accessed March 21, 2011, http://www.friends.ca/news-item/6903.

44 Grant Robertson, "Satellite Radio Deal Strikes Discord in Canada," *Globe & Mail*, February 21, 2007, accessed March 21, 2011, https://www.friends.ca/news-item/ 1332.

45 Grant Robertson, "U.S. Ruling Sets Clock Ticking on Radio Deal," *Globe & Mail*, March 25, 2008, accessed March 21, 2011, http://www.friends.ca/News/Friends_ News/archives/articles03250803.asp.

46 Vito Pilieci, "Sirius-XM Merger Could Ding CanCon," *National Post*, March 31, 2008, accessed March 21, 2011, http://www.friends.ca/news-item/2032.

47 *Broadcasting & Cable* Staff, "New Logic," *Broadcasting & Cable*, August 3, 2008, accessed March 22, 2011, http://www.broadcastingcable.com/article/94228- New_Logic.php.

10

WORLDSPACE SATELLITE RADIO AND THE SOUTH AFRICAN FOOTPRINT

BEN ASLINGER

The launch of WorldSpace's AfriStar satellite in October 1998 made three beams of up to eighty channels available to subscribers on the African continent. WorldSpace is a subscription-based satellite radio service founded in 1990 by Noah Samara designed for emerging markets in Africa and Asia. WorldSpace's AfriStar and AsiaStar satellites, launched in 1998 and 2000, respectively, each have three beams—East, West, and South—that transmit a mix of news, music, and entertainment to subscribers with branded World-Space receivers. WorldSpace is an interesting case study as a venture that attempted to find a middle ground between explicitly prodevelopment satellite and communications projects (influenced by modernization theory) and brazenly neoliberal commercial ventures such as Rupert Murdoch's Star TV. WorldSpace's activities illustrate tensions created by the company's attempts to reconcile the utopian visions of modernization theory that treat satellite communication technologies as a source of economic and cultural uplift with the rapidly evolving strategies of the culture industries in an age of unstable and capricious flows of transnational capital.

WorldSpace's offices in Johannesburg handle the technical and administrative issues for AfriStar's southern beam, and the South African context makes an excellent case study for analyzing how the satellite broadcaster chooses between global, national, and local radio broadcasters in creating its programming lineup.[1] South Africa's British and Dutch colonial histories, apartheid, the Truth and Reconciliation Commission, and the rise of a more inclusive parliamentary democracy with the election of Nelson Mandela have drawn international media attention to the country and have placed the nation at the center of debates surrounding African politics. WorldSpace's programming determines which musical, national, regional, and political cultures will become part of AfriStar's cultural footprint. The introduction and marketing of satellite radio in Africa gives scholars an opportunity to see the continuities and discontinuities between "old" and "new" media and examine which broadcasters are invited to participate in the new media economy of satellite radio and which broadcasters are reinscribed into terrestrial radio and regional, local, and/or language-based spaces. I argue that WorldSpace's South African programming evidences tensions between neocolonialism and pan-African politics and between regional, local, national, and transnational broadcasting interests.

For-Profit Development?

As WorldSpace began lining up broadcasters and entering various national markets, the company had to navigate the tension between the utopian rhetoric of its modernist development mission and its status as a transnational media company. WorldSpace's difficulties in reconciling development and economic neoliberalism raise the question of whether it would be possible for any for-profit enterprise to satisfy both stockholder desires and development objectives. These tensions surfaced early on in profiles of WorldSpace CEO Noah Samara. While some journalists treated Samara as a visionary and emphasized his personal stake in the company and his idealism, others treated him as a profit-driven media mogul.[2] Marina Bidoli opens her 1998 story on WorldSpace by asking if Samara is "Radio Gaga or the Rupert Murdoch of the Developing World?"[3] Certainly, the potential comparison to Murdoch likens Samara more to a defender of free flows and neoliberal economic policies than to a humanitarian, development-driven executive. Samara's mogul status was bolstered by the fact that he had a 91 percent stake in WorldSpace before the company's initial public offering (IPO); his 67.3 percent share of the company after the IPO helped fuel a kind of cult of personality surrounding the CEO.[4]

Samara's and WorldSpace's insistence that the company's main mission was to develop communication infrastructures in the developing world resuscitated a kind of 1990s technological boosterism in journalists. Gracia Hillman, president and CEO of WorldSpace Foundation, said, "The ability to widely disseminate information about the treatment and prevention of HIV/AIDS and other diseases is the very reason the WorldSpace system was created."[5] Anitha Soni, managing director of WorldSpace Southern Africa, stated in 2001 that the company's business efforts "are in line with our mission to create information affluence in those parts of the world where lack of appropriate infrastructure and prohibitive costs continue to work against access to information."[6] After presenting the WorldSpace technology to Ethiopian officials, including Foreign Minister Seyoum Mesfin and Education Minister Genet Zewde, Samara said, "If there is a computer that can give a solution to all questions of the problems of Africa, the solution would be WorldSpace."[7]

While headlines such as "Bridging the Digital Divide: The Delivery of Digital Radio, via Satellite, Promises to Reduce the Information Gap between the Developing and Developed Worlds" stressed WorldSpace's efforts at constructing communications infrastructures,[8] journalists such as Jamie Doward wondered if WorldSpace's goals were too ambitious. In 1998, Doward wrote in the *Observer*, "*Ambitious* is the word. The reason WorldSpace is not that well known here is that it aims to provide emerging countries with state-of-the-art radio even before it has become apparent the idea can take off in the West."[9] Noting the disparity between the numbers of radio stations per listener in industrialized nations and the developing world, Doward's report suggested that WorldSpace's goals, though admirable, were perceived by many analysts and investors as unattainable. African development agencies and educational institutions to which WorldSpace had promised receivers and specialized broadcasting content became increasingly skeptical as well. The *East African Standard* reported in 2003 that more than one year after educational broadcasts were supposed to go live to Kenyan schools, less than 1 percent of schools had been sent the needed receivers to listen to the programs, illustrating that WorldSpace was having trouble living up to its educational and social-justice mission.[10]

WorldSpace worked to place the transnational corporation at the center of global radio flows, sidelining the role that governments and nongovernmental organizations have played in global media governance and initiatives such as the World Summit on the Information Society (WSIS). In congressional testimony in May 2001, Samara twisted the digital divide from its critical political-economic roots and used it to rationalize the expansion of a U.S.-based transnational company. Samara said, "That Africa is on the wrong side of an 'information technology divide' should surprise no one. . . . Through most of Africa, the residents of most cities have only the most meager choice of news and entertainment via the companionable medium of radio." Piggybacking on the techno-utopian hopes and neoliberal aims of the Clinton administration's National and Global Information Infrastructures initiatives, Samara told congressional members, "To put it bluntly, the Internet does not currently serve all of Africa and will not be able to do so for years to come. WorldSpace does. . . . The only thing that stands between six hundred million people becoming a true market for American goods and services is information."[11] With a business model that combined leasing channel capacity, commercial advertising, and subscription fees, WorldSpace latched on to the same lifestyle marketing tropes that U.S. satellite firms such as XM and Sirius had used to brand their services; in WorldSpace's case, lifestyle marketing messages blended uneasily with rhetorical messages of African empowerment.[12]

Contesting the Footprint

As WorldSpace introduced channel lineups, pursued partners, and began broadcasting, conflicts began to emerge over which audiences and broadcasters mattered most and how WorldSpace conceived of potential listeners. While executives continually stressed that they were working to alter communications flows, WorldSpace's privileging of audiences such as American and British nationals, its embrace of "high quality" international broadcasters, and its failure to define a range of listeners point to conflicts over the cultural footprint of the AfriStar satellite.

Lisa Parks argues that understanding the footprint in only its technical sense ignores the ways that the distribution and use of satellites have much to tell us about the operation of cultural power. Parks writes, "We must imagine 'flow' and 'footprint' not as fixed schedules and closed boundaries but as zones of situated knowledges and cultural incongruities that may compel struggles for cultural survival rather than simply suppress them."[13] While AfriStar's signal covered the continent, WorldSpace determined which broadcasters would become international or regional players and which audiences would have access to international media content.

As WorldSpace devoted significant attention and marketing efforts to attract American and British expatriates—especially military personnel—the company's executives struggled to define the African user.[14] Agreements with the British Broadcasting Company (BBC) and Fox News Channel illustrated the economic value of American and British nationals.[15] Neil Curry, commissioner for the BBC World Service English networks, said, "I love that I can listen to the quality of the BBC programming anywhere in the world, and there are people out there who want the same thing. Marketing to niches means they can get it."[16] Curry's comment is one of the rare times an audience member gets to speak about the experience of listening in the coverage of WorldSpace in the trade press materials.

In an age of narrowcasting and an overproduction of marketing discourses and psychographic profiles of media audiences, WorldSpace seemed to have difficulty imagining a range of listeners or selling investors on the specific demographics of listeners. WorldSpace rarely reported the number of subscribers or actual listeners, preferring to think of its audience as everyone contained within the footprint. Sam Holt, senior vice president of content, told *Billboard*'s Frank Saxe in 2000, "We will succeed at the outset in our ability to be inclusive and offer people the opportunity to escape a niche and aspirationally become part of the bigger world."[17] The number of people and amount of demographic data on who would be able to "aspirationally become part of the bigger world" were up for debate, especially when considering the cost of receivers and subscription services. Tony Koenderman of the *Financial Mail* wrote in 1997, "WorldSpace's Noah

Samara has ambitious plans for Africans. But will they be able to afford them?"[18]

Word choices in stories stressed WorldSpace's continual need to assert the viability of its business model and highlighted the number of objections that potential investors and partners might raise. Simon Barber of *Business Day* writes, "Low-end models, WorldSpace promises, will go for about US$200. Samara insists that enough people in Africa can afford that to make the project viable."[19] By 2006, the *Motley Fool* openly contested WorldSpace's definition of the footprint as all-inclusive: "The real potential audience is very small compared with the 'billions' that WorldSpace tosses out when addressing the pie-in-the-sky potential audience for its planned roll-outs."[20] While WorldSpace asserted that its business model was workable and referred to the potentially vast scope of its listenership, journalists, investors, and analysts wanted more specific data on the number of subscribers and well-researched market constructions of listener demographics.

Access issues and definitions of the audience are significant elements that work to shape the cultural footprint of the AfriStar satellite, but determining which broadcasters became a part of the WorldSpace service reveals how firms extended their international reach. In 2006, after WorldSpace renewed a distribution agreement with CNN International, William Sabatini, the firm's vice president of global content, said, "We are proud to extend our relationship with CNN International and to continue providing subscribers with one of the world's most respected and ubiquitous English-language news sources. As we extend existing contracts and forge new relationships with high-quality content providers, we enhance WorldSpace's subscriber value and reinforce our position as the dominant international provider of satellite radio services."[21] Samara told *Satellite Week* in 1999, "When you're working with respected organizations such as CNN and Bloomberg, it brings players of a similar quality into the mix."[22] "High-quality content providers" seem to be either American or British broadcasters or broadcasters whose primary content echoed the subject matter and technical style of American and British broadcasting firms.

Kelechi Obasi, of the Lagos, Nigeria, paper the *News*, wrote in 1999 that many African broadcasters had mixed feelings about WorldSpace and the success of the AfriStar satellite. Chris Ubosi, director of operations for Steam Broadcasting Company, said, "It's a good development, a welcome one not only for Nigeria but for Africa, as it offers us the opportunity to be heard worldwide." But Obasi went on to note that Ubosi still believed many stations would remain local, given the cost of receivers, subscription fees, and equipment and airtime.[23] For Ubosi, the potential of being heard transnationally had to be balanced with the costs of uplinking and associating oneself with a transnational venture such as WorldSpace. Sud FM, based in Dakar, Senegal, became the first Francophone broadcaster to sign with WorldSpace, but only on the western beam.[24] Golfe FM of Benin and Radio Kledu of Mali also signed agreements with WorldSpace, but issues emerged regarding the beams on which signals appeared.[25] While BBC and CNN programming could be heard on all three of AfriStar's beams, African broadcasters were often confined to one beam. While this may have made many African broadcasters significant regional players, WorldSpace never worked to alter global communications flows to the degree that executives claimed.

WorldSpace and South African Broadcasters

Journalists writing for the South African business and popular press had mixed feelings about WorldSpace's entrance into the African mediasphere. This ambivalence may be due to the seismic shifts in South Africa's mediascape (such as increasing privatization and conglomeration of media interests), lingering questions about the role of media in the post-apartheid era, and the rapid influx of foreign capital into the South African mediasphere. Keyan G. Tomaselli and Ruth E. Teer-Tomaselli write, "in the mid-1990s, the African National Congress (ANC) led the way for media transformation, black political and economic empowerment, and an inclusive range of languages, political opinions, and populist genres, marking one of the most remarkable makeovers in international media history."[26] However, they also note that by 2007, four press companies—News24, Caxtons, Independent Newspapers, and Avusa—dominated news coverage.[27] While the vitality of South African media was a concern, Herman Wasserman argued that more debate was needed regarding "the role the media should play in the reconstruction of post-apartheid social identities and reconfiguration of social reality."[28] And Ruth Teer-Tomaselli noted with some worry the scale of media acquisitions by foreign firms: "From 1994 on, international capital began to acquire interests in local South African information and communications companies on an unprecedented scale."[29]

Samantha Sharpe observed that WorldSpace planned to invest more than 2.5 billion rand in Africa from 1998 to 2001,[30] and Lesley Stones noted that WorldSpace had signed a four-year deal with South African telecommunications company Telkom to base its pan-African service in Johannesburg.[31] While Sherilee Bridge noted the optimism surrounding the launch of the AfriStar satellite (valued at US$850 million) in 1998, she also explained that the cost of receivers and leasing channel capacity "could see [South African] radio stations lose out on the full potential of their investments."[32] Difficulties in establishing a workable business model in the South African market led to the drafting of the relatively inexperienced twenty-two-year-old new media executive Hamza Farooqui in 2004 to head up WorldSpace's South African offices.[33]

While WorldSpace pursued partnerships with South African and regional broadcasters, the combination of programming feeds from Fox News, CNN, Bloomberg, and the BBC and WorldSpace's lineup of branded channels left less space for national, regional, and local broadcasters. WorldSpace's branded music channels included UPop ("the station for globally focused pop music that knows no boundaries"), Riff ("presenting jazz to a global audience"), Bob ("offering new music with a new attitude, Bob serves a global audience with music from some of the world's best modern rock bands, including Green Day, Nirvana, Weezer, and Radiohead"), Radio Voyager ("playing today's hottest adult hits"), Flava (featuring bhangra, reggaetron, Kwaito, dancehall, and hip-hop), and Worldzone (world music).[34] In 2001, when WorldSpace acquired Radio Voyager, Sabatini said that Voyager's adult contemporary format "reaches the eighteen- to forty-nine-year-old demographic. This is a group that has a lot of buying power and a lot of influence in setting trends, and Radio Voyager is a great way to serve it."[35] In publicizing channels such as Bob and Radio Voyager, WorldSpace struggled to mention any artists who were not white, alternative rock, American or British artists whose careers began in the 1990s; this move added fuel to smoldering arguments that programming on the AfriStar satellite was yet another symptom of neocolonialism.

Tensions between global and local music flows came to the fore in corporate discussions of Flava, WorldSpace's hip-hop music channel. Sabatini said of Flava, "We are not just giving our subscribers premiere access to international hip-hop-flavored music, we are combining it with the freshest and latest in U.S.-based music of the genre and offering it to the world." Shawna Odour, programming director, said, "We are extremely excited about Flava and hope it will become a catalyst for the international hip-hop movement, creating a sense of community across the globe."[36] On the one hand, Sabatini as a high-ranking executive at WorldSpace hinted at the centrality of U.S. hip-hop and bordered on suggesting that genres such as bhangra, dancehall, and kwaito could only be understood in relation to American hip-hop styles. On the other hand, the programming director of the channel itself stressed hip-hop's status as a global cultural movement. But despite their somewhat oppositional stances, Flava's status as a WorldSpace branded channel worked to define global hip-hop cultures from above, potentially co-opting local musical styles.

South African radio broadcasters who did take advantage of the opportunity WorldSpace presented were primarily broadcasters who offered programming in English or Afrikaans and who were already major players in South African radio. Stations that signed agreements with WorldSpace included Kaya FM, an adult music and talk radio station directed to the growing black middle class, as well as Jacaranda and East Coast Radio, parts of the Kagiso

Media Group that broadcast adult contemporary music and news. Clive Barnett notes that Jacaranda and East Coast Radio were sold to private interests in the 1990s.[37] Barnett argues, "the reform and restructuring of a broadcasting and telecommunications sector previously tightly controlled by the apartheid state necessarily implies an increased role for private capital and the market."[38] Pat Dambe, managing director at Kaya FM, said, "I am excited that Kaya will soon be accessible throughout Africa with audiences receiving the station in digital sound as far north as Portugal, Cairo, and Turkey."[39] On the one hand, by linking with WorldSpace, these broadcasters expanded their footprint beyond the provinces in which their FM signals could be heard. On the other hand, what was publicized as national South African culture within southern Africa, with the exception of Kaya FM, was largely geared to listeners who were affluent, white, and cosmopolitan. The privatization and continental expansion of South African media interests bracketed the larger communication policy issue that Barnett raises, namely, "how liberalization can be regulated and made consistent with the promotion of nation-building, reconciliation, democratization and cultural diversity."[40]

Conclusion

In 2007, the law firm of Spector, Roseman, & Kodroff filed a class action lawsuit on behalf of stock purchasers. The suit alleged that WorldSpace had violated financial regulations by making misleading statements about subscriber numbers and revenues in both press releases and filings with the Securities and Exchange Commission. The suit also alleged that WorldSpace continued to count consumers who took advantage of a three-month promotional offer but then elected to cancel their subscriptions for three months beyond the cancellation dates.[41] The suit accused WorldSpace of breaches of corporate ethics and incidents of misconduct, but the case also highlights the need to avoid taking at face value the trade press, the relationships between reporters and the public-relations apparatuses of companies, or the "vaporware" that John Caldwell argues challenges the truth of stories and claims of journalistic credibility.[42]

In this chapter, I have used WorldSpace's construction of satellite radio programming and its design of services surrounding the AfriStar satellite in order to analyze the tensions involved in rolling out new technologies in emerging markets. It is easy to dismiss WorldSpace as another case of neocolonialism, treating executives such as Samara and Sabatini as villains. Judging by the allegations leveled at WorldSpace and its difficulties in imagining and constructing diverse audiences and programming types, there may be some truth to this line of thought. However, to simply demonize WorldSpace would ignore the ways in which larger regulatory and economic trends have handicapped even well-meaning firms' abilities to balance profit-making imperatives and forms of global citizenship. Also, to villainize WorldSpace's corporate leaders draws attention away from the subnational, national, regional, and global tensions surrounding WorldSpace's entry into markets such as South Africa.

NOTES

1 Due in part to the economic crisis, WorldSpace recently restructured its operations and renamed itself 1worldspace. This essay analyzes WorldSpace's activities from its origins until the end of 2008. Future research might investigate how the global economic crisis has reshaped this company's activities in Africa, India, and the Asia-Pacific regions.

2 Catherine Yang, "Media Mogul for the Third World," *Business Week*, June 30, 1997; Doug Abrahms, "Visionary Hopes to Bring Cheap Radios, Fed by Satellites, to the Developing World," *Washington Times*, August 8, 1994.

3 Marina Bidoli, "Reaching for the Skies," *Financial Mail*, November 6, 1998.

4 David Ehrlich, "IPO Outlook," *Deal*, June 28, 2005.

5 "Africa-at-Large: WorldSpace, Satellite Create First Public Health Channel," *Africa News*, May 22, 2000.

6 "Nigeria: WorldSpace Crosses the Digital Divide," *Africa News*, November 22, 2001.

7 "Ethiopia: Noah Samara Says Wants to Make a Proposal to Help the Transformation of Education," *Africa News*, June 21, 2002.

8 Richard Butler, "Bridging the Digital Divide: The Delivery of Digital Radio, via Satellite, Promises to Reduce the Information Gap between the Developing and Developed Worlds," *Telecommunications International Edition*, January 2002.

9 Jamie Doward, "Media: Will WorldSpace Come up Beaming?" *Observer*, May 31, 1998.

10 The East African Standard, "Kenya; Uncertainty over KIE Broadcast to Schools," *Africa News*, February 18, 2003.

11 House Committee on International Relations Subcommittee on Africa, *Bridging the Information Technology Divide in Africa*, May 16, 2001.

12 Sylvia Lyall, "WorldSpace Launches Sh30 Million Sales Campaign," *Daily Nation*, July 25, 2000.

13 Lisa Parks, *Cultures in Orbit: Satellites and the Televisual* (Durham, N.C.: Duke University Press, 2005), 62.

14 "WorldSpace Satellite Radio Launches Global Subscription Campaign; Initial Focus on American and British Expats," *PR Newswire*, March 3, 2004; "WorldSpace Satellite Radio Details Global Subscription Campaign," *Asia Pulse*, March 11, 2004.

15 "Fox News Channel to Air on WorldSpace Satellite Radio: American Expats and U.S. Military Anticipate Fox News Channel on WorldSpace Satellite Radio," *PR Newswire*, March 17, 2004.

16 "Reaching Niche Markets Key to Satellite Radio Success: Industry Experts Outline Opportunities to Engage and Cultivate Subscribers," *PR Newswire US*, November 20, 2006.

17 Frank Saxe, "WorldSpace Aims to Bring Satellite Radio to the Planet," *Billboard*, August 19, 2000.

18 Tony Koenderman, "Now for Radio by Satellite," *Financial Mail*, November 14, 1997.

19 Simon Barber, "Stars and Stripes: AfriStar Elegant in Theory, but Wait-and-See Policy Seems Best," *Business Day*, November 4, 1998.

20 Duncan McLeod, "Noah Samara's Horrible Year," *Financial Mail*, July 21 2006.

21 "WorldSpace and CNN International Extend Global Relationship: Three-Year Contract Renewal Ensures Delivery of Leading English-Language News Service to Europe, Middle East, Asia and Africa," *PR Newswire US*, August 14, 2006.

22 "WorldSpace Identifies Its 2 Primary Backers," *Satellite Week*, February 1, 1999.

23 Kelechi Obasi, "Africa-at-Large: Global Vibrations," *Africa News*, November 1, 1999.

24 "Senegalese Broadcaster Signs with WorldSpace to Transmit Digital Direct Radio Programming in Africa: Sud FM Will Offer French Language News, Entertainment Via the WorldSpace System," *PR Newswire*, September 1997.

25 "WorldSpace Signs Contract with Radio Kledu: Francophone Broadcaster to Increase Scope of Reach," *PR Newswire*, June 23, 1998.

26 Keyan G. Tomaselli and Ruth E. Teer-Tomaselli, "Exogenous and Endogenous Democracy: South African Politics and Media," *International Journal of Press/Politics* 13(2) (2008): 173.

27 Ibid.

28 Herman Wasserman, "Globalized Values and Postcolonial Responses: South African Perspectives on Normative Media Ethics," *International Communication Gazette* 68(1) (2006):81.

29 Ruth Teer-Tomaselli, "Transforming State Owned Enterprises in the Global Age: Lessons from Broadcasting and Telecommunications in South Africa," *Critical Arts: A South-North Journal of Cultural and Media Studies* 18(1) (2004):9.

30 Samantha Sharpe, "WorldSpace Takes on 10% in Radio Maker Baygen," *Business Day*, May 8, 1998.

31 Lesley Stones, "Radio Broadcast Deal Worth Millions," *Business Day*, August 26, 1998.

32 Sherilee Bridge, "Africa Home to World's First Satellite Relay," *Business Times*, October 25, 1998.

33 "WorldSpace Satellite Radio Names Hamza Farooqui Managing Director," *PR Newswire Europe*, March 29, 2004.

34 "WorldSpace Satellite Radio Unveils Audio Signature," *PR Newswire*, December 12, 2005.

35 "WorldSpace Acquires Radio Voyager: All English Radio Network Merges with Pioneer Digital Satellite Radio Company, Joins Roster of WorldSpace Unique Branded Music Channels," *PR Newswire*, October 29, 2001.

36 "International Hip-Hop Now Available 24x7 on WorldSpace's 'Flava': Channel Offers Freshest Hip Hop and Urban Music from around the World," *PR Newswire*, November 15, 2005.

37 Clive Barnett, "The Limits of Media Democratization in South Africa: Politics, Privatization, and Regulation," *Media, Culture, & Society* 21(5) (1999): 657.

38 Ibid., 650.

39 "WorldSpace Signs Contract with Kaya FM: Satellite Broadcast System Gives South African Broadcaster Reach throughout Africa," *PR Newswire*, September 15, 1998.

40 Barnett, "Limits of Media," 650.

41 "Spector, Roseman, & Kodroff, P.C. Announces Class Action Lawsuit against WorldSpace, Inc.," *Business Wire*, April 4, 2007.

42 John Thornton Caldwell, *Production Culture: Industrial Reflexivity and Cultural Practices in Film and Television* (Durham, N.C.: Duke University Press, 2008).

11

CONTENT VS. DELIVERY

THE GLOBAL BATTLE FOR
GERMAN SATELLITE TELEVISION

PAUL TORRE

Three days after the launch of its digital pay TV platform Premiere World, powerful German rights trader Kirch Group predicted that Rupert Murdoch's News Corporation and other competitors were permanently blocked from grabbing a significant chunk of the pay TV market in Germany.
—E. Hansen, *The Hollywood Reporter*, October 5, 1999

Getting content that will attract and hold subscribers is the primary critical success factor for the management of a digital platform. It is all about getting content before someone else spots its potential, and then tying up the rights exclusively.
—Alan Griffiths, *Digital Television Strategies*

In the era of globalization, influential policy actors are based not only in national governments but also in supranational bodies, regional and local administrations, as well as transnational and translocal networks and corporations.
—Paula Chakravartty and Katharine Sarikakis,
Media Policy and Globalization

Media expansion is a risky business, and the success of a new media venture depends on a number of factors. The challenges of competition, cooperation, and regulation are omnipresent considerations. Is it more important to control the media content, or is a successful launch dependent upon control of the delivery system? Within the media industries there are many examples of the basic conflict of content versus delivery, of television producers versus cable companies, or, in the context of emerging media, of gaming companies versus mobile providers. One entity may own the content while another owns the distribution system. In a number of cases, media cross-ownership, or consolidation across sectors, may bridge this bifurcation, as in the case of Disney producing television programming for its ABC television network and, even more directly, for the ABC stations that Disney owns and operates. Media industries typically function as competitors, even if cooperation and consolidation may smooth out many of the ostensible conflicts. Regulation of media industries often complicates media-sector expansion, with regulatory bodies deciding how best to arrange and control these fundamental conflicts over the production and delivery of media content. Media regulation at supranational levels involves governmental regulatory bodies that oversee the transnational or global interests of the media industries.

Media companies confront challenges of acquiring content, developing technologies, securing delivery systems, addressing regulatory demands, jockeying with competitors, and finding an audience. These challenges have been central to efforts to launch digital and satellite television in various contexts, as I explore in this analysis of the launch of satellite television in Germany. In his discussion of the launch of the Star and Phoenix television ventures in Asia, Michael Curtin considers the combination of infrastructure, politics, and brand management as a set of "sociocultural forces" confronting those companies launching satellite television. For instance, "within the infrastructural realm, government regulation and market forces significantly influence the configuration of delivery systems for satellite TV."[1] In this chapter I analyze the competitive arena, the regulatory environment, the political complexities, and the marketing challenges, both inside and outside German borders, as digital satellite television was developing in the territory.

Moving Toward Satellite Television

In the mid-1990s, with more than thirty million television households and a growing economy, Germany was considered the strongest market for television revenue, primarily via advertising. In 1994, the German media company the Kirch Group, the cable company Deutsche Telecom, the pan-European conglomerate Bertelsmann, and the French media company Canal Plus joined together in an attempt to launch the first digital television service within Germany; the service was called Media Service GmbH (MSG), but the consortium was blocked by the European Commission (EC).[2] The EC, citing its action against the proposed merger, noted that while it would "continue to favor agreements which promote technical progress, and which promote market entry, [it] would not allow agreements or mergers which have the effect of foreclosing markets or creating dominant positions, or giving the parties the possibility of denying access to new entrants." In this instance of supranational media regulation, the concern was to preserve competitive markets, to disallow an alliance that could have led to "excessive pricing, as well as a loss of innovation and product variety."[3]

In July 1996, however, the Kirch Group successfully launched Digitales Fernsehen 1 (DF1), the country's first digital and satellite television service, one that promised dozens of channels to satisfy multiple niche audiences. Bertelsmann, with major holdings in Germany, was offering the pay television service Premiere, and was planning its own satellite television service. Premiere initially utilized analog technology instead of digital, and employed cable delivery for most of its customers. At the outset, Kirch's DF1 used a single transponder on the Astra 1E satellite, originally at the 19.2°E orbital position. Both companies expanded their channel offerings on the Astra 1F satellite, with DF1's channels occupying ten transponders. These Astra satellites, also located at 19.2°E, were owned by Europe's first private satellite operator, the Luxembourg-based Société Européenne des Satellites (SES), and represented a break from state-managed deployment of satellite technologies and services.[4] Pay television and satellite programming contracts covered German-speaking Europe, which included Germany, Austria, German-speaking Switzerland, Luxembourg, Liechtenstein, and the Alto Adige region of Italy. Kirch's DF1 service was presented directly to Germany and Austria, and it was branded as "Teleclub" in Switzerland. DF1's revenue model included streams from subscriptions, pay-per-view, and advertising.

Both Kirch and Bertelsmann had been negotiating with the major studios in Hollywood in an attempt to lock in feature film programming rights for their respective services. In addition to securing American programming rights from the studios in the form of bulk packages called "output deals," they were also competing to carry branded channels from Discovery and National Geographic for their platforms.[5] These negotiations for American content were quite complex since satellite television had yet to be launched in many markets, and essentially these deals were speculative regarding the literal value of these programming rights alongside existing free-to-air channels.

Kirch was more successful in acquiring programming from Hollywood, and DF1's satellite offerings reflected these acquisitions. What did DF1 carry, and what audiences were targeted? Per the monthly guide *DF1 Magazine*, in 1998 the service included about thirty themed channels, including movie channels such as Star Kino (*Mission Impossible*), Cine Action (*Apollo 13* and *Twister*), and Cine Comedy (*The Cable Guy*). There were channels centered on westerns, crime stories, science fiction, romance, classical music, adult programming, and the Heimat-kanal, with German cultural fare. There were three each of children's and sports channels. There were familiar brands from the United Kingdom (Sky News and BBC Prime), and from across the Atlantic (MTV, VH-1, Discovery Channel, NBC, and CNBC). A double-page spread touts the "Top-Hits im April," spotlighting "Strong Women" on the pay-per-view service Cinedom, featuring Glenn Close's Cruella DeVille from *101 Dalmatians*; the members of *The First Wives Club* (Goldie Hawn, Bette Midler, and Diane Keaton); Barbra Streisand producing, directing, and starring in *The Mirror Has Two Faces*; and edgier fare with the women from *Bound* and *Set It Off*. Another double-page spread hyped Deutsches Sport Fernsehen's (DSF) live and commercial-free Formula One (F1) racing. Viewers could choose from six camera angles during the auto race: not only wide angles and close-ups but also the view from the pit and looking over the driver's shoulder. Clearly, this variety of programming was meant to appeal broadly—there was something for everyone—and DF1's content was arranged in several targeted packages. A Basic package was US$12 per month, a Movie package was US$20, a Sport package was US$17, and customers could receive all three of these packages for just US$25.

The development and dissemination of digital delivery (via both cable and satellite) allows for more channels and an increased ability to reach fragmenting audiences.[6] In his discussion of satellite television in the United Kingdom, Shaun Moores notes the characteristic shift in programming and scheduling initiated by satellite television, where "largely dispensing with the ethos of mixed programming, which is dominant in terrestrial television, satellite broadcasters prefer to deliver thematic packages." Moores notes, however, that increases in "diversity," which often lead to increased competition in the industry and increased consumer choice, may also lead to "a heated debate about 'quality' in broadcasting."[7] This debate raged in Britain, where Murdoch's Sky Television merged with British Satellite Broadcasting to form BSkyB in 1990. This may appear an odd progression, but diversity (even when referring to consumer choice) can lead to concerns about preserving the status quo. In her analysis of the nexus between presentation and perception, Charlotte Brunsdon contends that "the erection of a dish is also historically specific, a particular act, a concrete and visible sign of a consumer who has bought into the supranational entertainment space."[8] Moores notes some consumer resistance to satellite television in Britain, explaining that a common perception of satellite television revealed a "deep suspicion of American culture," and a "feeling of disgust [that] extended to many Hollywood movies on the film channels."[9] These are "textual" considerations that inform brand management, and it is telling that before launching Sony Entertainment Television (SET) in India, Sony first acquired a package of 400 classic Hindi films and distributed these prominently throughout its schedule until reaching its penetration benchmark.[10]

Within the German context, DF1 marketed its satellite television programming as a mix of local, regional, and imported content. As discussed above, DFI offered a broad set of themed channels, many of which combined American programming with Kirch's substantial library of programming rights to German and pan-European movies and series. Public perception of entertainment companies and their executives can also affect consumer reception. In the United Kingdom, Murdoch arrived into the marketplace from Australia and was widely considered a "marauding investor bent on building a global empire at the expense of local media enterprises."[11] In contrast, Leo Kirch was the son of a Bavarian vintner and intimately connected to German political and business interests. Kirch's rights trading was always a combination of high and low culture. He acquired rights to the Fellini classic *La Strada*, and also to the slapstick comedy of the Hal Roach library; he secured rights to live performances of opera singer Maria Callas, and also to Pamela Anderson's *Baywatch* and *V.I.P.* In the United Kingdom, Murdoch represented crass culture (in his tabloid and television holdings). Murdoch's entry into German satellite television was extremely controversial. Germany's digital sports programming featured Michael Schumacher, the only German to ever win the FI world championship, and the matches of the German Bundesliga balanced World Cup football. All of these factors served to counterbalance negative connotations of American content on German satellite television. In fact, beginning in 1999 Kirch's German free television network, ProSieben, organized its entire brand around a "Welcome to Hollywood" theme, and ProSieben continues to use the English-language tagline "We Love to Entertain You" to represent a programming mix that includes many American television series and blockbuster movies.

Germany and its television industry are organized into sixteen *Länder*, or federal states. German television is extremely competitive. There are two powerful public broadcasters (ARD and ZDF) that control more than one-third of the traditional television market share, numerous state-based regional channels, and government-subsidized Deutsche Telecom, which holds a monopoly on cable delivery. There are more than thirty free-to-air channels, including two major network families, and other specialized channels. In the mid-1990s, the typical media costs for each household included a mandatory public television fee of about US$20 per month, with another US$25 per month for cable. In this context, the additional costs of a DFI or Premiere Digital and Satellite television package (not including additional pay-per-view costs) were considered prohibitive by many, and DFI's early adopters were required to pay US$600 for the satellite receiver (set-top) box.[12] Thus one key explanation for the lackluster reception of satellite television on either platform was the high cost of basic media. In 1996, Germany had thirty million TV households. During that first year of the German digital and satellite experiment, Bertlesmann's Premiere Pay TV acquired only 300,000 new customers, or 1 percent of television homes. Kirch's numbers were ten times worse, with DFI counting only 30,000 new subscribers, for a national penetration rate of 0.1 percent.[13]

After months of fierce competition, and after assessing the risks each would face individually, Kirch and Bertelsmann agreed to join forces with German cable network Deutsche Telecom in a new digital and satellite television venture. Merger proceedings began in July 1997. After signing billion-dollar deals with nearly all of the studios, ordering one million digital boxes from Nokia, and buying out Canal Plus from Premiere for US$575 million, Kirch was especially grateful to have a partner to help share the financial risk. Both digital satellite television providers were glad to have realized the costs of independence early on. One crucial question remained, however: would the EC approve of the newly merged and rebranded service, to be known as Premiere World?

Premiere World and the European Commission

The first response came from the national media regulators. Antitrust authorities, including the German Commission on Media Concentration (KEK), were concerned about the two largest German media companies turning their duopoly into a de facto monopoly in the pay-television sector.[14] In a second and more significant development, the EC also decided to intervene once again. Given the trust-busting sentiments expressed by the commission mentioned above, this would not seem surprising, except for the fact that the commission had specifically recognized the necessity of media alliances when launching new media initiatives. The commission claimed to understand "the reality that few, if any, of today's market players will have the skills or resources to straddle the whole of the value chain within a converged environment, and the emergence of major players in the sectors affected by convergence will inevitably rely on partnering to varying degrees."[15] Nonetheless, the commission had several concerns, including dominance of program rights, variety of programming, and anticompetitive controls over the set-top box.[16]

Kirch and Bertelsmann attempted to make the necessary concessions to convince the commission that they could diffuse any monopolistic tendencies. For instance, Kirch offered to sell off 25 percent of its film rights, and was willing to open up the architecture of its set-top box to companies other than its partners, Nokia and Deutsche Telecom. The EC stood firm, however, claiming that the two media companies would still retain "control of the box," in addition to control of the content and the delivery system.[17] Bertelsmann decided that many of the concessions would not allow for profitability, and in the end, Bertelsmann appears to have understood the obstacles pay television would face in Germany better than Kirch.[18] Kirch fought the EC's decision, even going so far as to file legal action before the European Court

of Justice. Kirch contended that there was little chance of the merger domi-
nating the market, claiming that "Premiere will not be able to prevent entry
by new competitors in the area of digital pay television. On the contrary:
through building the necessary infrastructure and subscriber base, Premiere
will create the preconditions of competition, and through intense competition among more than thirty existing commercial and public broadcasters
the scope of Premiere is controlled effectively."[19]

Of course, Kirch was not planning on building a pay television infrastructure in order to share it with anyone, and yet this was the posture it
needed to adopt in front of the court. As for Kirch's second point, that the
competition between Germany's preexisting thirty channels would constrain
and limit the "scope of Premiere," that projection should have caused the
company to reconsider seriously the entire enterprise. Germany's competitive media environment, its "channel clutter," was regularly cited as a significant reason for the snail-like pace of the adoption of pay television within
Germany.[20]

Pan-European political realities contributed to the failure of the merger.
The EC's shutdown of the Bertelsmann-Kirch merger was interpreted as part
of an EC campaign to push for an accelerated process of liberalization of a
number of industries in Germany. Given the long-standing alliance between
Leo Kirch and the then chancellor Helmut Kohl, Kirch was caught in the
middle of a larger battle over Kohl's reluctance to sharply reduce governmental subsidies that were propping up several key German industries,
including support for public service television in the media sector. Kirch was
seen as collateral damage in the fight with the EC's competition commissioner, Karel Van Miert.[21]

Through the efforts of his American acquisitions executives, Kirch had
acquired far more pay television rights than Bertelsmann, and so it seemed
natural for Kirch to resolve the regulatory stalemate by offering to buy out
Bertelsmann, rather than vice versa.[22] In the spring of 1999, Kirch negotiated
a deal to pay US$1 billion for Bertelsmann's stake, and Premiere World was
officially relaunched on October 1 of that year.

Almost immediately thereafter, and even after receiving US$1.5 billion in
loans from Chase and several German banks, Kirch's operating costs forced the
company to consider accepting financial partners.[23] By the end of 1999, Kirch
had agreed to let Rupert Murdoch buy into Premiere, giving BSkyB a toehold
in the German media environment.[24] Murdoch's contribution of an additional US$1.5 billion, in cash and BSkyB shares, secured him a 22 percent share
of KirchPayTV. The EC allowed this investment to go forward. In addition to
the entrance of Murdoch's BSkyB, three companies held small stakes: Prince
al-Waleed bin Talal's Kingdom Holdings, Los Angeles–based Capital Research
Management, and Lehman Brothers Merchant Banking (see Figure 11.1).

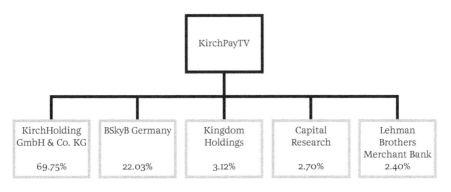

—————— Figure II.I Kirch PayTV investment structure, 2000. ——————

At the end of 1999, more than three years into the satellite television venture, Premiere had barely 2 million subscribers, a mere 6 percent of the German television market. Kirch projected a jump up to 2.9 million subscribers by the end of the next year, but the tally at the end of 2000 was 2.2 million—only 200,000 subscribers had joined during the course of the entire year.[25] Murdoch invested in Premiere with the understanding that the recently united service would be able to grow in Germany without the threat of competition, and he had limited patience with Premiere's slow growth in 2000. By June 2001, six months prior to the deadline for substantial subscriber growth, Murdoch adjusted the terms of his investment in Premiere. He granted a six-month extension (until the summer of 2002), but required that if the growth benchmark was not met, the put option repayment of his US$1.5 billion investment must be made entirely in cash.[26] A study by Zenith Media forecast that by 2002 more than half of Europe's 182 million TV homes would be pay-television subscribers.[27] But Premiere was struggling in Germany, even as Spain's Via Digital and Quiero were suffering from sluggish subscriber growth and persistent losses. Compared to Premiere's 6 percent adoption rate in Germany, France's CanalDigital was seeing steady growth at 25 percent, and Murdoch's BSkyB had reached more than 30 percent of the television households in the United Kingdom (see Figure II.2).

Is Content King?

Kirch responded to the subscription crisis at Premiere with renewed attempts to acquire desirable, high-profile programming. If blockbuster movies from each of the major studios in Hollywood were not impressive enough, then perhaps major sporting events would prove more attractive.[28] Kirch acquired 75 percent of the F1 auto racing circuit for more than US$2 billion. The company was already spending more than US$350 million annually for Germany's national soccer league, the Bundesliga. Soccer ratings, like

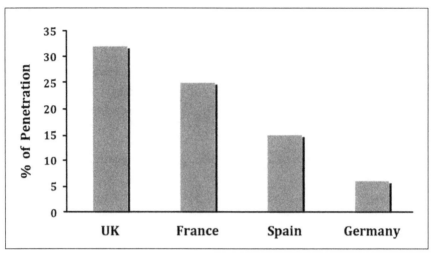

Figure II.2 Satellite TV adoption in Europe, 2002.

those for FI, were huge, averaging market shares of 20 to 30 percent for the Bundesliga and 40 to 60 percent for the World Cup matches. But Kirch may have overestimated the value of certain rights. For instance, in 1998 soccer rights holder FIFA sold the worldwide broadcasting rights for US$80 million, and yet Kirch was willing to pay US$800 million for the same rights just four years later, in 2002.[29] Kirch found that its free-to-air and satellite broadcaster clients were not able to pay their asking prices for the rights to FI and World Cup soccer. After such huge outlays for sports rights, Kirch became embroiled in separate controversies with the race-car manufacturers and FIFA. In both instances, the case was made that the rights to FI and soccer could not be set aside for pay television, but that they needed to be made available to the population at large—that each was in some way a public trust and needed to be available for all audiences.[30] Of course, Kirch had secured these rights for the express purpose of hyping unique, exclusive product on its pay television service.

The Kirch Group's relationships with the Hollywood studios were strained by the high cost of the feature films and television series that it had licensed, especially in relation to digital and satellite television, where optimistic projections were not matched by healthy revenues. Kirch used the realities of sluggish subscriber growth to attempt to renegotiate contracts signed in the midst of a bidding war with Bertelsmann. The American studios typically colluded to adopt a "Most Favored Nation" strategy, where one studio would only agree to a change if all other parties agreed as well. Nonetheless, Kirch representatives began talks with the Hollywood studios in mid-1999. They proposed various adjustments to Kirch's overall television programming contracts, asking for reductions in the number of TV movies, and un-restricted multiplexing (multiple reruns during a twenty-four-hour period). A Kirch executive indicated that "it should be the primary aim of all parties involved to make the platform business in Germany a lasting success, not let it crumble under the weight of historical lunacy," observing that all the con-tracts were negotiated in 1996 and 1997 and that "German historians refer to this two-year period as the 'Mother of All Pay-TV Wars.'"[31]

Nearly every studio was responsive to KirchMedia's entreaties over the next two years, and each made limited concessions that amounted to a total savings of US$300 to US$400 million over the first five-year term of a ten-year contract. The savings represented by these concessions were approximately 10 percent of the overall cost of the various arrangements, and, while signif-icant, such adjustments were far from adequate. As 2002 began, Kirch's satel-lite division was losing nearly US$2 million per day.[32]

Kirch v. Universal and *Universal v. Kirch*

> *The close-knit international TV fraternity has just been fractured.*
> *For the first time ever, one of Hollywood's foreign TV clients—in fact, its*
> *No. 1 client—has slapped a megabuck lawsuit on one of the studios. . . .*
> *[T]he real concern in Hollywood is that the Teutonic tiff could broaden—*
> *and that other Eurocasters will want to review their own deals.*[33]

Over the course of these renegotiations, one Hollywood studio repeatedly refused to adjust its digital satellite television output arrangement. The Kirch Group had entered into a pay-TV output deal with Universal Studios in July 1996 that included feature films, television series, and television movies, but for pay television only, as Universal had a free-to-air output deal with Kirch's rival, Bertelsmann-controlled RTL. At that time, Canada's Seagram's owned Universal; it was subsequently acquired by France's Vivendi, and most recently by General Electric. One unique aspect of the content arrangement involved Universal working with Kirch to launch two Universal-branded channels, 13th Street and Studio Universal, on the Ger-man digital satellite platform. For Kirch, an incentive for the arrangement was that Universal promised to devote special efforts to produce higher-quality television programming for the channels on the platform—since it would be highly visible and representative of the Universal brand. Kirch even paid a healthy US$65 million "inducement fee" earmarked for develop-ing this quality programming. However, the business relationship between the companies deteriorated to the point where Kirch filed a lawsuit against Universal, and Universal reciprocated with a countersuit.

Conflicts between Kirch and Universal over their contract can be traced back to early 1998, less than two years after the deal began. When Kirch and Bertelsmann went before the EC to seek approval for the DFI-Premiere merger, Blair Westlake, chairman of Universal Television, was the only top U.S. television executive to challenge both the request for a merger and Kirch's later unilateral acquisition of Premiere.[34] As mentioned, the commission blocked this merger. As Kirch prepared to take over satellite operation, Universal's Westlake insisted, as a stipulation for the takeover, that the EC require Kirch adhere not only to the Universal-Kirch deal but also to all of the previously negotiated output arrangements with each of the studios. Westlake presented his position as antimonopolistic—he wanted to hold the Kirch Group to deals that Kirch had negotiated while in competition with Bertelsmann. Nonetheless, Kirch was able to secure adjustments from the other studios, but Universal was not similarly accommodating. Kirch noted this episode of Westlake's involvement when filing a US$2 billion lawsuit, accusing Universal of "improperly intervening" with the European Union's Commission for Competition Policy.[35]

As with a similar, simultaneous lawsuit lodged by Kirch against Paramount Studios, the dispute with Universal centered on the lack of feature films and the poor quality of the television programming. Kirch claimed that Universal had drastically reduced the number of blockbuster films included in its pay-TV output during the first years of the agreement. In 1996, the year the agreement was signed, Universal had produced thirty films, but in the years that followed, Universal produced only ten to fifteen films per year. In a number of cases, Universal also engaged in "split-rights" deals, where it would produce blockbuster films with another studio, splitting the foreign rights with it instead of passing along the films to Kirch for German pay television.

Just as Kirch had accused Paramount of doing, Universal combined dramatically reduced feature-film output with television product of questionable quality. Soon after the output deal was signed, and as part of corporate-parent Seagram's new entertainment approach, Universal essentially spun off its television production division. The Home Shopping Network (HSN) acquired a major stake, and Universal authorized HSN chairman Barry Diller to increase production of syndicated programming, such as *Hercules* and *Xena*, and decrease production of quality primetime network series, such as the *Law & Order* franchise.[36] The Kirch lawsuit claimed that "Universal failed to provide all of its output of movies, failed to use a US$65 million inducement payment for high-quality television and motion-picture programming, and transferred its television production business to USA Network, which did not produce quality shows with appeal to German-speaking Europe." The Kirch complaint also alleged that KirchMedia was entitled to terminate the contract "because of governmental rulings blocking the merger with Bertelsmann and constituting *force majeure* events."[37]

In September 2000, however, the presiding judge threw out the lawsuit without allowing Kirch to present the "extrinsic evidence" of the negotiations leading up to the output contract.[38] Kirch officially stopped payment to Universal in October 2000, owing more than US$100 million for a year's worth of output. Universal stopped delivery of all programming materials and dubbing elements in April 2001 and filed a countersuit. Trial dates were postponed all the way up to Kirch's bankruptcy filing in April 2002. Universal kept up the pressure, however, and the matter was still in litigation, even when the reconstituted German company signed a new multiyear output arrangement with Universal in August 2002.[39]

Throughout 2001, Kirch was also embroiled in a lawsuit with Paramount and was contemplating legal action against Sony. The primary argument advanced by Universal and Paramount was that Kirch was suffering from buyer's remorse and was making after-the-fact excuses to avoid payment of legitimate obligations.[40] There was speculation that these lawsuits were an attempt to buy time and mask the Kirch Group's increasing inability to manage its substantial debt load. The conventional wisdom, however, was that "a combination of high Hollywood film prices and a failure to attract subscribers turned Premiere into a financial disaster for Kirch."[41] Indeed, Premiere recorded losses of more than US$1 billion in 2001, and the pay television service ended that year with roughly 2.3 million subscribers, only 100,000 more than the previous year. By the end of 2001, Kirch was facing multiple put options from Murdoch and other investors, and Kirch's banks were becoming reluctant to extend credit.

Conclusion

This chapter began by underscoring several key factors affecting the launch of a digital satellite television service. Media companies must wrestle with competitors, navigate the political terrain, and negotiate with regulators, and they must construct an effective marketing plan that will encourage consumers to sign up for new content delivered via new technologies. Kirch confronted these issues and obstacles in launching satellite television in Germany, and Rupert Murdoch faced a similar mediascape in launching satellite television in the United Kingdom, and more indirectly via his investment in Germany. Murdoch was also pursuing satellite opportunities in Asia, where he encountered the difficulties of presenting digital content to multiple cultures across multiple time zones, particularly in the emerging markets of China and India. After initial setbacks in each country, Murdoch was forced to retrench and recalibrate his entry posture. Hernan Galperin notes that despite "the expanding jurisdiction of intergovernmental bodies, nations retain key instruments to direct the evolution of their media sector, whether in terms of market structure, technology, or content."[42] Murdoch acknowledged a key miscalculation in constructing and presenting China's Star as "an English-language, pan-Asian platform aimed at upscale house-holds across the continent."[43] Kirch's media mix for DF1 and Premiere achieved a balance that had eluded Murdoch: how to acquire and program Hollywood content without leaving aside locally produced and culturally situated programs that would reach a variety of audiences. For Murdoch's second iteration of satellite television, Phoenix, alliances with savvy, politically connected locals produced a branding campaign that sought to represent the service as something that was brand new but also profoundly Chinese.[44]

For Kirch, the pursuit of the perfect programming mix was prohibitively expensive. Contracts were signed to send US$6 billion to the Hollywood studios, and yet given the nearly insurmountable obstacles presented by regulations, competition, technologies, and marketing, the content simply was not worth the price. As discussed, business strategies predicated upon acquiring and controlling both content and the means of delivery can lead to the intrusion of national and supranational media regulators intent on preserving open markets. And yet when German media companies sought alliances in their quest to launch new media technologies, including satellite television, the EC extended its intergovernmental regulatory reach to manage the evolution of the German media sector, effectively launching a global battle for German satellite television. At the turn of the millennium, Germany's foray into satellite television—the pursuit of content, the development of new technologies, the joint ventures with competitors—was hindered by transnational regulatory bodies (the EC) and transnational corporate entities (News Corp. and Universal). Instead of acting as a bulwark against the intrusion and dominance of Hollywood, the EC effectively weakened European alliances that might have defended the Continent against further encroachment. Contrary to Kirch's brash claim that he had "a lock on German TV," and that his launching of Premiere World would prevent cross-border invasions, Rupert Murdoch recently doubled his investment in satellite television in Germany, and on July 9, 2009, Premiere World was formally rechristened "Sky Deutschland."

NOTES

1 Michael Curtin, *Playing to the World's Biggest Audience: The Globalization of Chinese Film and TV* (Berkeley: University of California Berkeley Press, 2007).

2 Jeffrey A. Hart, *Technology, Television, and Competition: The Politics of Digital TV* (Cambridge: Cambridge University Press, 2004).

3 European Commission, "Green Paper on the Convergence of the Telecommunications, Media and Information Technology, and the Implications for Regulations," COM(97)623, December 3, 1997, accessed March 22, 2011, http://ec.europa.eu/avpolicy/docs/library/legal/com/greenp_97_623_en.pdf

4 Jean K. Chalaby, *Transnational Television Worldwide: Towards a New Media Order* (London: I. B. Taurus, 2005).

5 Paul J. Torre, "Block Booking Migrates to Television: The Rise and Fall of the International Output Deal," *Television and New Media* 12(2) (March 2011): 501–520; Debra Johnson, "Race for Digital Fuels Sales at MIP," *Broadcasting and Cable*, April 29, 1996, 19.

6 David Page and William Crawley, *Satellites Over South Asia: Broadcasting, Culture, and the Public Interest* (Thousand Oaks, Calif.: Sage, 2001).

7 Shaun Moores, *Media and Everyday Life in Modern Society* (Edinburgh: Edinburgh University Press, 2000).

8 Charlotte Brunsdon, *Screen Tastes: Soap Opera to Satellite Dishes* (London: Routledge Press, 1997).

9 Moores, *Media and Everyday Life*.

10 Curtin, *Playing*; Page and Crawley, *Satellites Over South Asia*.

11 Curtin, *Playing*.

12 David A. L. Levy, *Europe's Digital Revolution: Broadcasting Regulation, the EU, and the Nation State* (London: Routledge Press, 1999).

13 Jonathan Annells, "EC Slaps Down Digital Deal," *Hollywood Reporter*, May 28, 1998, 1.

14 Serge Robillard, *Television in Europe: Regulatory Bodies* (London: John Libbey, 1995).

15 European Commission, "Green Paper/COM(97)623."

16 Mark Wheeler, "Supranatural Regulation: Television and the European Union," *European Journal of Communication* 19(3) (2004): 349–369.

17 Miriam Mils, "Digital War Heating Up: EC Says 'No' to Megamerger, Ices Plans," *Variety*, June 8–14, 1998, 15.

18 Miriam Mils, "CLT-UFA Draws Digital Line," *Variety*, May 28, 1998, 16.

19 KirchGruppe Press Release, "Kirch Group Takes EU Commission to the European Court of Justice: Premier Merger Project Wrongfully Prohibited," August 3, 1998.

20 R. Davies, "Sleeping Giant: The German Market Offers Huge Potential for Pay-TV," *Television Business International*, December 20, 2000.

21 A. Von Gamm, "Quo Vadis Kirch? After Being Black-balled from the German Market Two Years Ago, Rupert Murdoch Once Again Is Casting His Eyes to Deutschland," *European Media Business and Finance*, August 10, 1998, 1–2.

22 Michael Williams, "Pay TV Tug of War: Bertelsmann Selling Premier Stake to Kirch," *Variety*, February 24, 1999, 11.

23 Ed Meza, "Pay TV Cash Flash," *Variety*, November 30, 1999, 12.

24 Stylianos Papathanassopoulos, *European Television in the Digital Age: Issues, Dynamics and Realities* (Cambridge: Polity Press, 2002).

25 John Hopewell and Emiliano de Pablos, "Euro Pay TV Doesn't Pay," *Variety*, April 23–29, 2001, 7.

26 P. Davies and R. Major, "Murdoch Says Deal with Kirch Has Changed," *New Media Markets*, June 22, 2001, 14.

27 Zenith Media Report, "TV Trends in Europe," August 1999.

28 Petros Iosofidis, Jeanette Steemers, and Mark Wheeler, *European Television Industries* (London: BFI, 2005).

29 Jennie James, "Has the Sports Bubble Burst?" *Time Europe*, March 11, 2002, 59.

30 Erik Kirschbaum, "ARD Attacks Pay TV: German Pubcaster Calls FeeVee Unfair," *Variety*, May 21, 1996, 10; Dan Knutson, "The Engine of Change: European Automobile Association May Start New Automobile Racing Events," *Auto Racing Digest*, September 2010, 14.

31 W. Hahn, memorandum to Hans Seger, *Premier World*, February 5, 2001.

32 James, "Sports Bubble."

33 Elizabeth Guider and Liza Foreman, "Put Out Over Output: Germany's Kirch Takes U to Court Over TV Deal," *Variety*, January 3–9, 2000, 10.

34 Ibid.

35 Janet Shprintz, "Kirch Sues U Over Pay TV," *Variety*, December 16, 1999, 34.

36 D. Harris and Eric Hansen, "$2 Bil. Kirch Suit Says Uni Didn't Deliver Pay TV Fare," *Hollywood Reporter*, December 16, 1999, 25.

37 First Amended Complaint, KirchMedia v. Universal Studios, Los Angeles Superior Court, January 25, 2000.

38 Elizabeth Guider, "Judge Rejects Kirch Suit," *Variety*, September 8, 2000, 11.

39 Elizabeth Guider, "U Expands Kirch Suit," *Variety*, August 28, 2002, 14; Ed Meza, "Pay TV Phoenix Premiere Rises from Kirch's Ashes," *Variety*, September 1, 2002, 12.

40 Scott Roxborough, "Media Event: Kirch Watch '02," *Hollywood Reporter*, December 28, 2001, 3.

41 Meza, "Pay TV Phoenix."

42 Hernan Galperin, *New Television, Old Politics: The Transition to Digital TV in the United States and Britain* (Cambridge: Cambridge University Press, 2004).

43 Curtin, *Playing*.

44 Ibid.

ORBITAL MATTERS

12

WHEN SATELLITES FALL

ON THE TRAILS OF COSMOS 954 AND USA 193

LISA PARKS

Thousands of satellites and space objects have fallen back to Earth since the space age began. More than forty years ago, in May 1968, the Nimbus B-1 weather satellite plummeted into the Pacific Ocean just off the coast of Santa Barbara, California, where I live. The fallen satellite was recovered, intact, from the bottom of the Santa Barbara channel, but its failure caused an enormous stir because of the four pounds of plutonium it had on board. The event prompted an investigation that moved from the depths of the ocean to the launchpad at Vandenberg Air Force Base out to orbit and back down to Earth. The fall of Nimbus B-1 in California's coastal backyard serves as a provocative reminder of the environmental risks, high costs, and unique materialities and spatialities of satellite technologies.

When satellites fall, they draw attention to the vertical stretch of world history and events—a space that extends up from Earth's surface through the atmosphere, stratosphere, and ionosphere; into the multiple orbital paths; and out to the edges of the supersynchronous or "parking" orbit, where satellites go to die. As much as this stretch may be conceived as a space of flows, waves, and flight-paths, it is also a space of capital accumulation filled with metallic hardware, synthetic materials, and toxic waste. This massive aeroorbital space, trafficked by aircraft, rockets, satellites, and signals, is a dynamic field of technologized movement and electromagnetic activity as well as a giant graveyard where dead satellites float. It is a space most of us have never visited—a space that only astronauts have seen—and yet it is a space that we cannot afford to overlook.

Both James Hay and Christy Collis examine dimensions of this aeroorbital domain in their respective chapters of *Down to Earth*. Hay explores how air space, outer space, and cyberspace were invented within the discourses of liberal governance as "free" and "open" spaces for exploration and enterprise, on the one hand, and as extensions of sovereign territories subject to regulation and securitization, on the other. Collis hones in on the plethora of legal discourses that historically have been mobilized to define, order, and regulate the space of the geostationary orbit. This chapter interlinks with these critical geographies of aero-orbital space in the sense that it considers historical events that have transpired within this domain, technical objects that occupy it, and human attempts to visualize it.

After five decades of satellite launches and space probes, there are now thousands of objects in orbit. According to a 2010 report from the U.S. Space Surveillance Network and the National Aeronautic Space Administration's (NASA's) Orbital Debris Program Office, there are 15,550 satellite-related objects currently orbiting Earth.[1] All of these objects are 10 cm in diameter or larger, and they range in size from a tiny paint chip to a massive rocket fuel tank. Only 3,333 of them are functioning satellites.[2] The rest are considered orbital debris. In addition to being filled up with debris, orbital space is trafficked primarily by objects deployed by the former Soviet Union, the United States, and China. Of the 15,550 satellite-related objects in orbit, 14,045 are from these three countries, and only 2,615 are satellites.[3] To draw public attention to the problem of orbital debris, NASA has released a series of visualizations over the years.[4] The 2009 graphic shown in Figure 12.1 spotlights the heavy concentrations of debris in low Earth orbit, posing risks to satellites in the vicinity as well as to the planet's atmosphere and surface.[5] In 2010, the Obama administration announced a new space policy that identified curbing the growth of orbital debris and preserving the space environment as top national priorities.[6]

With the increasing number of objects in orbit comes a variety of financial, political, environmental, and security concerns. Satellite owners are concerned about whether such debris will damage or interfere with their functioning satellites and compromise their investments. Environmentalists are concerned about the fact that orbit is polluted with floating space junk. And state officials are concerned about whether objects in orbit will plummet to the planet and threaten populations or valuable resources. When satellites fall back to Earth, typically they incinerate as they reenter the atmosphere, but sometimes fragments survive the extreme heat and fall to the planet's surface. In this chapter I focus on two incidents when malfunctioning satellites fell back to Earth and became high-profile media events. The first involved a Soviet radar satellite, Cosmos 954, in 1978, and the second involved a U.S. spy satellite, USA 193, in 2008.

Although satellite failures can be catastrophic in terms of their environmental impacts, they also can arouse new critical curiosities and ways of

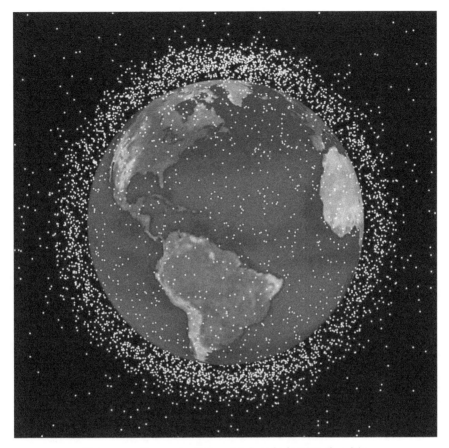

Figure 12.1 Concentrations of debris in low Earth orbit.
Source: NASA's Orbital Debris Program Office, 2009.

thinking about orbital matters. This chapter begins with a discussion of the failures of Cosmos 954 and USA 193 that emphasizes the importance of using mapping and visualization in efforts to comprehend the materiality and positioning of satellites, draws attention to the moment of technological failure as a site for increased knowledge production and public awareness of satellites, and foregrounds the astronomical cost and precariousness of satellite technology. The chapter culminates in a more analytical discussion that suggests satellite failures necessitate a rethinking of *capital* and *capitalism* in relation to the unique conditions of orbit and challenge us to consider the satellite as an object of power and agency, and as part of a satellitarian world order.

Trail One: The Collision of Cosmos 954

On January 24, 1978, a Soviet radar satellite known as Cosmos 954 plunged into the Great Slave Lake area of the Northern Territories in Canada. The satellite was launched from the Baikonur facility in Kazakhstan on September 18, 1977. It was one of sixteen RORSATs (ocean radar satellites) launched into low Earth orbit by the Soviet Union, and was designed to track deep-running American nuclear submarines.[7] By October 29, 1977, monitors at the North American Aerospace Defense Command (NORAD) revealed that Cosmos 954 had drifted out of orbit and predicted it would reenter Earth's atmosphere sometime in April 1978. The primary concern about Cosmos 954's tumble back to Earth was the nuclear reactor it had on board. Because the satellite was carrying 110 pounds of enriched uranium 235, some officials predicted Cosmos 954's crash could result in the "worst nuclear contamination since Hiroshima and Nagasaki."[8] Since U.S. and Canadian officials were uncertain where Cosmos 954 would land, they decided not to issue a public announcement detailing the nuclear concern. When it became clear, however, that Cosmos 954 would fall months earlier than anticipated, on January 18, 1978, the U.S. State Department relayed a secret message to its North Atlantic Treaty Organization (NATO) allies and to Australia, Japan, and New Zealand informing state diplomats about the matter.

By January 24, 1978, the satellite's fall back to Earth had become public knowledge, and the world's news agencies deployed reporters to the icy tundra near Yellowknife to investigate the "killer satellite," a forty-six-foot-long vehicle weighing more than five tons. Seemingly overnight, a satellite the public had never known to exist became the object of urgent searching, scrutiny, and media spectacle. Journalists reported on the massive size of the debris field, the fragments that had been recovered, and the complex nature of the retrieval mission. Since the satellite fell in remote lands, the press also highlighted the Canadian government's attempts to communicate with Inuit and Chippewa communities living in the vicinity of the crash, whose water and food supplies were in danger of exposure to radiation.

Covering the massive (30,000 square miles) debris field was challenging for journalists and recovery workers alike. Maps showed an enormous swath across the Northwest Territories, stretching from Great Slave Lake up to Baker Lake. CBC camera crews were invited aboard Hercules aircraft outfitted with radiation sensors as they flew over the frozen tundra looking for "hot spots."[9] From the air, recovery personnel found ten to twelve larger fragments and more than 3,000 smaller, windblown particles. Tens of millions of pepper-flake-size radioactive particles were also scattered over a 124,000 square km crash footprint.[10]

While cartography was used to represent the huge scale of the debris field, there was also a great deal of visual attention devoted to the fragments themselves. Photos of the larger fragments of Cosmos 954 appeared in press coverage and reports on the recovery effort. These images emphasized the satellite's materiality by drawing attention to the texture, shape, color, and size of the ruins. One of the biggest fragments was discovered by two hikers on an expedition through the icy tundra. Mike Mobley and John Mordhorst were tracing the footsteps of historic explorers in the Wardens Grove area when they sighted what they thought was a rack of caribou antlers in the distance on the ice of the Thelon River. They approached, only to discover a piece of Cosmos 954 melting the surrounding snow. Another photograph shows fragments described as an office "wastepaper basket" that had been pieced together. The caption explains that most of the parts were metallic, but some were composed of fiber-filled plastic.[11] Yet another photograph showed a Canadian penny in the snow next to a tiny piece of particulate from the satellite; the caption indicates that some such particles "were highly radioactive and proved to contain both uranium 235 and fission products resulting from the operation of the nuclear reactor."[12]

While photos drew attention to the smoldering remains of an unknown satellite and sparked questions about its history, cost, uses, and environmental impacts, maps visualized the satellite's trajectory through time and space as it fell from orbit to Earth. These visualizations helped both to materialize the satellite and make it matter. In doing so they demarcated the increasingly important domain between Earth and outer space, a domain where geopolitics have been extended to become "astropolitics." Building on the work of Gerard O'Tuathail and others, Fraser MacDonald uses the term "astropolitics" to "develop an agenda for a critical geography of outer space," suggesting that critical geopolitics "must lift its gaze to the politics of the overhead."[13] Images of satellite ruins and maps of the recovery field draw attention to the politics of the overhead by intimating the politically charged space between orbit and Earth—a space where secret satellites can become public spectacles, technological failures can become incidental weapons, and state agencies can become planetary monitors.

Canada spent C$14 million on the recovery and debris clean-up effort.[14] In the months following the crash, the Canadian government sought compensation from the Soviet Union in the amount of more than C$6 million under the 1972 Convention on International Liability for Damage Caused by Space Objects. This international law holds satellite owners liable for the damages caused when their satellites or spacecraft fall back to Earth. The Soviets fought this case, claiming that Cosmos 954 had broken up by the time it fell to Earth and thus could no longer be recognized as a "satellite" when it landed in the Northwest Territories. After three years of negotiations, however, the Soviets paid Canada C$3 million in damages.[15]

The fall of Cosmos 954 not only established an occasion to test satellite liability law, but Operation Morning Light—the name given to the first phase of the search for Cosmos 954 debris—became a prototype for future nuclear recovery missions and satellite reentries. Cosmos 954 is one of a small handful of recovered satellites that have been used to study and develop models to prepare for other satellite reentries and mitigate future damages. Later generations of satellites, in fact, have been carefully designed with their termination in mind, and are now programmed to ensure a particular breakup scenario, whether to thrust into a higher (super-synchronous) parking orbit, burn up more quickly upon reentry, or fall into the ocean. Whatever the case, satellites, like other technologies, have specific life cycles, and their material composition and orbital position cause unique challenges for their disposal.

Trail Two: The Interception of USA 193

Almost exactly thirty years after Cosmos 954 fell, another satellite drifted out of orbit and began to move toward Earth. This time it was a secret U.S. satellite rather than a Soviet one. USA 193 was a classified spy satellite (also known as NROL-21) that had been launched on December 14, 2006, from Vandenberg Air Force Base in California. Communication with the satellite failed shortly after its launch. Rather than allow USA 193 to fall to Earth's surface as Cosmos 954 had in 1978, the U.S. government devised an elaborate scheme to intercept and destroy the satellite with an SM-3 missile. U.S. officials expressed concerns about the 1,000-pound tank of hydrazine fuel onboard USA 193 and claimed it could form into a toxic cloud the size of two football fields if the satellite were to crash on Earth, thereby posing a serious public health risk. Many were skeptical of this claim and speculated instead that the United States did not want this classified satellite to fall into foreign hands—especially since future fleets of U.S. spysats were slated to use similar technologies. Still others interpreted the U.S. satellite shoot-down as a geopolitical showdown in which the United States set out to demonstrate its anti-satellite (ASAT) missile capabilities forcefully, following a controversial and high-profile test the Chinese had conducted in 2007. Whatever the explanation, USA 193's interception became a media event that revealed satellite termination could have an almost surgical precision. It also exposed the systems of risk management that had solidified in the thirty years since Cosmos 954 crashed.[16]

Like Cosmos 954's plunge into the Northwest Territories, the interception of USA 193 became a high-profile media event. Reporters who covered USA 193's failure emphasized the risky nature of the satellite shoot-down, used Google Earth to predict and map where the fragments would land, and offered speculations about the public health risks.[17] The event drew attention to a cohort of amateur satellite trackers, including Marko Langbroek and Trevor Paglen, who had already been tracking and photographing the secret satellite (along with many others) since its 2006 deployment.[18] Their distant and abstract views of the secret satellite work in subtle ways to give these unknown and classified objects form and intelligibility. While these shots differ quite dramatically from the close-up images of the Cosmos 954 ruins, they serve as significant traces or intimations of technologies that citizens pay for but are not supposed to know about.

On February 21, 2008, an SM-3 missile launched from the U.S. Navy vessel USS *Lake Erie* blasted into USA 193 as it passed over an area west of the Hawaiian Islands in the Pacific Ocean. After the strike, the U.S. Department of Defense held a press conference and released a video showing the missile as it struck USA 193, shot from the perspective of an aircraft hovering nearby, as the satellite turned into an incandescent gaseous blob.[19] According to the Department of Defense, "no parts bigger than a football survived the strike," but others speculated that there may have been more fragments and greater environmental impact than the U.S. military reported.[20] T. S. Kelso of the Center for Space Standards and Innovation plotted the debris left over from USA 193's destruction and by February 2008 had cataloged forty-five pieces.[21] More than a year later, in May 2009, Kelso reported that his CelesTrak system, which monitors orbital debris, was tracking 174 pieces of debris from USA 193.[22] That the satellite was carefully intercepted rather than allowed to plummet to Earth did not mean it had no environmental impact: USA 193 remains as a scattering of parts and residues floating in orbit. In this sense, the satellite's interception can be understood as part of a longer history of the U.S. military's use of extraterritorial domains, whether the air, oceans, deserts, or islands, for weapons testing and toxic waste dumping.[23]

In addition to linking the interception of USA 193 to a history of pollution, it is worth pointing out the astronomical cost of the satellite. Much information about USA 193 remains classified, but it is known that the satellite was part of a design scheme called Future Imagery Architecture, involving Boeing and Lockheed Martin, for which the U.S. government paid more than US$10 billion.[24] The shoot-down operation alone cost American taxpayers US$40 to US$60 million.[25] Since USA 193 was a secret satellite, it is unclear how and whether it was insured. Nevertheless, the knowledge of its costly production and destruction compels us to begin recognizing and investigating the elaborate systems of risk management that have been put into place to guard such costly and strategic investments.

Satellites and Dandelions

We live in an age in which extremely expensive machines are made and installed in orbit without public knowledge, only to be spectacularly blown away and become total losses before our eyes. Given such scenarios, the study of satellite failures, finances, and futures remains a vital path for further investigation. Cosmos 954 and USA 193 are just two of hundreds of satellites that have failed since the late 1950s. Although these were both secret projects, a number of public and commercial satellites have also been destroyed in recent years. In January 2007, the NSS-8 satellite owned by Netherlands-based SES New Skies was destroyed during its liftoff from the ocean-based launchpad known as Sea Launch.[26] Later that year the JCSAT-11 satellite fell back to Earth when a Proton rocket failed to enter its second stage after launching from the Baikonur facility in Kazakhstan. In February 2009, Russia's Cosmos 2251 collided with a U.S. Iridium 33 satellite over northern Siberia traveling at more than 15,000 miles per hour. This was the first time that two satellites collided in orbit. The collision generated enormous debris clouds and has posed serious risk-management challenges for satellite operators.[27] Later that month NASA's US$280 million Orbiting Carbon Observatory satellite, designed to conduct global-warming tests, tumbled fatefully into the ocean near Antarctica.

The satellite industry is interwoven with one of the most complex and expensive insurance industries on the planet. For private satellite owners, insurance premiums are typically an operator's second-largest cost. Any given satellite can have ten to fifteen large insurers and twenty to thirty smaller companies issuing policies for different phases of the satellite's development, transport, launch, in-orbit operation, and termination. In 2003, a basic premium for a satellite worth US$250 million cost between US$40 million and US$55 million.[28] Intelsat paid nineteen companies to insure eight of its satellites for US$1.5 billion in 2007. With a fleet of fifty-two satellites, Intelsat also "self-insures" by manufacturing back-up satellites rather than purchasing costly policies.[29] Satellite operators also regularly maneuver or adjust the positions of orbiting satellites in order to avoid debris and mitigate damages.[30]

Satellite failures and collisions not only represent hundreds of millions of dollars in losses; they are symptomatic of the "dandelion economics" that underpin the satellite industry. Just as the capital to manufacture, launch, and operate a satellite accumulates and the technology takes shape, the object can be blown away in an instant, its fragments either darting cataclysmically toward Earth's surface or floating into the oblivion of space. In this sense, a satellite is as precarious as a dandelion that vanishes with a single blow. The concept of dandelion economics accounts for the instant annihilation of meticulously designed and extremely costly technologies, technologies that took years of state and commercial funding, labor, materials, and knowledge to produce. When Cosmos 2251 and Iridium 33 collided in February 2009, each of the satellites became a total loss in a flash, and their collision created enormous debris fields containing thousands of particles. The U.S. Space Surveillance Network cataloged 521 pieces of debris (23 of which have already decayed from orbit) associated with Iridium 33 and 1,267 pieces of debris (50 of which have decayed) associated with Cosmos 2251.[31] Both of these satellites remain in orbit, but now exist as scatterings of debris rather than as functioning satellites. Hence their value instantly shifted from that of multimillion-dollar satellites to that of multimillion-dollar liabilities. Blowing on a dandelion may lead to pollination or wish fulfillment; blowing up a satellite in orbit only proliferates risk and negative value.

Just as scientists used photographs to spotlight pieces of Cosmos 954 after it fell to Earth and video to document USA 193's interception, they have developed visualizations of the Cosmos 2251–Iridium 33 collision. An animation created by Analytic Graphics, Inc. (AGI) reveals the two satellites crashing and creating immense debris fields, represented in red. The red debris clouds move in opposite directions and become scatterings in low Earth orbit where hundreds of other functioning satellites are located.[32] AGI also released a series of static visualizations. One illustrates the relative positions of the satellites prior to their collision as city lights on Earth's surface appear below, implying the proximity of the collision to populated areas on Earth. Another tries to predict debris trajectories and shows the "new debris" as scatterings of red dots intermixed with all other "existing space objects" shown in green, implying the high level of risk to functioning satellites.[33] T. S. Keslo, who tracked the debris of USA 193, also monitored the ruins of this collision and presented findings on his CelesTrak website.[34] Finally, a layer modeling the Cosmos 2251–Iridium 33 collision appeared in Google Earth so that users could "fly" through the debris fields, check the altitude of each fragment, and grasp the proximity of these satellite ruins to Earth's surface.[35]

Such visualizations are used not only to monitor debris trajectories and prevent collisions but also to mitigate damages and manage risks.[36] They have become the primary means by which satellite operators and insurers identify liabilities and prove damages. One satellite operator describes these "risk-visualization tools" as being just as important as insurance.[37] In this context, these images can be understood as instrumentalizing the visualization of orbit—their production is motivated by the need to identify, count, track, and evaluate thousands of objects of value, whether of use or risk value, so that current and future capital investments on Earth and in orbit can be protected. And yet these visualizations need not be tethered so tightly to the agencies or agendas of those who generate them.

Such images can be repurposed and used to articulate struggles and contestations over the meanings and uses of satellites and orbital space. I interpret these visualizations of satellite ruins as powerful symbols of the precariousness of capital and capitalism in orbit. They not only remind us that enormous accumulations of capital can turn into negative value in an instant; they reveal imploded technology and investments where no one is there to witness them. In this sense, they reverberate symbolically with the desert explosions in the finale of Antonioni's 1970 film *Zabriskie Point*, during which capital accumulations (refrigerators, electronics) implode against a smoky blue sky of unforgettable pyrotechnic display. One of the most intriguing representations of the Cosmos 2251–Iridium 33 collision appeared in *National Geographic*: artist Stefan Morrell tried to imagine and represent what the crash may have looked like up close. Morrell's rendition shows the two satellites thrashing into one another at great speed, signaled by a blurring effect, as a dispersion of parts sprays out into orbit.[38] The visualization is useful because it models satellite capital in ruins while simultaneously creating a position for viewing satellites as objects in close-up and for thinking through the unique conditions of orbital space and matter, including congestion, speed, heavy metals, pollution, and risk.

Some might understand satellite failures in relation to the phenomenon of "creative destruction,"[39] especially since companies have already begun to capitalize upon the problem of orbital debris by developing special satellites designed to capture and slow the velocity of debris particles so that they will not cause damage to satellites.[40] However, I want to link these moments of satellite failure and destruction to what J. K. Gibson-Graham refers to as "the end of capitalism (as we knew it)." Gibson-Graham's book of the same title complicates the totalizing and unified ways in which "capitalism" has been invoked in critical discourse, and sets out to inscribe *difference* within the study of capitalism and catalyze the study of noncapitalist projects.[41] Such an intervention, Gibson-Graham suggests, is vital to creating a world beyond capitalism. As she writes, "[H]ow do we begin to see this monolithic and homogeneous Capitalism not as our 'reality' but as a fantasy of wholeness, one that operates to obscure diversity and disunity in the economy and society alike?"[42] She continues, "My intent is to help create the discursive conditions under which socialist and other noncapitalist construction becomes a 'realistic' present activity rather than a ludicrous or utopian future goal. To achieve this I must smash Capitalism and see it in a thousand pieces. I must make its unity a fantasy, visible as a denial of diversity and change."[43]

Spinning this idea in a slightly different direction, what happens to capitalism when a satellite breaks into a thousand pieces? Can this moment be used to fracture and diffract the meanings of capital and capitalism so that orbital matters are discursively broken down, differentiated, and better understood? Mediated moments of satellite failure serve as important discursive sites because they trigger questions about key terms such as "value," "loss," "risk," and "responsibility" and, in so doing, can provoke a reassessment of the meanings of "capital" and "capitalism" as well. Instances of satellite failure challenge us to imagine and recognize the unique materials, operations, costs, locations, scales, distances, speeds, and durations of satellites, and to create new political, economic, and cultural concepts and theories tailored to orbital conditions. The concept of dandelion economics is a small step in this direction. It questions what it means to have a satellite (representing immense capital accumulation, as a satellite is worth more than the gross national product [GNP] of many nation-states) instantly lose all value, whether by chance or by force. It recognizes the lingering, dispersed, and unpredictable material effects of satellite failure or destruction. And it challenges us to think about the liabilities and negative value of satellite ruins and orbital debris. In so doing, it begins to inscribe an orbital layer within the discursive fields of "capital" and "capitalism," and pushes us to consider how the matters of orbit might alter critical thought.

While critical theorists regularly engage with questions of property, ownership, and value in relation to parcels on Earth, fewer have considered how the value of a satellite is calculated, how it is insured and by whom, and which orbital slots are most valuable and why.[44] Information about real estate and property values, insurance, and traffic patterns on Earth abounds, but what about the equivalents in orbit? How much does it cost to use a transponder on a satellite, and who uses them the most? After fifty years of satellite use, such information should be more widely available, yet information about the business of satellites is often proprietary and costly, and is sometimes classified. Rethinking capital and capitalism in orbit involves expanding the kinds of knowledges about satellites that circulate in public, and considering the agency and power of satellites as well.

Satellites as Actors

If satellite failures can lead to more nuanced understandings of capital and capitalism, they also can be used to foreground the agency of objects. In an effort to recognize the significant roles of objects in the modern world, Bruno Latour and others set out to develop a social theory that would account for the relations between human and nonhumans, the animate and the inanimate.[45] Actor-network theory can be used to explore and articulate the material-semiotic associations and relations of human and nonhuman actors that make up a social network. Of particular relevance is Latour's suggestion that humans "delegate" force, values, duties, and ethics to nonhuman actors.[46] "Parts of a program of action," he explains, "may be delegated to a human or to a nonhuman."[47] Thus as much as a satellite can be understood as a capital investment, it can be thought of as an actor in a social network. While the satellite is designed, built, and remotely controlled by humans on Earth, it is imbued with particular "programs of action" or "delegated intentionalities." Whether the relay of signals, the imaging of Earth, or the maintenance of a geopolitical order, the satellite carries out particular "programs of action" and as such plays an active role in the formulation, mediation, and (de)stabilization of social, cultural, and political relations.

Cosmos 954 and USA 193 were deployed to monitor parts of Earth during and after the Cold War. These satellites carried out programs of action in relation to different national security agendas. Cosmos 954 and USA 193 could "see" Earth from the perspective of a physical position that no human can. They were part of social networks that included nonhuman Earth stations and human intelligence analysts assigned to watch for particular activities or signs of particular activities. The "delegated intentionalities" of these satellites came into relief especially during the moments of their failure and destruction. These moments led to intensive investigation, reconstruction, and publicity of these satellites' programs of action as forensic experts set out to understand what these objects were intended to do and what went wrong. This process led to a kind of retrospective or posthumous blueprinting in which the satellites' programs of action were sketched out and made intelligible in public only after they were destroyed. The crash site visits, the piecing together of remains, and the mapping and simulation of debris all worked to convey the actions of these objects not only by reassembling them in public but also by drawing immense attention to their loss.

According to Latour, one of the best ways to determine an object's actions is to contemplate what the situation would be if the object were not there.[48] He writes, "every time you want to know what a nonhuman does, simply imagine what other humans or other nonhumans would have to do were this character not present. This imaginary substitution exactly sizes up the role, or function, of this little character."[49] Conducting this imaginary substitution and thinking of the multitude of actions that satellites perform is a monumental task. Satellites have been delegated so many functions and have become so fundamental to social networks that the more interesting question may be how they came to take on so many functions and have so much force. To name just a few of these, satellites acquire meteorological data, circulate financial flows, uplink and downlink television signals, deliver navigational data, monitor natural resources, and target sites for bombing. They are actors in networks of weather forecasting, international trade, broadcasting, mapping, Earth sciences, and warfare.

By delegating so many duties to the satellite, humans have arguably created a *satellitarian* order in which a large share of daily activities and transactions—whether commerce, automobile navigation, television viewing, or policing—occur via satellite.[50] This techno-social (satellitarian) order is manifest not only in a host of regularized satellite-based transactions and activities, or *satellite actions*; it can also be detected in the (re)territorialization or satellitization of space on Earth and in orbit. Where satellite installations have led to the production of orbital paths and positions or slots, the satellite transmission of audiovisual signals has transformed Earth's surface into a crisscrossing of mediated fields or "footprints." The idea of a satellitarian order is meant to recognize the power-bearing actions and structural effects of satellite technologies without casting them as autonomous, omnipotent, or unchangeable. Satellites are, after all, one among many actors in a multitude of networks.

The power of the satellite as object is grounded also in its potential for failure. Since satellites are nonhuman actors in so many social networks, this potential is significant. Satellite failure can disrupt one network or shut down an entire techno-social order. In an effort to avoid such scenarios, satellite inventories are conceptualized as "fleets" or "constellations" so that if one satellite fails, there is another to provide emergency coverage. The failures of Cosmos 954 and USA 193, as well as those of other satellites mentioned, suggest that even the actions of dead satellites matter. When satellites malfunction, move adrift, or fall back to Earth, they can have powerful effects. Satellite failures have enabled us to think about relations and actions between Earth and orbit, from micro to macro scales, and between humans and nonhumans. Latour's insistence that nonhumans are actors is crucial because it challenges humans to find a new language, system, or method for understanding and communicating with satellites, one that will not continue to put Earth and its orbital environs in jeopardy.

Conclusion

It is ironic that humans have delegated so many actions to satellites in the area of security and yet, in the process, have managed to put the planet at great risk. Both Cosmos 954 and USA 193 were installed to advance global security in the midst of and after the Cold War, but both ended up failing, triggering geopolitical crises and harming the global environment. Fallen satellites expose the thresholds of human control over technology and call the dream of planetary management into question. They also point to three directions for further research.

First, fallen satellites suggest the need for further research on the intersections of satellite technologies and environmental studies. Satellites require expanded thinking about what the "environment" is and a recognition that this domain extends contiguously from below Earth's surface, up through the atmosphere, and out through orbital space. We must continue to investigate, document, theorize, and historicize through this vertical stretch (as Hay, Collis, MacDonald, and others have done), and begin to delineate the environmental impacts of satellite technologies. In the age of global warming, there has been intensified attention directed to the atmosphere and its fragility. We might direct further attention to the issue of orbital debris given the thick concentrations of heavy metals, toxic gases, and hazardous waste that now surround our planet. If nothing else, a fallen satellite compels us to ask, "What *else* is up there?!"

Second, there has been an important shift in the visual representation of fallen satellites during the past thirty years. While changes in imaging technologies have, of course, played a role in setting the possibilities for the representation of these satellites, it is also evident that there has been a trend toward greater remoteness, abstraction, and simulation when satellites appear in media culture. In the case of Cosmos 954, citizens could examine and scrutinize the minute fragments of the fallen satellite; these images were made available in press photos. In contrast, USA 193, in its brief twilight of publicity, was relatively invisible, seen primarily in the moments of its launch and annihilation, or in the form of visual sketches designed for the news media. Satellites seem to be able to "see" everything (think about Google Earth), but we cannot seem to see *them*. What this reveals is that technical knowledge about satellites is as highly guarded a resource as the thing itself. Information about USA 193 was deemed to be so valuable that the U.S. government decided to destroy it—so that if it fell back to Earth, information about it would not become "known." The visibility of satellites, whether seen close-up or from afar, is important, then, because it points to an entire orbital assemblage that is shrouded in secrecy yet paid for by publics, and that is culturally pervasive and yet heavily laden with risks.

Finally, then, we need further research on what I have called "dandelion economics" and on the elaborate systems of risk management that have emerged to protect enormous investments. In other words, by studying satellite failures we can learn much about the excesses and obscenities of capitalism and state power. The giant debris fields created by the collision of Cosmos 2251 and Iridium 33 mark the current stage of satellite history—a stage in which orbital congestion threatens to produce ever more losses and environmental threats. Rather than continue to keep satellites at arm's length, we need to bring them down to Earth and develop materialist analyses that approach these objects as having a life cycle and history that begins with the extraction of heavy metals (whether aluminum, beryllium, or titanium) from deep underground and may end with the satellite's deliberate destruction, slow decay in a parking orbit, or instantaneous combustion while falling back to Earth. When satellites fall, we can glimpse their altered states and inscribe them within a vertical field of world history, critical inquiry, and orbital matters.

NOTES

1 "Satellite Box Score," *Orbital Debris Quarterly News* (July 2010): 12.

2 Ibid.

3 Updates of this data are published regularly in *Orbital Debris Quarterly News*, accessed March 23, 2011, http://orbitaldebris.jsc.nasa.gov/newsletter/newsletter .html.

4 See the Orbital Debris Graphics section of the NASA Orbital Debris Program website, accessed March 23, 2011, http://orbitaldebris.jsc.nasa.gov/photogallery/beehives.html.

5 Space.com Staff, "Space Junk Problem Detailed," Space.com, September 12, 2009, accessed March 24, 2011, http://www.space.com/news/090912-space-junk-images.html.

6 "New U.S. National Space Policy Cites Orbital Debris," *Orbital Debris Quarterly News* (July 2010): 1.

7 "Cosmos 954: An Ugly Death," *Time*, February 6, 1978, accessed March 24, 2011, http://www.time.com/time/magazine/article/0,9171,945940,00.html.

8 Colin A. Morrison, *Voyage into the Unknown: The Search and Recovery of Cosmos 954* (Stittsville, Ont.: Canada's Wings, 1982), 7.

9 Michael Bein, "Star Wars & Reactors in Space: A Canadian View," 1986, accessed March 24, 2011, http://www.animatedsoftware.com/spacedeb/canadapl.htm.

10 Atomic Energy Control Board (AECB), "AECB Publishes Summary Report on Cosmos 954 Satellite Crash," News Release 80-21, October 22, 1980, cited in Bein, "Star Wars," accessed March 30, 2011, http://www.animatedsoftware.com/spacedeb/canadapl.htm.

11 *Operation Morning Light: Northwest Territories, Canada, 1978—A Non-Technical Summary of U.S. Participation* (Las Vegas: U.S. Department of Energy, 1978), 17, 49.

12 Ibid., 56

13 Fraser MacDonald, "Anti-*Astropolitik*: Outer Space and the Orbit of Geography," *Progress in Human Geography* 31(5) (2007): 592, 610. Gerard O'Tuathail is also known as Gearóid Ó Tuathail or Gerard Toal.

14 Bein, "Star Wars."

15 Anatoly Zak, "Dangerous Space Reentries of Spacecraft," *Space.com*, June 20, 2000, accessed July 6, 2009, http://www.space.com/news/spacehistory/dangerous_reentries_000602.html.

16 A post on *Physics Today* asked whether the satellite may have been powered by a plutonium 238 power source or solar power. The National Reconnaissance Office (NRO) denied to *Physics Today* that there was a plutonium power source on board; see "More Doubts Surface over Pentagon's Explanation for Shooting Down Spy Satellite," *Physics Today*, February 18, 2008, accessed November 4, 2009, http://blogs.physicstoday.org/newspicks/2008/02/experts_query_pentagons_explan.html.

17 "US 193 Satellite Shootdown KMZ File," posted by Knobee, Google Earth Community Forum, February 19, 2008, accessed March 24, 2011, http://bbs.keyhole.com/ubb/ubbthreads.php?ubb=showflat&Number=1117483&site_id=1#import.

18 "USA 193," Sattrackcam Leiden Station (B)Log, December 26, 2007, accessed March 24, 2011, http://sattrackcam.blogspot.com/2007/12/usa-193.html; John Schwartz, "Satellite Spotters Glimpse Secrets and Tell Them," *New York Times*, February 5, 2008, accessed March 24, 2011, http://www.nytimes.com/2008/02/05/science/space/05spotters.html?_r=2&ref=science&oref=slogin; Trevor Paglen, *Blank Spots on the Map: The Dark Geography of the Pentagon's Secret World* (New York: Dutton, 2009), 97–125.

19 Armed Forces Press Service, "Navy Missile Hits Decaying Satellite over Pacific Ocean," *Defenselink*, accessed March 24, 2011, http://www.defenselink.mil/news/newsarticle.aspx?id=49024.

20 Gerry Gilmore, "Navy Missile Likely Hit Fuel Tank on Disabled Satellite," *Defenselink*, February 21, 2008, accessed March 24, 2011, http://www.defenselink.mil/news/newsarticle.aspx?id=49030.

21 Leonard David, "US Spysat Orbital Debris Tracked," *Live Science*, February 29, 2008, accessed July 8, 2009, http://www.livescience.com/blogs/2008/02/29/us-spysat-orbital-debris-tracked/.

22 T. S. Keslo, "USA 193 Post Shoot-down Analysis," CelesTrak, May 26, 2009, accessed March 24, 2011, http://celestrak.com/events/usa-193.asp.

23 For instance, for a listing of known satellites launched with nuclear materials on board, see Regina Haven, "Nuclear Powered Space Missions: Past and Future," November 8, 1998, accessed March 24, 2011, http://www.space4peace.org/ianus/npsm3.htm. Some of the satellites on this list are still in orbit; others exploded in the atmosphere; and others returned to Earth.

24 Noah Schactman, "Rogue Satellite's Rotten, $10 Billion Legacy," *Wired*, February 20, 2008, accessed March 24, 2011, http://www.wired.com/dangerroom/2008/02/that-satellite/.

25 Jamie McIntyre and Mike Mount, "Attempt to Shoot Down Spy Satellite to Cost Up to $60 million," CNN, February 15, 2008, accessed March 24, 2011, www.cnn.com/2008/TECH/02/15/spy.satellite/index.html.

26 Justin Ray, "Sea Launch Rocket Explodes on Pad," *Spaceflight Now*, January 30, 2007, accessed March 24, 2011, http://spaceflightnow.com/sealaunch/nss8/.

27 Marc Jones, "Satellite Collision May Set Coverage Precedent," *Allbusiness.com*, February 23, 2009, accessed March 24, 2011, http://www.allbusiness.com/insurance/aviation-insurance-spacecraft-satellite/11870631–1.html.

28 Andrea Maleter, "Strategies to Mitigate High Satellite Insurance Premiums," *Satellite Finance* 64 (December 10, 2003): 46.

29 Lisa Daniel, "Satellite Insurance: Operators Returning to Outside Providers," *Satellite Today*, November 1, 2007, accessed March 24, 2011, http://www.satellitetoday.com/via/features/Satellite-Insurance-Operators-Returning-To-Outside-Providers_19461.html.

30 Space.com Staff, "Space Junk."

31 T. S. Kelso, "Iridium 33/Cosmos 2251 Collision," Celestrak.com, March 5, 2009 (updated April 28, 2010), accessed March 24, 2011, http://celestrak.com/events/collision/.

32 "Iridium 33—Cosmos 2251 Collision," Analytic Graphics, Inc. website, February 12, 2009, accessed March 24, 2011, http://www.agi.com/media-center/multimedia/current-events/iridium-33-cosmos-2251-collision/.

33 Ibid.

34 Keslo, "Iridium 33/Cosmos 2251 Collision."

35 "When Two Satellites Collide, in Google Earth," *Barnabu*, February 12, 2009, accessed March 24, 2011, http://www.barnabu.co.uk/when-two-satellites-collide-in-google-earth/; a follow-up post, "Knee Deep in Satellite Debris," March 1, 2009, accessed March 24, 2011, showed an increase in the number of debris pieces being tracked, http://www.barnabu.co.uk/knee-deep-in-satellite-debris/.

36 G. B. Valsecchi, A. Rossi, and P. Farinella, "Visualizing Impact Probabilities of Space Debris," *Space Debris* 1(2) (2000): 143–158, accessed March 24, 2011, http://www.springerlink.com/content/m43u1o1p26655461/.

37 Daniel, "Satellite Insurance."

38 Stefan Morrell's illustration is included at the outset of Michael D. Lemonick's article, "Clearing Space," *National Geographic*, July 2010, 30–31, accessed March 24, 2011, http://ngm.nationalgeographic.com/big-idea/12/space-trash.

39 Joseph Schumpeter, *Capitalism, Socialism and Democracy* (1942; reprint, New York: Harper, 1975). I thank Sandra Braman for encouraging me to consider this connection.

40 Debra Werner, "ATK Proposes Satellite to Fight Space Debris," *Space News*, August 9, 2010, accessed March 24, 2011, http://spacenews.com/civil/100809-atk-satellite-fight-space-debris.html. Thank you to James Schwoch for sharing this article with me.

41 J. K. Gibson-Graham, *The End of Capitalism (As We Knew It): A Feminist Critique of Political Economy* (Minneapolis: University of Minnesota Press, 1996), 1–5.

42 Ibid., 260.

43 Ibid., 263–264.

44 For work that does take on these issues, see Christy Collis's chapter in this book.

45 Bruno Latour, *Reassembling the Social: An Introduction to Actor-Network Theory* (Oxford: Oxford University Press, 2005).

46 Bruno Latour, "Where Are the Missing Masses: The Sociology of a Few Mundane Artifacts," in *The Object Reader*, ed. Fiona Candlin and Raiford Guins (London: Routledge, 2009), 234.

47 Ibid., 235.

48 Latour looks at "accidents" as "a way of hearing what the machines silently did and said." As an example, he points to the meticulous detail generated about the Space Shuttle *Columbia* after it exploded. Ibid.

49 Ibid., 232.

50 I am grateful to the participants of the Satellite, Border, Footprint Workshop held at Hartware MedienKunstVerein in Dortmund, Germany, in August 2010, for offering their thoughts about these satellites in relation to their artwork. I am especially grateful to Anselm Bauer, David Halbrock, and Franziska Windisch, members of the Urban Research Performance Collective based in Cologne, Germany, and to Frances Hunger for organizing the workshop. For further information, see *Satellite, Border, Footprint* (Dortmund, Germany: Hartware MedienKunstVerein, 2010).

13

AFP-731 OR
THE OTHER NIGHT SKY

AN ALLEGORY

TREVOR PAGLEN

On February 28, 1990, the Berlin Wall was crumbling, and the Cold War was thawing. In the weeks before, Germany had agreed on a plan for reunification, the Central Committee of the Soviet Communist Party pledged to give up its monopoly on power, and the first McDonald's had just opened in Moscow. In Cape Canaveral, Florida, the Space Shuttle *Atlantis* sat on a movable launchpad with its bone-white delta wings lit with floodlights from below.[1]

STS-36 was going to be an odd mission. There were no high-minded press releases about the wonders of space travel or the scientific instruments aboard *Atlantis* that evening, no media kits for the journalists assigned to the launch, and no talk of the payload. One of the few reports about the upcoming launch came from the pages of *Aviation Week and Space Technology*: "A secret Pentagon shuttle mission set for Feb. 16 will carry a 37,300-pound advanced reconnaissance satellite to be used by the Central Intelligence Agency and the National Security Agency. Designated AFP-731, it is a 'combination' spacecraft carrying both digital imaging reconnaissance cameras and signal intelligence receivers."[2]

After the shuttle *Columbia*'s first few missions in the early 1980s, the spacecraft's activities had become rather ho-hum affairs, virtually unnoticed outside specialized aerospace journals such as *Aviation Week and Space Technology* and *Florida Today*, Cape Canaveral's hometown newspaper. The Space Shuttle program's low profile had, of course, been rocked by the January 28, 1986, *Challenger* explosion carrying teacher Christa McAuliffe aboard. It took the National Aeronautics Space Administration (NASA) almost three years to launch the *Discovery* on the first post-*Challenger* mission, STS-26, which launched on September 29, 1988.

By 1990, shuttle missions had once again receded into the dim corners of the public imagination. It was supposed to be that way. These missions were supposed to be reliable and routine—and to a great extent, they were supposed to be secret.

The Space Shuttle was, in a sense, designed as a secret spacecraft. It conjoined two Cold War space races. On one hand was the public race to demonstrate state virility and generate nationalistic zeal by putting men in orbit, landing on the moon, and all the rest. But there was another space race, a "black" space race conducted in secret but with similar vigor. The "black" space race took place among the Cold War's shadows and secrets. It was a race to launch ever-more-powerful reconnaissance satellites, develop anti-satellite weapons, and control the strategic "high ground" that space represented. Where John Glenn, Alan Shepard, and Neil Armstrong became national icons, stories from the "black" space race remain largely untold. If NASA was the iconic organization behind the "white" space race, the National Reconnaissance Office (NRO) spearheaded the secret race to the stars. The very existence of the NRO, the nation's "other" space administration, was classified. Although it enjoyed the largest budget of any agency in the intelligence community, the NRO was a "black" agency.[3]

The relationship between NASA and the NRO, which had been forced to cooperate in the program, was shaky at best. To be sure, the Space Shuttle had some strong supporters within the U.S. Air Force (which, on paper, oversaw the Space Shuttle's "military requirements," as the words "National Reconnaissance Office" were at the time top secret), who imagined a space plane emblazoned with the letters "USAF" on its fuselage. But within the intelligence community, the Space Shuttle also had strong critics. To its detractors, the shuttle was known as the "turkey"; at classified briefings, its critics circulated cartoons underlining the point.[4]

First conceived in the 1960s, the Space Shuttle was meant to serve as a cost-efficient, reliable way to achieve low Earth orbits. But the Space Shuttle had a catch. To make up for the shuttle's development, maintenance, and operational costs, the spacecraft only made economic sense if it put rockets out of business. It was only affordable if it had a monopoly on American spaceflight. In turn, this meant that the shuttle would have two varieties of missions: in addition to the public displays of American space exploration, the shuttle would be tasked with launching all of the United States' "black" spacecraft. The Space Shuttle would have to be a joint effort between NASA and the NRO. NRO requirements meant significant design changes to what NASA had envisioned: the shuttle would have a much larger payload bay (for holding colossal spy satellites) than NASA required, and would be designed to land at a launch site at Vandenberg Air Force Base on the California coast after a single orbit.[5]

The NASA logo on the shuttle's fuselage would be a kind of Barthesian myth: the truth, but not the whole truth. For a number of shuttle missions, the NASA logo served as a cover story.

Cover Stories

In this sense, the Space Shuttle was nothing new. It had always been impossible to hide titanic rocket launches from Americans living within hundreds of miles from the launch sites at Vandenberg and the Cape. Moreover, to Soviet observers, an unannounced launch might look like the opening salvo of a nuclear war. Thus, beginning with the first classified satellite launches, "black" space missions were conducted under the guise of elaborate cover stories.

The legacy of cover stories began on January 22, 1958, when the National Security Council issued Action Memorandum 1846. With this document the Eisenhower administration, embarrassed by the launch of Sputnik, made the development of a reconnaissance satellite the nation's highest technical intelligence priority.[6]

Classified satellite development had been in the works for years. From 1955, the air force had been working on a satellite reconnaissance system code-named "Pied Piper," but had made little headway. One of the many problems with Pied Piper was how to get reconnaissance images down from space. The air force wanted to relay a television signal from the satellite down to Earth, broadcasting the reconnaissance "take" in real-time down to ground-based interpreters. But the idea was far ahead of its time—in August 1957, RCA (the company responsible for the system) informed the air force that such a television signal would provide such poor resolution that it was not worth the effort to design and develop. The alternative was an ejectable film payload. When a reconnaissance satellite had exposed a requisite number of images, it would drop a film canister that could be recovered as it parachuted down to Earth. To the air force, the idea of scurrying around trying to snatch film canisters midair was ridiculous. It chose not to pursue the idea.[7]

After Sputnik, Eisenhower publicly canceled the Pied Piper program, unleashing a torrent of anger from congressmen who interpreted the move as Eisenhower pinching pennies where the nation could least afford it. "We of course couldn't tell anyone that the air force program was being replaced by a bigger one," the Central Intelligence Agency's (CIA's) Richard Bissell would later recall. The point of "canceling" the programs was to hoodwink both the Soviet Union and the American media into thinking that the United States had given up on the idea of space-based reconnaissance. Of course, the air force programs were not canceled at all—the lead role in developing reconnaissance satellites was given to the CIA. The agency's new space reconnaissance program would hide under the cover story of Discoverer.[8]

Ostensibly an air force project to conduct biomedical experiments in space, Discoverer's true purpose was to provide a public explanation for the new launchpad at Vandenberg Air Force Base on the California coastline and to hide—in plain sight—all of the attention-grabbing activities that go into putting a satellite in orbit. General Electric, which was building the camera capsule for Corona (the program's real code name), went as far as publishing a pamphlet on the Discoverer program, explaining how the satellite would ride an Agena rocket into space, and how its capsule containing "scientific data" would be recovered.

The Space Shuttle would continue the tradition of using "scientific experiments" as cover stories, of wrapping secret missions in public images of civic science and national progress.

On April 12, 1981, the first shuttle mission (STS-1) lifted off from Cape Canaveral with all the fanfare associated with a great leap in the advancement of public science. An IMAX film called *Hail Columbia!* documented the spacecraft's debut, and the rock band Rush put a song called "Countdown" on their 1982 *Signals* album, whose liner notes contained a tribute to the first *Columbia* crew. A year later, the shuttle's "other" patron would launch its first payload.

STS-36

STS-36 was going to be one of these "other" Space Shuttle missions. Its February 1990 launch window was classified; its weight was classified; its mission was classified; its payload was classified. That alone told aerospace-industry buffs that there was something special about STS-36. Another clue came from the résumé of the mission's pilot, a man named Colonel John H. Casper.

Casper, like scores of other astronauts going back to the Mercury program, began his career as a test pilot. According to the pilot's NASA biography, Casper began his aviation career as a combat aviator in Vietnam before becoming an air force test pilot. Upon entering the world of flight tests, his NASA résumé says that Casper was named "Operations Officer and later Commander of the 6513th Test Squadron, where he conducted flight test programs to evaluate and develop tactical aircraft weapons systems."[9] Translation: Casper flew secret airplanes.

Nicknamed the "Red Hats," the 6513th Test Squadron was one of the air force's most unusual units. The unit's patch sported a large brown bear wearing a red bowling hat climbing over the top of a globe. Above the bear, there was a collection of six red stars. The top of the patch read "Red Hats"; at the bottom is the unit's slogan: "More with Less." To military insiders, the symbolism was unmistakable. The Red Hats flew a squadron of covertly acquired

Soviet MiGs. The collection of stars on the patch referenced the unit's operating location. Six stars is the sum of five and one: Area 51, a "black" flight test center deep in the Nevada desert. The unit had been formed after the United States acquired several MiGs from Israel in the late 1960s. Israel provided the United States with two MiG-17Fs captured during the Six-Day War and a MiG-21F provided by an Iraqi defector.[10]

After completing his assignments with the Red Hats, Casper went on to become the deputy chief of the Special Projects Office at the air force headquarters in the Pentagon, where he continued to work on "black" projects. In 1984, NASA asked Casper to join. The following year, Casper became an astronaut. STS-36 would be Casper's first shuttle flight.[11]

Among defense-industry analysts and aerospace journalists, common knowledge held that STS-36 would deploy another Key Hole–class (KH) reconnaissance satellite, a highly classified but nonetheless somewhat run-of-the-mill imaging craft. The KHs are essentially classified versions of the Hubble Space Telescope, but instead of pointing out toward the edge of the universe, they are pointed down at Earth. The name "Key Hole," however, is a bit of a misnomer: in 1990 the air force and the NRO had stopped using the code name to designate its imaging satellites, but in common parlance the name had persisted.[12] According to a leak in *Aviation Week and Space Technology*, STS-36 would launch a payload whose official designation would be AFP-731 (Air Force Project 731), a randomly chosen designation.[13]

Although the exact launch time was classified, we know that Space Shuttle *Atlantis*, commanded by John Casper, rumbled off the Cape Canaveral launchpad at exactly 2:50:22 a.m. (EST). Almost immediately, a strange sequence of events began to unfold.

The first clue that something was amiss came from the characteristics of the STS-36 orbit. It went into a 62-degree inclination, the steepest inclination of any shuttle mission before or since. This was unique, but the truly bizarre part of the mission would occur weeks later, after STS-36 had completed seventy-two orbits and touched down on the dry lake at Edwards Air Force Base at 10:08 a.m. (PST) on March 4, 1990.[14]

On March 16, 1990, the Soviet news agency Nostovi reported that AFP-731 had blown up nine days earlier, on March 7—three days after STS-36 landed. Two days later, the *New York Times* picked up the story: "A new American spy satellite has apparently malfunctioned after less than a month in orbit, and parts of it are expected to be destroyed re-entering the atmosphere within a month, United States and Soviet military space officials say." The *Washington Post* quoted unnamed intelligence and congressional sources calling the reported explosion a "serious setback" and a "major concern." An article in *Aviation Week* explained that the "apparent failure of the US$500 million AFP-731 imaging reconnaissance satellite launched by the space shuttle *Atlantis*

Feb. 28 is a serious setback in the U.S. strategic intelligence program," and that "the apparent failure of the AFP-731 imaging spacecraft was the third in a series of major Western space failures in the last month."[15]

For its part, the Pentagon did little to confirm or deny the Russian report. President George H. W. Bush said nothing; the same held true for Dick Cheney, Colin Powell, and the rest of the administration. There were no congressional investigations, no name calling or posturing in the halls of Washington. The political uproar following AFP-731's multibillion-dollar boondoggle seemed strangely restrained. After the initial reports of the explosion, the matter simply went away.

It was as if AFP-731 had been nothing more than a ghost, seen like the residue from a bright light, fading from behind closed eyelids.

Enter Ted Molczan.

The Other Night Sky

In the threadbare apartment where he lives with his two cats, Rusty and Sparky, Ted Molczan tells me that he's always been a "space nut." His desk is neatly stacked with star atlases. Wil Tirion's *Uranometria 2000* sits atop the *Atlas Eclipticalis*, the Czech astronomer Antonín Bečvář's beautifully rendered 1958 guide to the heavens. Across the room is a bookshelf crammed with more space books: Patrick Moore's *Men of the Stars*, Robert Divine's *The Sputnik Challenge*, Desmond King-Hele's *A Tapestry of Orbits*, and William Burrows's classic tome on secret satellites, *Deep Black*. Perched on a tripod next to the bookshelf is a pair of Celestron binoculars with lenses the size of two-liter Coke bottles. I've brought along my own small pair. In a good-natured way, Molczan tells me they might work OK at the opera.[16]

Ted Molczan is one of the world's leading figures of a very peculiar version of amateur astronomy. Most amateur astronomers content themselves to stare deep into the cosmic wonders sprinkled through the night sky: the Triangulum in the Orion Nebula, the gaseous blue clouds surrounding the Pleiades, the globular clusters in Sagittarius, the Andromeda Galaxy near Cassiopeia, or the great craters of Copernicus and Plato on the surface of the moon. At the high end of the hobby, amateur astronomers employ sophisticated automated systems to help university researchers search for extra-solar planets, collect gamma ray outbursts, and generate observational data used by professional researchers. Amateurs have a virtual monopoly on discovering brighter novae. In astronomy, the line separating backyard amateurs from their cousins at major research universities can be blurry and indistinct.[17]

But Ted Molczan is a different breed of amateur astronomer. Molczan's specialty is a host of celestial objects even more obscure than the galactic cluster in Hercules or the dark nebula in Aquila. Using his pair of high-powered binoculars, a stopwatch, and a self-fashioned computer program called "Obsreduce," Molczan works with a handful of people around the world who observe and keep detailed records on nearly 200 classified American satellites in Earth orbit. Molczan tries to understand a very different night sky than the starry night most of us see when we stare into the heavens on a clear night. Molczan maps the other night sky.

The other night sky is a landscape of fleeting reflections: of glints, glimpses, traces, and flares. Of unacknowledged moons and "black" spacecraft moving through the predawn and early evening darkness, where the rising and setting sun lights up their stainless steel bodies, and they blink in and out of sight as they glide through the backdrop of a darkened sky hundreds of miles below. In most cases, the reflection is all we get. The other night sky doesn't want to be seen: even full-time defense industry journalists and aerospace historians have a hard time knowing exactly what's what.

On the faintest end of the spectrum are the geosynchronous satellites—spacecraft like the Milstar constellation (originally for "Military Strategic and Tactical Relay"), perched 22,241 miles above Earth's surface so that they can "cover" about half the planet. Milstar 5, for example, is parked on the equator over eastern Africa, about halfway between Nairobi and Mogadishu, while Milstar 6 is perched on the other side of the globe near the Galapagos Islands off the coast of Ecuador. Then there are the Mercury, Mentor, Magnum, and Advanced Orion eavesdropping satellites, purported to look like umbrellas the size of football fields, sitting in geosynchronous orbits to vacuum up communications, telemetry, and electromagnetic signals emanating from below. Even among NRO and National Security Administration (NSA) insiders, the code names of these geosynchronous satellites are said in hushed tones. All but invisible to ground-based observers, and containing some of the most highly classified systems and charged with collecting the planet's most sensitive data, they are Earth's most secret moons.

The illuminated hulls of reconnaissance satellites follow the inverse-square law: objects closer to Earth are exponentially brighter. Just visible to the unaided eye at an altitude of about 1,100 km are the Naval Ocean Surveillance Satellites (NOSS) (code-named Parcae, after Zeus's three daughters), whose mission is to track naval vessels by eavesdropping on shortwave and other transmissions. The NOSS satellites cruise across the night sky in formations of twos and threes, appearing as points of light moving across the sky in a triangular formation. In other words, they look exactly like late-generation UFOs, or delta-winged aircraft using a cloaking device. In fact, they are *so* easy to mistake for UFOs or "black" aircraft that UFO researchers have come to realize that a number of "black triangle" sightings can be explained by the Parcae constellations.[18]

The kings of the night sky, however, are the imaging satellites. If you've ever looked up at the sky just after twilight or just before dawn and have seen a satellite moving across the sky, there's a good chance that you've seen one of these. They are some of the brightest objects in the sky. The size of school buses and dwelling in precariously low orbits, their polished hulls light up like meteors when they reflect sunlight toward Earth below. The imaging satellites come in two basic "flavors": those that use photographic imaging methods, known as KHs, and those that use something called "Synthetic Aperture Radar," known by the code names Lacrosse and Onyx. And right about here, our knowledge of the other night sky's denizens starts to run out.

"One way to look at it," Molczan told me about his unusual hobby, "is that it's a form of science-based investigation, but more like detective work. I guess that it's like modern detective work, which is also based on science." Molczan's pastime might sound unbelievably complicated, and to a certain extent it is, but in another sense it's pretty straightforward if you know what you're doing. The hobby is possible, Ted explains, because no matter how many security classifications and code words the NRO uses to hide its secret satellites, the agency can't classify Kepler's three laws of planetary motion. The tools of satellite observing are so simple that they seem almost anachronistic: a good pair of binoculars, some star charts, and a stopwatch. That's it, really. Molczan does it all from the balcony of his apartment.

With these simple tools, Molczan can generate an incredible amount of detail about "black" satellites and secret moons. By synthesizing his own observations with those of his fellow observers such as Russel Eberst in Scotland, Pierre Neidrick in France, and Mike McCants in Austin, Texas, Molczan can predict where a spacecraft will be with astonishing accuracy. "I have a fanatical devotion to accuracy," he chuckles.

Molczan's and the other observers' hobby shares origins with the various conflations of public science and underlying militarisms that have long characterized the space race. In the mid-1950s, it was clear that both the Soviet Union and the United States would begin populating Earth orbit with artificial moons in the near future, but the United States lacked the capacity to track them. To solve the problem, the Smithsonian Astrophysical Observatory (SAO) created Operation Moonwatch, a national (and later international) program to support satellite observing as a popular hobby. Inspired in part by the Ground Observer Corps, an early Cold War program designed to teach ordinary citizens how to spot Soviet bombers and to act as a national "early warning system" against potential Soviet attacks, the SAO encouraged legions of ordinary people to form local satellite-spotting clubs. While Moonwatch was billed as a large-scale citizen-science project, the military was an active sponsor and patron of the program. Officials at the air force and navy provided Moonwatch teams with telescopes and training support, and encouraged Moonwatch teams to set up shop on air force bases.[19]

Within a few years, Operation Moonwatch's popularity waned. The culture of McCarthyism, which motivated people to act as vigilant citizens on the lookout for Soviet machinations, began to fade. So, too, did the initial novelty of seeing artificial satellites in the night sky. While many Moonwatchers dropped out, those who stuck with the hobby became exceptionally good at it. By the time Moonwatch closed its doors for good in 1975, its remaining members were amateurs in name only. Observational data collected by the remaining Moonwatchers was often more accurate than data collected by their professional counterparts.[20]

Throughout the Moonwatch program, amateur observers enjoyed a symbiotic relationship with NASA. The space agency supplied amateurs with the lists of orbital elements it maintained for all the spacecraft it tracked. Amateurs, in turn, refined these orbital elements using data from their own observations. Even after Moonwatch ended, NASA's Goddard Space Flight Center continued doling out reams of orbital data to the amateur observers who'd signed up for the service. Data from the Goddard Center contained elements for all the spacecraft it was tracking, including American reconnaissance satellites. Although the classified objects were not identified as such in the data set, it was relatively simple for amateurs to tell which elements corresponded to spy satellites. Until 1983.

That year, the Reagan administration abruptly stopped publishing orbital elements for American military and reconnaissance satellites. Ironically, this move gave amateur satellite observers a renewed sense of purpose. Russel Eberst, a former Moonwatcher based in Scotland, explained that the abrupt classification "unwittingly set a challenge to the amateur network of observers." The new game was "to see if they could maintain reliable orbits for these 'secret' objects."[21] Amateur observers rose to the challenge: faxes, telephone calls, and the mail were replaced by bulletin board systems (BBS) boards, the precursor to the World Wide Web. Observations were shared, orbits refined. The amateurs developed a highly accurate guide to the blank spots in the official satellite catalog. They monitored existing satellites for changes in behavior, and updated their catalog with every classified object the NRO put into orbit.

When Ted Molczan (a self-taught observer) heard about the Space Shuttle STS-36 mission in 1990, and read that the shuttle would deploy something dubbed AFP-731, he assumed that the payload would be another KH-class object, a standard optical reconnaissance satellite. Molczan also assumed that tracking the spacecraft would be a relatively straightforward problem, "but being impatient and wanting as much data as I could get, knowing launch date and time, it was easy to determine that it wasn't going to be in range of most active observers. We were out of luck for weeks." Because the skies above Toronto would be in the sun's shadow when Molczan expected the spacecraft to fly overhead, there would be no reflection. The payload wouldn't be visible from his Toronto balcony. "I needed to find observers in the North." So with a bit of "inspired phone calling," Molczan recruited teams of agreeable amateur astronomers at Yellowknife in the Northwest Territories, at Whitehorse in the Yukon, and in Alaska, then held a series of informal training sessions to teach the astronomers the basics of satellite observing, and supplied the groups with "look angles" where they could expect to see the classified spacecraft.

"Things went really well," Molczan explained. After the Space Shuttle deployed the unidentified satellite, "it went into [the] orbit that it was supposed to be in from *Aviation Week*. The whole idea of tracking it was to refine the estimated orbits that I'd predicted. That all worked like a charm." The amateurs based in the Far North reported back that AFP-731 was nice and bright in the night sky—about Mag -1, the same brightness as Jupiter—and supplied Molczan with useful observations. In the United Kingdom, Russel Eberst made additional observations. "A few days of this, and the guys up North asked if I had all the data I needed. Remember that it was about 40 degrees below [zero] outside."

With the data in hand, Molczan decided that he had what he needed, even though it would be weeks before he'd actually be able to see the object from his home. Then something unexpected happened: "About a week goes by and a press release comes out from the Soviets saying that the satellite may have blown up."

AFP-731 had disappeared. The last elements Molczan had for it looked like this:

USA 53 (AFP-731) 18.0 4.0 0.0 4.1

1 20516U 90019 B 90309.99079700 -.00002298 00000-0 -95528-3 0 03

2 20516 65.0200 194.0588 0009734 214.9671 144.9440 14.26241038 04

Interlude

On the twenty-third floor of his downtown Toronto high rise, Molczan is teaching me how to get the maximum accuracy from a satellite observation. We're sitting at his desk looking at a graphic on his computer screen: a line representing a predicted satellite pass bisects a pair of stars we plan to use as fixed reference points to measure the pass of a peculiar object launched in 1999 called USA 144Deb.[22]

"So it should be about 60 percent down from this star to the next," I say. "Try to forget that number," Molczan tells me. "You don't want to bias your observation." One of the dangers in observing satellites holds true for all strictly empirical work, he explains: you have to see what's there, not what you want or expect to see. To get accurate data, you have to be as objective and unbiased as possible. Trying to make oneself a "reliable witness" requires a tremendous amount of self-discipline.

Minutes later, I'm looking through Molczan's binoculars at the two stars we just saw on his computer screen. A point of light, shimmering like a diamond on blue-black velvet, glides into my field of view and bisects the pair of stars. I click the button on Molczan's stopwatch as it passes. The USA 144Deb object appears to be exactly where Molczan predicted it would be: about 60 percent down from the top star to the bottom star. Then a moment of self-doubt: did I see it there because it actually was there, or because I expected it to be? I hadn't been able to forget the number. Molczan and I step back in the apartment. There's an hour until the next predicted satellite pass.

"I can't speak for everyone involved in this hobby," Molczan once told me about his pastime, "but for me, this is about democracy. There are elements out there who want to keep everything secret. I try to put pressure in the other direction. I try to put checks on that power. When people ask me about what gives me the right to make these decisions, I say 'Citizenship in a democracy gives me the right to make these decisions.' I don't break in, I don't steal stuff. I assert my right to study the things that are in orbit around the earth and study them with the belief that space belongs to all of us. I exercise my right to know what's there."

The more we talked, the more I couldn't help but to start thinking of Molczan as a kind of latter-day Galileo, measuring reflections from secret moons in ways that recall Galileo's measurements of Jupiter's moons in *Sidereus Nuncius*, the book based on his initial experiments with the telescope. Galileo's observations helped open up the Pandora's box of classical empiricism. His critics claimed that his self-fashioned telescope was feeding him illusions. The church told him to disavow the Copernican implications of his work—or else.

Molczan's work echoes the work of classical empiricists—the likes of Kepler, Galileo, and Newton—and their contemporary descendants in the hard sciences. His work sits atop two much-maligned epistemological assumptions and methods: that abstracting space into something fixed and calculable can reveal (as opposed to produce) an underlying truth and that reflections can serve as an index of verisimilitude. Molczan's predictions rely on the notion of space as something physical, fixed, and calculable, the notion of absolute space most often associated with Newtonian physics and much aligned as a far too limiting account of space in contemporary spatial thought, and a powerful technology of domination.[23] Furthermore, Molczan's measurements of the other night sky stem from a second outmoded epistemological assumption: that signs are parts of things rather than representations of things; that signs do not necessarily conjure practical reality into existence, but can instead function as ciphers to decode underlying truths about reality *as it is*; that reflections exist prior to observation and naming. For Molczan, the truth is indeed "out there" for anyone willing to look with open eyes and an open mind.

For theoretically inclined social scientists, humanities scholars, and even a number of our colleagues in the hard sciences, Molczan's methodological assumptions might seem as epistemologically anachronistic as his pencil-and-stopwatch approach to collecting data. One of the critical tradition's most consistent projects is, of course, a thorough debunking of any easy link between reflection and correspondence. Critical scholars have come to see *vision*, and by extension, *truth*, like the NASA logo on the Space Shuttle or the cover story for a classified satellite. Vision and reflection easily function as myth and fetish: appearances easily mask far more than they reveal, naturalizing the contingent, frangible, and historical.[24] There is no underlying universal truth to reveal, and no rational subject or "modest witness" to disinterestedly report whatever might be "out there."[25]

But the more I thought about it, the more profound Molczan's quip about democracy seemed. Almost 400 years earlier, the Enlightenment tradition had held out the promise of a world in which emperors and inquisitors could not dictate what would be true for all. The notion of democracy Molczan found by studying the other night sky was, of course, a liberal democracy, one rightly criticized for too easily producing the darkness of slave trades, satanic mills, genocides, and arms races. But this tradition also animated countless other movements—from abolition to contemporary antitorture campaigns—whose aims were to counter those dark spaces. Of course, the notion of critique is itself embedded in the Enlightenment tradition.

I wanted to believe in Molczan's methods.

In a world of Abu Ghraibs, Guantanamo Bays, renditions, waterboardings, state secrets, wiretappings, and "black" sites, it seemed as if the critical de-linking of reflection and correspondence and the denigration of absolute space was no guarantee of a more equitable society. The point was famously underlined when an official in the George W. Bush administration told journalist Ron Suskind, "We're an empire now, and when we act, we create our own reality. And while you're studying that reality—judiciously, as you will—we'll act again, creating other new realities, which you can study too, and that's how things will sort out. We're history's actors . . . and you, all of you, will be left to just study what we do."[26] Wholeheartedly embracing the radical instability of signs could also produce nightmares: a world in which torture was not torture and disappearances "weren't happening." Not trusting one's own eyes can also have consequences: it can make it that much easier for authority to trump reason in dictating truth.

And in this context, Molczan's work reminded me of another antiquated notion from the liberal democratic tradition: in a democracy you're supposed to have a right to your own opinion, but you're not supposed to have a right to your own facts. Truth wasn't always supposed to be the woman Nietzsche famously denounced as a temptress and a tease. Molczan reminded me that perhaps truth is sometimes like a point of light in the evening sky, the sun's reflection against something authorities say isn't there. Singular. Visible to anyone who bothered looking through a telescope. Insisting on the verisimilitude of reflections could be a radical gesture. As Winston Smith wrote in his secret notebook, freedom can mean insisting on two plus two making four.[27]

Or perhaps I was being seduced. Perhaps I was, ironically, doing the thing that a classical faith in empiricism forbids above all else: allowing myself to see something because I wanted to see it. Soon, Molczan would come up against his own "observer effect," the general notion that the act of observation has a tendency to change the thing being observed, an effect that undermines any easy distinction between the observer and the observed. Molczan would come to learn that the other night sky was also paying attention to him as he went out on his balcony each night when the clouds cleared enough to see the stars.

March 14, 1990

Two days before AFP-731 appeared to explode, Teledyne Industries filed a rather unusual collection of documents with the U.S. Patent Office. Patent 5345238 described a solution to a problem that had plagued military and reconnaissance satellites throughout their entire history: they were too easy to track. "Oftentimes space based weapons systems are looking out into the non-reflective background of outer space," noted the patent authors. "This makes the tracking of the target easier, because there is no background radiation or other noise background in the sensor's view. The satellite, which is a radiation source and a radiation reflector, is very evident in this radiation-free background." Reflections, in other words, were a problem. They revealed the location of a satellite to anyone looking to track it. Moreover, once an observer or "detection threat" acquired a few observations, he or she could predict a satellite's path with remarkable accuracy: "Once a satellite or other space object is in orbit, they follow very precise orbital tracks. Therefore, once a satellite's position is accurately determined and tracked, predictions of the future location of this satellite are very accurate." Teledyne Industries was proposing a solution to these pesky consequences of satellites' materiality. The patent was for a "satellite signature suppression shield for camouflaging a satellite's location from ground based and airborne tracking and detection systems," whose purpose was to "suppress the laser, radar, visible and infrared signatures of satellites."[28]

The Teledyne patent seemed to solve a problem that the NRO had been worrying about for as long as it had been conducting classified space operations. As early as 1963, there were proposals within the NRO for a "Covert Reconnaissance Satellite," an alternative to the bus-size Key Hole imaging birds that shone so bright in the night sky. A "covert system" proposed in a 1963 memo to the NRO deputy of technology would "rely, above all, on concealment," in a dual sense: to facilitate bureaucratic secrecy, the covert system would have a "separate and tight security system," and "simplified check-out and handling procedures, requiring a minimum of personnel." Physically hiding the covert system could be achieved through "covert and at least portable launch and recovery, preferably mobile" and through a "reduction of radar and optical cross-sections below the detection threshold."[29] The NRO had, in other words, long sought to develop a "stealth satellite" that could appear and disappear at will, and that would thwart "active and passive detection systems."

The stealth system described in Teledyne's 1990 patent was in essence a giant movable mirror designed to reflect the blackness of space toward predetermined "threat detection sites" on Earth. Its major limitation was that the threat detection sites had to be programmed into the satellite's control software. The mirror would not make a satellite invisible in general; the spacecraft could only hide from specifically programmed sites. The satellite suppression shield had to be "oriented in the direction of the threat."[30]

October 1990

Six months after AFP-731 "exploded" and disappeared, Ted Molczan came across something unusual in the sky. "I get this message from Russel Eberst in Scotland and others," said Molczan, "and I realize we've seen precise unknowns on the same nights." The world's most accomplished spy-satellite observers had all seen an especially bright object. None of them could identify it. "Russell refined the orbits, and we expected it to match something already up there," Molczan explained. Figuring that the unknown object would end up being a wayward rocket body or a forgotten Soviet spacecraft, he checked the numbers against publicly available satellite catalogs and his own records of classified orbits. Molczan couldn't ID the spacecraft, so he turned to another set of records: objects that he'd tracked and subsequently lost. Using the observations of the unknown object, Molczan precessed the orbits back in time. "Lo and behold, they lined up on the seventh of March— the day the Russians said the satellite exploded." The satellite "had been exceedingly bright and was still bright, especially considering its height of about 800 kilometers [497 miles]."

The "explosion" of AFP-731 appeared to have been a ruse, something akin to an embattled submarine shooting oil, lifejackets, and debris from its torpedo tubes to create the impression that it had been blown up. But Molczan's calculations showed that the bright object the amateurs found in October could only be AFP-731, alive and well. The Soviet press report could have been part of an international game of deception: perhaps the Soviets hadn't been fooled at all, but had wanted to indicate to the United States that they had.

Molczan and the other satellite observers continued tracking AFP-731 for several weeks, reporting their sightings and refining their data on a BBS board. With a few weeks, a reporter named Todd Halvorson at *Florida Today* caught on to what was happening and wrote an article about the sightings for Cape Canaveral's local paper. A month later, the *New York Times* added a tidbit at the bottom of an article about a different shuttle mission: "Amateur satellite trackers in Canada and Europe reported they had spotted the spy satellite in a higher orbit than anyone suspected—an indication that the craft was not only working but also highly maneuverable."[31]

Autumn clouds obscured the night sky over Canada and Europe in the few days after Halvorson's article was released. And when the skies opened up, AFP-731 was once again nowhere to be found.

NOTES

1 Much of this essay develops ideas from Trevor Paglen, *Blank Spots on the Map* (New York: Dutton, 2009).

2 "Secret Mission," *Aviation Week and Space Technology* 132(4): 23.

3 For the NRO, see Jeffrey T. Richelson, "The NRO Declassified," *National Security Archive Electronic Briefing Book No. 35*, September 27, 2000, accessed March 25, 2011, http://www.gwu.edu/~nsarchiv/NSAEBB/NSAEBB35/; for the NRO budget, see Commission on the Roles and Capabilities of the United States Intelligence Community, "The Cost of Intelligence," chap. 13 in "Preparing for the 21st Century: An Appraisal of U.S. Intelligence," website dated February 13, 1996, accessed March 25, 2011, http://www.fas.org/irp/offdocs/report.html.

4 Dwayne A. Day, "The Spooks and the Turkey: Intelligence Community Involvement in the Decision to Build the Space Shuttle," *Space Review* (November 20, 2006), accessed March 25, 2011, http://www.thespacereview.com/article/748/1

5 Day, "Spooks."

6 William E. Burrows, *Deep Black* (New York: Random House, 1986), 104.

7 Ibid., 84, 90–91.

8 Ibid., 107.

9 John H. Casper's NASA online biography, accessed March 25, 2011, see http://www.jsc.nasa.gov/Bios/htmlbios/casper.html.

10 For the "Red Hats" and history of purloined MiGs, see Curtis Peebles, *Dark Eagles* (Novato, Calif.: Presidio, 1995), 217–244; for patches, see Trevor Paglen, *I Could Tell You but Then You Would Have to Be Destroyed by Me* (Brooklyn, N.Y.: Melville House, 2008).

11 Casper, biography.

12 Jeffrey T. Richelson, *America's Secret Eyes in Space* (New York: Harper and Row, 1990), 231.

13 "Secret Mission," *Aviation Week and Space Technology* 132(4): 23.

14 For launch details on STS-36, see NASA's web page about the mission, accessed March 25, 2011, http://science.ksc.nasa.gov/shuttle/missions/sts-36/mission-sts-36.html.

15 Warren Leary, "Problems Are Reported with New Spy Satellite," *New York Times*, March 18, 1990.

16 All Molczan quotes come from two days of interviews in Toronto, Canada, in July 2008, and from a year's worth of e-mail correspondence with the author.

17 For amateur astronomy and professional science, see Timothy Ferris, *Seeing in the Dark* (New York: Simon and Schuster, 2002).

18 See Anthony Eccles, "UFOs and the NOSS Problem," *Anomalist*, accessed March 25, 2011, http://www.anomalist.com/features/Noss.html; "Recent Australian UFOs Were Just U.S. Navy Satellites," Listserv post under Reader Feedback, accessed March 25, 2011, http://www.ufoinfo.com/roundup/v08/rnd0804.shtml.

19 Patrick McCray, *Keep Watching the Skies! The Story of Operation Moonwatch and the Dawn of the Space Age* (Princeton, N.J.: Princeton University Press, 2008), 108–109.

20 Ibid., chap. 7.

21 Quoted in Patrick Radden Keefe, "I Spy: Amateur satellite spotters can track everything government spymasters blast into orbit. Except the stealth bird code-named Misty," *Wired* 14(2), February 2006, accessed March 25, 2011, http://www.wired.com/wired/archive/14.02/spy.html.

22 For the full story of the USA 144Deb object, see Paglen, *Blank Spots on the Map*.

23 For some of the classic critiques of absolute space, see Henri Lefebvre, *The Production of Space*, trans. Donald Nicholson (London: Blackwell, 1990); Neil Smith and Cindi Katz, "Grounding Metaphor: Towards a Spatialized Politics," in *Place and the Politics of Identity*, ed. Michael Keith and Steve Pile (London: Routledge, 1993), 67–83.

24 The literature on these questions is far too vast to address here. A few books that have been particularly helpful to me in thinking through these ideas include: Michel Foucault, *The Order of Things* (New York: Vintage, 1994); Alexander Koyré, *From the Closed World to the Infinite Universe* (Charleston, S.C.: Forgotten Books, 2008); Neil Smith, *Uneven Development: Nature, Capital, and the Production of Space* (Athens: University of Georgia Press, 2008); and David Harvey, "Space as a Keyword," in *David Harvey: A Critical Reader*, ed. Noel Castree and Derek Gregory (Oxford: Wiley-Blackwell, 2006), 270–294. In his essay, Harvey notes the danger of ignoring absolute space: "there is a serious danger of dwelling only upon the relational and lived as if the material and absolute did not matter" (147).

25 For example, Michel Foucault, *The Foucault Reader*, ed. Paul Rabinow (New York: Vintage, 1984), 59–60.

26 Ron Suskind, "Without a Doubt," *New York Times*, October 17, 2004.

27 Winston Smith is the protagonist in George Orwell's 1949 classic novel *1984*.

28 U.S. Patent No. 5345238, Eldridge et al., "Satellite Signature Suppression Shield," filed March 14, 1990, granted September 6, 1994, accessed March 25, 2011, http://www.gwu.edu/~nsarchiv/NSAEBB/NSAEBB143/nph-Parser.htm.

29 "Memorandum for Deputy for Technology/CSA, Subject: A Covert Reconnaissance Satellite," April 17, 1963, declassified November 26, 1997.

30 U.S. Patent No. 5345238.

31 Warren Leary, "Space Shuttle Lifts Off with Secret Military Cargo," *New York Times*, November 16, 1990.

14

MICROSATELLITES

A BELLWETHER OF CHINESE AEROSPACE PROGRESS?

ANDREW S. ERICKSON

Central to China's rise in space—no less important than its becoming the third nation to test an anti-satellite weapon (on January 11, 2007) and the third to orbit an astronaut (on October 15, 2003)—is its rapid development of microsatellites. Microsatellites (weighing 10 to 100 kg, or far less than the average satellite) are believed by both Western and Chinese analysts to represent the key to improving space capabilities by lowering the cost of establishing a robust presence in space with built-in redundancy to ensure system continuity. They do so by enabling mass production and modularization, and through their flexible use in multisatellite constellations for applications such as communications. For these reasons, microsatellites will be the focus of this chapter. To be sure, microsatellites cannot be separated entirely from their larger counterparts in function and significance. This chapter will consider small satellites (those weighing up to 500 kg) and their microsatellite "cousins" (which weigh less than 100 kg). The development of small-satellite technologies played a role in the cultivation of microsatellites somewhat later. Satellites weighing more than 500 kg, such as the 2,200 kg Beidou navigation satellites, are beyond the scope of this study.

China's increasingly sophisticated microsatellites are a vital element of Beijing's overall aerospace development, but to what end? Small/microsatellite production offers Beijing three major benefits: support for national development, lucrative and geostrategically relevant foreign sales, and potential military space control applications. The first two benefits have provided a major motivation for Chinese microsatellite development thus far, and may well be the most important current benefit. In foreign sales, for example, Chinese satellites (albeit of larger ones, thus far), components, and

——————————— Table 14.1 Chinese Satellite Categories ———————————

Type	Weight Range	Chinese Name	Pronunciation
large satellite	>=500 kg	大卫星	dàwèixīng
small satellite	<500 kg	小卫星	xiǎowèixīng
minisatellite	10–200 kg	微小卫星	wéixiǎowèixīng
microsatellite	10–100 kg	微型卫星	wéixíngwèixīng
nano-satellite	<110 kg	纳卫星	nàwèixīng
pico-satellite	<1 kg	皮卫星	píwèixīng
femto-satellite	<0.1 kg	飞卫星	fēiwèixīng
LightSat	DARPA blanket designation for small satellites	轻卫星	qīngwèixīng

launch and training services have performed relatively well by giving developing nations otherwise unaffordable access to space.

China today has only a fraction of the overall space capability of the United States, has major gaps in coverage in every satellite application, and relies to a considerable extent on technology acquired through nonmilitary programs with foreign companies and governments. But China is combining this new knowledge with increasingly robust indigenous capabilities to produce potent advances of its own. China's satellite developers are experimenting with a new workplace culture that emphasizes modern management, standardization, quality control (including ISO 9000 management initiatives), and an emerging capacity for mass production—part of a larger trend in China's dual-use military-technological projects. For a complete categorization of Chinese satellite designations, see Table 14.1.[1] Note that there is some overlap between these categories, and they are used somewhat differently in different publications.[2]

History: A National Development Imperative

Long before the concept of microsatellites was considered to be a key development trend in either the West or China, satellite development writ large was considered critical to furthering China's national interests. In January 1958, Qian Xuesen, the father of China's space program, initiated Project 581[3] to build China's first satellite when, with other scientists, he drafted a satellite development program and designated a working group.[4] Following the Soviet launch of Sputnik III in 1958, Project 581 became a top national priority. "We too should produce man-made satellites," Mao declared to his fellow leaders on May 17, 1958.[5] Premier Zhou Enlai later added, "We should try our best to develop our own meteorological satellites."[6]

On April 29, 1965, China's Defense Science and Technical Commission submitted the "Plan for the Development of China's Artificial Satellites," which called for launching China's first satellite in 1970–1971.[7] On August 10, 1965, Zhou Enlai formally approved the plan, which directed "that the satellite should be visible from the ground and that its signals should be heard all over the world."[8] In May 1966, Qian and his scientific colleagues solidified the plans for China's first satellite launch, agreeing on a name (Dong Fang Hong [East is Red]-1 [DFH-1]), a launcher (CZ-1), and a deadline (the end of 1970).[9] DFH-1's successful launch on April 24, 1970, from China's Jiuquan launch facility made China only the fifth country to launch a satellite.[10] DFH-1's mission was political: its sole function was to broadcast the Chinese national song, "The East Is Red."[11]

Satellite development has been a consistent priority for China since the 1960s. During the past three decades, the country's satellite development and testing have gradually increased in volume and sophistication. China developed and launched the DFH series of large satellites, the Shijian (SJ), or "Practice," series of small satellites, and Da Qi-1 and 2, a pair of atmospheric research balloons.[12] SJ-1 through SJ-4 performed a wide variety of scientific experiments. Some Western experts have speculated that SJ-2 and Da Qi (DQ)-1 and DQ-2 also performed electronic intelligence (ELINT) missions.[13] They are listed in Chinese sources as having "succeeded" as "technological probes" and as "scientific survey balloons, used to research the upper atmosphere."[14]

Initially produced to demonstrate China's national capacity and promote key technological advances, satellites were determined to have vital military significance as well as great potential to support national development. Satellites are regarded as key to China's strategy of efficient investment, a strategy grounded in the notion that a nation can leapfrog traditional stages of technological development.[15] Over the past three decades, satellites have provided China with tremendous benefits in land survey, crop monitoring, forestry, hydrology, geology and petroleum exploration, archaeology and cultural preservation, meteorological observation, natural disaster response and mitigation, oceanography, space environmental exploration, communications and broadcasting, and scientific and technical experiments.[16] These functions are regarded as vital for national modernization given China's vast, largely mountainous territory; complex terrain; and imbalanced economic development.[17]

By the end of the Cold War, China, according to one Chinese analyst, had "become one of the few countries in the world with an ability to launch all categories of satellites with her own launching vehicle; control and manage satellites with her own TT&C [tracking, telemetry, and control] communications network, with services for launching and TT&C of foreign satellites starting to be provided."[18] Chinese analysts believed their nation's technologies of satellite telemetry and recovery and their ability to launch geostationary satellites have been on a par with the most advanced countries in the world.[19] Beijing also became more involved in international satellite conferences and cooperation by the end of the Cold War. Development had progressed steadily and appeared to be gaining momentum.

For most of the Cold War, the United States, the USSR, and other leading space powers sought to build ever larger and more sophisticated satellites. Because of technological and manufacturing limitations, the Soviet Union produced a greater volume of satellites with simpler construction and shorter mission lives than their U.S. counterparts. Here China may have benefited from its relative limitations. Resource constraints meant that most of China's satellites were of relatively low mass and complexity, though there was a slowly emerging group of high-performance or mission-specific satellites. A national focus on developing the civilian economy meant that many of China's satellites were dual-use and multifunctional in nature. Though few people on either side of the Pacific realized it at the time, the stage was being set for China to align itself with a powerful technological trend.

Chinese development of small and microsatellites began with the 863 Program/National Defense Basic Science Research Program, which Beijing launched in 1986. Microsatellites therefore represent the next frontier of People's Republic of China (PRC) aerospace development, one that is receiving increasing priority. According to one PRC analyst, "Developing our country's small satellite technology, [which is] leading our country's remote sensing spaceflight enterprise, has risen to [the level of] an important impetus."[20] While China has been prioritizing its satellite programs since the 1960s, the country experienced limited success until quite recently. In the 1990s, however, with significant increases in technology access and funding, progress accelerated markedly (as it did in other technological development sectors).

Development work with foreign partners has been central to this progress. Nearly every PRC satellite in recent years has benefited significantly from foreign technology (for example, from the United Kingdom, the European Space Agency, and Brazil). PRC advances in microsatellites would have been limited without these contributions. While this international development assistance is a trend in the satellite industry overall, it appears also to represent an important aspect of China's technology development strategy. This suggests that, for the foreseeable future, China's satellite development will exhibit significant foreign influence. But China has been careful to diversify its development partners, and there is no chance of a repeat of the Sino-Soviet split, in which rapid withdrawal of Soviet advisers in 1960 severely limited Chinese aerospace development for years. Moreover, China is cultivating a new generation of extremely talented engineers who are learning from foreign partnerships while developing their own capabilities. The indigenous development and production capabilities that China has already accrued should not be underestimated.

This early history offers yet another example of China's capacity to achieve its national military-technological goals through an all-out effort (as with nuclear weapons and ballistic missiles, albeit on a smaller scale). But now a potent combination of world-class technological competence and commercial dynamism is propelling microsatellite development in ways that China's government could not achieve alone. Today rapid advancement and significant breakthroughs are possible in the field of microsatellite development for two main reasons: the programs are no longer solely reliant on government prioritization, and there is now consistent access to high-level personnel, funding, and technology. Other sectors may well benefit from similar dynamics, but the minimal regulation and unique aspects of satellites suggests that this will continue to be a leading sector.[21]

Chinese Development and Production Facilities

A wide variety of facilities support Chinese satellite development efforts. Beijing Satellite Manufacturing Factory (BSMF), subordinate to China Aerospace Science and Technology Corporation (CAST) since February 27, 1968, is a state-owned enterprise. Formerly the Chinese Academy of Sciences' Beijing Scientific Instrument Factory, BSMF has been involved in the assembly, integration, and testing of a wide variety of satellites since before 1970. Today it has an assembly workshop, five professional laboratories, seven small producing workshops, and the Beijing Xingda Technology Development Company (which develops commercial products). Another supporting facility, CAST, has developed several models of small satellites; the Chinese Academy of Sciences (CAS) has conducted research on payloads for small and nanosatellites; the Shanghai Space Administration has developed small satellite propulsion systems; and the Harbin Institute of Technology has established a small satellite research center and is running some small satellite projects related to the 863 program. Finally, microsatellites such as Shijian 5 and Haiyang 1 have been tested at the CM series magnetic test facility.[22]

The most sophisticated and cost-effective microsatellite development seems to be taking place at China's foremost research universities and major satellite production companies. These appear to be organizationally lean, employ young technical talents, and operate on quasi-market principles. This recipe for success, applied unevenly elsewhere in China's defense/science and technology industry, is bearing significant fruit. China has also received considerable technological assistance (and potential managerial and organizational influence) from many foreign entities, most prominently Surrey

Satellite Technology Limited (SSTL), a leading U.K.-based supplier of small satellites.[23]

Tsinghua University ("China's MIT") has clearly been recognized as a core PRC microsatellite development center.[24] Tsinghua's role is hardly coincidental. As one Chinese writer observes, microsatellites rely on "machinery, mechanics, electronic information, optics, heat energy, material, [and] control, [and] can drive multi-disciplinary synthesis development, making it a perfect domain for colleges and universities to enter."[25] On September 16, 1998, Tsinghua established the Aerospace Technology Research Center/ Tsinghua Space Research Center (TSRC) to pursue (in part) microsatellite development, and, as of 2003, the center already had "a 300 square meter super-clean room with ten thousand–level cleanliness, a 1200 square meter research and test base," a number of laboratories, and fifty graduate students.[26]

In June 2000, Tsinghua University Enterprise Group joined China Aerospace Science and Industry Corporation (CASIC) and Tsinghua Tongfang Limited Company to fund and jointly establish Aerospace Tsinghua Satellite Technology Company, Ltd. In September 2001, China Yintai Investment Company became the fourth shareholder by providing venture capital.[27] Aerospace Tsinghua is among China's first satellite development and manufacturing companies to establish itself in accordance with modern industrial practices. It shares many features with typical Western businesses, including accounting and logistics management, research and development flowsheets, and an ethos of standardized management; countless documents outline everything from purchasing to quality control to technical specifications; and the firm passed its ISO 9000 review in 2003. As a result of these efforts, the firm claims that "there has not been any quality problem during any satellite launch process of the company."[28]

Also reminiscent of best practices in the West, Aerospace Tsinghua employees are carefully selected and mentored. Their average age in 2004 was under thirty-one. Employees can choose a department in which to work, based on their personal interests and aspirations, and they are given promotions with reasonable frequency if they perform well. Based on its core principles of market orientation, economies of scale, and integrating market demand with technological development, Aerospace Tsinghua has reportedly "developed small satellites and related products on the basis of international and domestic market needs."[29]

Founded by CAST and its parent company, CASC, in August 2001, Aerospace Dongfanghong Satellite Company, Ltd. (DFH Satellite Company) is China's foremost satellite manufacturer. The firm is engaged in the research and design of small satellites and microsatellites, and the system design and production R&D of satellite application projects. DFH Satellite Company works closely with China Space Technology Research Institute and the Fifth Academy. In 2003, General Manager Li Zuhong, a leader of the small satellite development group and formerly vice president of the Fifth Academy, reported that the company had more than 120 personnel, 80 percent of whom held postsecondary degrees. According to *China Aerospace News*, DFH Satellite Company has implemented innovative corporate management methods.[30] If true in practice, this human resource approach indicates both unusual personnel efficiency for China (a breakthrough also reported in the Shenzhou piloted spaceflight program) and high project priority. Such horizontal coordination represents a potent model that might be gradually introduced throughout China's defense-industrial sector. In a key example of the commercial aspect of Chinese satellite development, DFH Satellite Company is also said to have established a joint venture with a Japanese high technology enterprise based in China, and intends to "gradually seize [a portion of] the small satellite market."[31]

In late 2004, DFH Satellite Company completed construction of the China National Engineering Research Center on Small Satellites and Applications in northwest Beijing.[32] Billed as the world's largest small satellite facility, the center has the capacity to build six to eight microsatellites per year. Given the current trajectory of China's satellite construction, this target seems to be readily attainable. According to *People's Daily*, the center "will strengthen the cooperation with foreign and Chinese institutions, [thereby] promoting the industrialization of microsatellites."[33]

Microsatellite Projects

This section surveys some of the latest microsatellites that China has developed. (Appendix 14.1 contains a complete list with technical details of all Chinese small satellite projects.) Chuangxin-1 (CX-1), or "Innovation," was successfully launched aboard a Long March 4 rocket from Taiyuan on October 21, 2003. During the first fifty-five days of its orbit, it successfully endured "two solar gales and . . . securely resisted twenty-three single-particle motions."[34] China's first modern microsatellite with a weight of less than 100 kg is also its first low-orbit digital communications satellite. It stores and rebroadcasts data through two-way communication with ground terminals in Shanghai, Beijing, Xinjiang, and Hainan. The results of various experiments "showed that the entire system operated well and met the requirements to enter the user trial phase."[35] The microsatellite was produced by the CAS, which initiated its development in 1999 as part of its Knowledge Innovation Project. The satellite's objective consists of electronic memory, transmission, and communications experiments. The payload has electronic memory, processing, and transmission (store and forward) functions.[36] It has been credited with achieving "breakthrough progress in low-orbit communications technology, monitoring, and control; design of shared communications channels for communications businesses; satellite in-orbit forecasts; satellite self-management and operation; and miniaturized satellite communications terminals."[37] CX-1 was launched with the second China-Brazil Earth Resources Satellite (CBERS 2).[38]

Naxing-1 was launched with SY-1. At 25 kg, Naxing-1, or "Nano-satellite," made China the fourth country (after Russia, the United States, and the United Kingdom) to launch a satellite approaching nano-satellite designation (10 kg or less). Hailed as an important breakthrough by the Chinese press, it enabled China to enter the international arena of small satellite research. NX-1 was produced by Tsinghua University and its subsidiary, Aerospace Tsinghua Satellite Company, and was designed to conduct high-technology experiments. NX-1's "primary missions" included using a CMOS camera to conduct image formation experiments. Other experiments involved miniature inertia survey, orbit maintenance and change, software upload procedures, and partial primary devices. As this technology matures, it will be used in applications such as optical image formation and environmental and meteorological observation. NX-1's designers claim that it is not only the smallest three-axis satellite in the world, but that it is also "China's first satellite having software uploading capabilities."[39]

On September 27, 2008, China launched, monitored, and controlled a satellite from a spacecraft for the first time.[40] The Shenzhou 7 spacecraft spring-launched the Banxing ("Companion") microsatellite (also called BX-1). The Shanghai Institute of Technical Physics, working under the CAS, developed and delivered BX-1 in less than three years.[41] Reportedly related to predecessor Chuangxin 1 (launched in 2003), the 0.4 m, cube-shaped satellite has a payload of less than 10 kg, and includes systems to support three missions: the in-orbit release experiment; photography of the Shenzhou 7; and a subsequent change of orbit to "chase" the orbit module. After being launched, the BX-1 flew around the Shenzhou 7 and photographed it with CCD cameras before moving 100 to 200 km away under control of a ground tracking station. This ground tracking station "measure[ed] the relative distance between the companion satellite and the orbit module." After the astronauts returned to Earth on September 28, the researchers at the ground flight control center were able to direct the small companion satellite to "chase" the orbit module, catch up to it, and enter into an elliptical orbit around it.

Zhu Zhencai, BX-1's chief designer and a researcher at the Shanghai Microsatellite Engineering Center, stated that "the working life that was originally expected is three months at the least. Therefore, we need to perform some more technical experiments, including Earth observation and further or long-term evaluation of the orbit."[42] Future experiments aside, the BX-1

performed admirably in its first journey, surpassing overall mission requirements (for example, making upward of twenty revolutions around the orbit
module, when just three revolutions would have qualified as a "success"),
and a Chinese news report concluded, "The success of [BX-1] lays a foundation for in-orbit troubleshooting and support for large spacecraft. The functions and applications of spacecraft can be extended and broadened and . . .
will also provide useful experience for the rendezvous and docking of spacecraft in the future."[43]

China's Small Satellite Buses: Indicators of Mass Production Ability

In a development that mirrors Western efforts to reduce costs and enhance
reliability, satellite buses or standardized platforms will constitute the backbone of China's future microsatellite efforts. China is developing at least five
variants of three major small satellite buses: CAST968A, B, and C; CAST2000;
and CASTMINI (for true microsatellites). By analyzing the performance
parameters of China's small satellite buses, future researchers may discern
for which combination of capabilities and missions China's new generation
of microsatellites have been optimized. In turn, this will provide insights into
China's true space interests and intentions.

CAST968. DFH Satellite Company has developed CAST968, China's first
small satellite bus.[44] Already applied to several satellites, its operating performance has reportedly been extremely stable and reliable.[45] According to
CAST, the bus "has strong expandability suitable for various payloads" and
can be configured for "different kinds of missions from [low and medium
Earth] orbits, as well as single satellite, multisatellite, and piggyback launch
missions."[46] These missions include Earth and ocean observation, space science, communications, engineering tests, reconnaissance, and surveillance.
CAST968's payload mass is roughly 30 to 60 percent of the platform mass. It
is designed with a network-based system to manage and control multiple
missions and resources simultaneously, and is able to support a variety of
attitude control modes. In accordance with international standards, it
employs a USB measurement and control system. CAST968 is powered by a
combination of solar and battery power. Its primary thermal system is passive, while its secondary system is active. DFH Satellite Company provides
technical support and states that CAST968 will be used successfully with a
variety of new satellites. The CAST968 bus is divided into three variants—A,
B, and C—each more advanced than the last. The CAST968A bus was the first
created by CAST, in August 1996. The mission of the CAST968 bus program
was to use mature technology and equipment, integrated systems, and computer software to create a platform that could support technological development and advanced experiments. Progress in these areas was important:

decision makers wanted to raise performance requirements and avoid catastrophic system failure. The successful deployment of the SJ-5 small satellite's payload in 1999 determined that the CAST968A bus had satisfied performance targets and user requirements.

The CAST968B bus was created to satisfy Chinese customer demands and to achieve true interchangeability, seriation, and "combinationalization."[47] These concepts, which are also gaining popularity in other areas of China's defense sector, are critical to the dual-use approach to modern defense-industry development in China. CAST968B was developed primarily to support HY-1, and, accordingly, CAST968B's propulsion system was enlarged to satisfy HY-1's payload work requirements. This enabled HY-1 to achieve orbit change and maintenance capability. Attitude control precision was increased, as were solar cell area, "segment transfer efficiency," and platform output power.[48] CAST968B abandoned its predecessor's driven-type solar array in favor of a more advanced version capable of automatically tracking the sun.[49] The GPS location system and satellite orbit self-stabilization capability were also augmented.

Analysis of Chinese and foreign markets for small satellites, and ongoing demonstration work concerning various kinds of small satellite bus requirements, determined that CAST968A and B could still not meet comprehensively high performance payload application demands. For instance, Earth observation satellites need to carry such diverse payloads as high-spectrum and high-resolution cameras, synthetic aperture radar (SAR), and microwave/optical remote sensing equipment. For this reason, the CAST team decided once again to revamp their product to meet the demands of current end-users; CAST968C was developed with the goal of meeting all types of high-resolution, high-performance, and multiuse payload requirements. China apparently plans to use this bus in future advanced satellites. HJ-1A and 1B, the initial members of China's first satellite constellation, all reportedly use the CAST968 bus.

CAST2000. DFH Satellite Company has already started to develop the CAST2000 bus to achieve small satellite volume production capability.[50] According to the China National Aerospace Information Center (CNAIC), CAST2000 was used in the SY-2 small satellite launched on November 18, 2004; if this is true, SY-2 would be the first satellite to use the recently developed CAST2000 bus. The bus enabled SY-1 to demonstrate cutting-edge technology, including high accuracy control, integrated management, highly effective power sourcing, multipurpose structural technology, enhanced attitude and control precision, standardization, network capabilities, high-speed information exchange, new power sources, and new control-system technology.

The CAST2000 bus reportedly delivers improvements in antimagnetic, vibration, and radiation protection; system expansion ability; and multimission adaptability. In another indication of commercial motivations being behind much of China's microsatellite development, CNAIC emphasizes that CAST2000 is world-class, meets international standards, uses commercial data-storage components extensively, and is intended for eventual export. As indicated earlier in this study, the buses represent a breakthrough in technological development under the aegis of a completely new management model for China's strategic technological industry. Organizationally, DFH Satellite Company is divided between the CAST968 and CAST2000 satellite buses, with separate but parallel chains of command.[51]

Conclusion

While I have found that Chinese publications often use florid expressions and exaggerate capabilities by emphasizing relatively minor achievements out of context and overlooking deficiencies, the overall scale and form of China's microsatellite developments suggest a significant increase in satellite capabilities. This should surprise no one: China's microsatellite development and production appear to be part of a broader pattern of defense-industry development that serves both commercial and military purposes. China has demonstrated key successes, albeit with substantial foreign assistance. In doing so, Beijing is demonstrating the value of a development paradigm that differs significantly from that of the United States. Many Chinese microsatellites appear to be dual-use in capability, whereas U.S. satellites are more strictly segregated (the U.S. military's extensive use of commercial satellite imagery being a significant exception). Whereas China encourages technology transfer from Europe, Israel, Brazil, and Russia to its benefit, America has alienated even some of its closest allies, and forfeited key sales opportunities, with its unyielding approach to export controls. A great beneficiary of these transfers is the European Space Agency, which is in high demand as a partner around the world, and particularly in China. By prohibiting the use of U.S. components in foreign satellites not cleared through the stringent International Traffic in Arms (ITAR) process, Washington gives Europe and China powerful incentives to develop their own. This raises the distinct possibility, at some point in the future, of Europe (and perhaps even, someday, China) setting international standards to its advantage.

Those who maintain that the U.S. political system precludes a more effective dual-use approach and that national security concerns make export controls nonnegotiable should nevertheless consider a third area, over which the United States wields undisputed control—its own satellite development approach. While U.S. satellites are unquestionably more sophisticated than their Chinese counterparts (and are likely to remain so for some time to come), the U.S. tendency to create ever larger and more expensive satellites surely has drawbacks. With the proliferation of space debris and anti-satellite weapons, it is unwise to concentrate so much economic and strategic value in a single location with little possibility of timely replacement. For at least some space applications, the use of some configuration of small/ microsatellites (with replacements perhaps cycled through more frequently, even in nonemergency situations) could have significant benefits. In addition to the obvious security and potential cost savings, this approach could help satellites fall less far behind technologically during long in-orbit lives, and even reach orbit with a better technology freshness-to-cost ratio thanks to more extensive use of commercial off-the-shelf (COTS) technologies.

Now is the time to act, before mounting concerns about threats to satellites trigger a major increase in expensive and cumbersome countermeasures that offer only marginal security benefits. Perhaps best of all, the resulting standardization, increase in development tempo, commercial interrelation, and ability to take risks could make the U.S. satellite industry far more competitive and versatile technologically. This process holds inherent challenges, with corresponding improvements in launchers being an important component. But in an era in which the U.S. government may not be able to afford a purpose-built designer solution to every space application, it may be the only truly sustainable way forward. China's microsatellite program has taken many lessons from U.S. programs; it would be the height of conceit and folly for Washington to imagine that it has nothing to learn from Beijing.

NOTES

The views expressed in this essay are those of the author alone and do not represent the policy or estimates of the U.S. Navy or any other element of the U.S. government. The author thanks Lyle Goldstein, Joan Johnson-Freese, William Martel, Anthony Mastalir, Oriana Mastro, and Kathleen Walsh for their incisive comments.

1 It is worth noting that while both Western and Chinese sources categorize satellites by weight, they often use terms in an overlapping fashion and even interchangeably. Unless otherwise specified, this study will use the term "microsatellite" as a collective term meaning any satellite weighing less than 100 kg. Because satellites are categorized by weight, some of China's earliest and least sophisticated satellites are still considered to be "microsatellites." PRC satellites are often given multiple names, and sometimes renamed, which has led to inconsistency in Western reports. This study will use designations currently used by U.S. government analysts and attempt to resolve ambiguity wherever possible.

2 Lin Laixing, Beijing Institute of Control Engineering, "Study on Microsatellite Application in Space Attack and Defense Overseas," *Journal of the Academy of Equipment Command & Technology* 17(6) (December 2006): 47–49 (original in Chinese); Chen Yi, "What Is the Meaning of 'Microsatellite'?" *Outer Space Exploration* (October 2003).

3 This program has also been referred to as Project 651. See Stephen J. Isakowitz et al., *International Reference Guide to Space Launch Systems*, 4th ed. (Reston, Va.: American Institute of Aeronautics and Astronautics, 2004), 261.

4 Iris Chang, *The Thread of the Silkworm* (New York: Basic Books, 1995), 225.

5 Yu Yongbo et al., *China Today: Defense Science and Technology* (Beijing: National Defense Industry Press, 1993), 1:30; "DFH-1," accessed March 27, 2011, http://www.globalsecurity.org/space/world/china/dfh-1.htm.

6 Fang Zongyi, Xu Jianmin, and Guo Lujun, "The Development of China's Meteorological Satellite and Satellite Meteorology," in *Space Science in China*, ed. Hu Wenrui (Amsterdam: Gordon and Breach Science Publishers, 1997), 239.

7 See, for example, Tu Shancheng, "Space Technology in China: An Overview," in Hu, *Space Science*, 15; also see Chang, *Silkworm*, 226.

8 Isakowitz et al., *International Reference Guide*, 261.

9 The Chinese designations are 东方红一号 and 长征一号, respectively. CZ-1 was developed from the DF-4 missile. Ibid.

10 Yu Yongbo et al., *China Today*, 1:98; Roger Cliff, *The Military Potential of China's Commercial Technology* (Arlington, Va.: RAND, 2001), 28. Previous nations were the Soviet Union, the United States, France, and Japan.

11 Shi Jianzhong, "The Song 'East Is Red' Was Transmitted from Space: Recording the Process of Developing 'Dong Fang Hong' 1's Shortwave Transmitter" (original in Chinese), *Aerospace Industry Management* 5 (2005): 12–13.

12 For further information on DQ-1A and DQ-1B, see Hu Wenrui, "Space Science in China: Progress and Prospects," in Hu, *Space Science*, 5.

13 "SJ-2," accessed March 27, 2011, http://www.globalsecurity.org/space/world/china/sj-2.htm; "DQ-1," accessed March 27, 2011, http://www.globalsecurity.org/space/world/china/dq-1.htm.

14 Gu Songfen et al., eds., *The History of World Space Science Development* (original in Chinese) (Zhengzhou: Henan Science and Technology Press, 2000), 272; Editorial Committee, *The Soaring Journey of Chinese Spaceflight* (original in Chinese) (Beijing: China Literature and History Press, 1999), 418.

15 Min Guirong, "Review of Chinese Space Programs," in *Devotion to Spaceflight in the Benefit of Humanity: The Collected Works of Academician Min Guirong* (original in Chinese) (Beijing: China Aerospace Publishing House, 2003), 189.

16 See, for example, Guo Huadong, ed., *Radar Remote Sensing Applications in China* (New York: Taylor & Francis, 1999); Min Guirong, "Spin-off from Space Technology in China," paper presented to the forty-third Congress of the International Astronautical Federation, IAF-92–0180, essay in Min, *Devotion*, 253.

17 Zhuang Yaoli and Wang Hualong, "Modern Microsatellite Communication," *Modern Communication* (February 2002).

18 Yu Yongbo et al., *China Today*, 1:176.

19 Ibid., 2:889.

20 Li Zhizhong, Wang Yongzhang, and Xu Shaoyu, "Microsatellites' Earth Observation and Application Prospects," *Territory, Natural Resources, and Remote Sensing* (April 2004).

21 Unlike Chinese aircraft, for instance, satellites do not have passenger safety requirements and can be usefully tested at any stage of development or sophistication.

22 See, for example, Chiang Yuen-p'ing, "The ROC's Countermeasures to the PRC's Satellite Development" (original in Chinese), *Navy Studies Monthly* (April 2005), OSC# CPP20070402312004; and Huang Bencheng and Ma Youli, eds., *Spacecraft Space Environment Test Technology* (original in Chinese) (Beijing: National Defense Industry Press, 2002), 235.

23 For more information, see Surrey Space Technology Limited's website, accessed March 27, 2011, http://www.sstl.co.uk/.

24 "Why Must Tsinghua University Research and Develop Microsatellites?," *Aerospace China* (January 1999).

25 Sun Diqing, "'Aerospace Tsinghua-1' Microsatellite," *Modern Physics Knowledge* (March 2001).

26 Gong Ke, Vice President, Tsinghua University, "Bring Cooperation into Bloom, Stride Forward with Development—Spaceflight Science and Technology at Tsinghua" (original in Chinese), in *China Spaceflight Moves into the World: Commemorating the Tenth Anniversary of the Founding of China National Space Administration*, ed. Luo Ge (Beijing: China Space Navigation Press, 2003), 257.

27 Xia Guohong, You Zheng, Meng Bo, and Xin Peihua, "An Attempt by China's Spaceflight Enterprise System to Blaze New Trails—Aerospace Tsinghua Satellite Technology Company" (original in Chinese), *Aerospace China* (August 2002): 11–13.

28 Unless otherwise specified, this and the following four paragraphs are derived from "Hangtian Tsinghua Satellite Co. Approaches Satellite Market," *Zhongguo Hangtian Bao* (May 28, 2004): 4, OSC# CPP2004061000201.

29 Xia Guohong et al., "Attempt."

30 Unless otherwise specified, data in this and the following paragraph are derived from Zhao Shanshan, "Four Small Satellites Will Be Launched in the Next Ten Months" (original in Chinese), *China Aerospace News*, December 24, 2003, accessed March 27, 2011, http://www.chinaspacenews.com/News/news_detail.asp?id=7670; Zhao Can, "What Will Hangtian Dongfanghong Satellite Company Rely On to Grow Into Adulthood," *Zhongguo Hangtian Bao* (December 24, 2003): 4, OSC# CPP2004010900125. *China Aerospace News* is published by CAST and CASIC.

31 Zhao Shanshan, "Four Small Satellites"; Zhao Can, "What Will Hangtian Dongfanghong Satellite Company Rely On to Grow into Adulthood?" *Zhongguo Hangtian Bao*, December 24, 2003, 4, OSC# CPP2004010900125.

32 Zhao Can, "Adulthood."

33 "Small Satellites a Big Deal in China," *CNN Science & Space*, December 15, 2004.

34 "'Chuangxin 1' Satellite Has Been Operating for 55 Days," *Jiefang Ribao* (December 15, 2003), OSC# CPP20031215000149.

35 Sun Zifa, "PRC 'Innovation-1' Minisatellite Enters Users Trial Phase," *Zhongguo Xinwen She* (February 19, 2004), OSC# CPP20040219000127.

36 Li Bin, "Chinese Academy of Sciences Has Established a Research and Manufacturing Base for Modern Microsatellites," *Xinhua Domestic News Service* (February 19, 2004), OSC# CPP20040219000126.

37 Sun Zifa, "PRC 'Innovation-1.'"

38 Stephen Clark, "Earth Monitoring Satellite Launched by China and Brazil," *Spaceflight Now* (October 21, 2003), accessed March 27, 2011, http://www.spaceflightnow.com/news/n0310/20cbers2/.

39   "Hangtian Tsinghua Satellite Co. Approaches Satellite Market," *Zhongguo Hangtian Bao* (May 28, 2004): 4, OSC# CPP2004061000020I.

40   Unless otherwise specified, data in this section are derived from/corroborated with the most extensive report on the subject yet published, David Wright and Gregory Kulacki, "Chinese Shenzhou 7 'Companion Satellite' (BX-I)," Union of Concerned Scientists, October 21, 2008, accessed March 27, 2011, http://www.ucsusa.org/assets/documents/nwgs/UCS-Shenzhou7-CompanionSat-10-21-08.pdf.

41   "China's Shenzhou-7 Launches Small Monitoring Satellite," *Xinhua* (September 27, 2008), OSC# CPP20080927968233.

42   Gao Lu, "Special Report: Shenzhou-7 Flight Partner Unveiled," *Xinhua* (September 27, 2008), OSC# CPP20080928338004.

43   "Accompanying Satellite of Shenzhou-7 Achieves Orbiting of Orbital Module," Military Report newscast, CCTV-7, October 6, 2008, OSC# CPM2008II28017008; see also Military Report, CCTV-7, September 28, 2008, OSC# CPM2008I0I00 5I004.

44   In Chinese, the CAST968 bus is referred to as "小卫星公用平台" or as "卫星系列平台."

45   "From 100 to 1000 kilograms, Our Country Is Moving toward Small Satellite Manufacturing Seriation" (original in Chinese), November 24, 2003, accessed March 27, 2011, http://www.spacechina.com/index.asp?modelname=nr&recno=6574.

46   Unless otherwise specified, information in this paragraph is derived from "Small Satellite Platform," Chinese Academy of Space Technology, www.cast.cn/en/ShowArticle.asp?ArticleID=85. CAST's English-language website, accessed March 27, 2011, can be found at http://www.cast.cn/CastEn/index.asp.

47   Unless otherwise specified, information in this paragraph is derived from "CAST968B Platform Synopsis" (original in Chinese), China Academy of Space Technology, accessed March 27, 2011, http://www.cast.cn/cpyyy/cp.htm.

48   The original Chinese term for "solar transfer efficiency" is "电池片转换效率."

49   The original Chinese term for "solar array" is "太阳帆板."

50   The Chinese designation for the CAST2000 bus is CAST2000 "小卫星公用平台."

51   Zhao Shanshan, "Four Small Satellites," http://www.china-spacenews.com/News/news_detail.asp?id=7670; and Zhao Can, "Adulthood."

— Appendix 14.1 PRC Small Satellite Projects —

Abbreviation/ English Designation	Chinese name	Equipment/Function	Weight (kg)	Manufacturer	Launch Date (PRC time)	Launcher, Site	Orbit
DFH-1 (Dong Fang Hong 1) East is Red 1	东方红一号 dōngfāng hóngyīhào	Radio transmitter	173		4/24/1970	Long March (LM)/ CZ-1 Jiuquan	439×238/ 68.5/114
SJ-1 (Shi Jian-1) Practice-1	实践一号 shíjiànyīhào	Measure space environment parameters, e.g., high-altitude magnetic field; magnetometer, particle detectors • cosmic and x-rays	221		3/3/1971	CZ-1 Jiuquan	266×1826/ 69.9/106
SJ-2 (Shi Jian-2) Practice-2	实践二号 A shíjiànèrhào	Detect solar activities, charged particles in near-Earth space, infrared & UV radiation background of Earth & atmosphere; beacon transmitter • ionosphere study	257	Beijing Institute of Spacecraft Systems Engineering	9/20/1981	FB-1 Jiuquan	237×1622/ 60/103 or 232/1598, 59.5?
SJ-2A (Shi Jian-2A)	实践二号甲 shíjiàn 2 èrhàojiǎ				9/20/1981	FB-1 Jiuquan	237×1622/ 60/103 or 232/1608, 59.4?

(continued)

Appendix 14.1 PRC Small Satellite Projects *(continued)*

Abbreviation/ English Designation	Chinese name	Equipment/Function	Weight (kg)	Manufacturer	Launch Date (PRC time)	Launcher, Site	Orbit
SJ-2B (Shi Jian-2B) Practice-2B	实践二号乙 shíjiàn 2 èrhàoyǐ	Passive radar calibration test • measure atmospheric density	28		9/20/1981	FB-1 Jiuquan	237×1622/ 60/103 or 232/1608, 59.4
STTW-2 (Shiyong Tongbu Tongxin Weixing) **DFH-2A** Zhongxing 2		Operational geostationary communications satellite	441		3/7/1988	CZ-3	
SY-1 (Shiyan 1)	试验卫星一号 shìyànwèixīng yīhào	Communications performance and new technology tests • "partially successful"	461		1/29/1984	CZ-3 Xichang	474×6480/ 36/161
SYTXW (Shiyan Tongxin Weixing)	试验通信卫星 shìyàntōngxìn wèixíng	After completing communications experiments, provided applications from fixed position above the equator at 125 degrees east longitude	461		4/8/1984	CZ-3 Xichang	35599× 35792.8/0.6 2/1431.5
SYTXGBWX-1 (Shiyong Tongxin Guangbo Weixing 1) **DFH-2**	实用通信 广播卫星一号 shíyòngtōngxìn guǎngbōwèixīng yīhào	On February 20, fixed position above the equator at 103 degrees east longitude	433		2/1/1986	CZ-3 Xichang	35783× 35792/ 0.09/1436

Name	Description	Mass	Date	Launch	Orbit
SYTXW-2 (Shiyong Tongxin Weixing 2) 实用通信卫星二号 shiyòngtōngxìn wèixīngèrhào	Operational geostationary communications satellite; on March 23, fixed position above the equator at 87.5 degrees east longitude	441	3/7/1988	CZ-3 Xichang	35786.4× 35862.6/ 0.07/1438
SYTXW-3 (Shiyong Tongxin Weixing 3) DFH-2A Zhongxing 3 实用通信卫星三号 shiyòngtōngxìn wèixīngsānhào	Operational geostationary communications satellite; on December 30, fixed position above the equator at 110.5 degrees east longitude	441	12/22/1988	CZ-3 Xichang	35782.5× 35790.2/ 0.56/1436.1
SYTXW-4 (Shiyong Tongxin Weixing 4) DFH-2A 实用通信卫星四号 shiyòngtōngxìn wèixīngsìhào	Operational geostationary communications satellite; on February 14, fixed position above the equator at 98 degrees east longitude		2/4/1990	CZ-3 Xichang	35783.3× 35797.8/ 0.11/1436.3
DQ-1A Da Qi-1 大气一号 dàqìyīhào	Research balloon • atmospheric density measurement	2.6	9/3/1990	CZ-4A Taiyuan	789/811, 99
DQ-1B Da Qi-2 大气二号 dàqìèrhào	Research balloon • atmospheric density measurement	3.3	9/3/1990	CZ-4A Taiyuan	596/629, 99
SJ-4 (Shi Jian-4) 实践四号 shíjiànsìhào	Acquired space environment parameters @ 20–36,000 km altitude, a PRC first • cosmic ray detection	400	2/8/1994	CZ-3A Xichang	200× 36000/ 28–28.5? or 36092/28.2

(continued)

Appendix 14.1 PRC Small Satellite Projects (continued)

Abbreviation/ English Designation	Chinese name	Equipment/Function	Weight (kg)	Manufacturer	Launch Date (PRC time)	Launcher, Site	Orbit
SJ-5 (Shi Jian-5) Practice-5	实践 5 号 shíjiànwǔhào	Magnetosphere research, space-charged particle measurement, S-band high-speed data-link transmitter tests, large-capacity solid-state storage test, fluid science experiments • first PRC small scientific experiment satellite designed based on common bus (CAST968A)	300/ 298*	Joint development with Brazil Aerospace by East is Red Corporation (est. 2001) of China Academy of Space Technology (CAST); and China Academy of Sciences	5/10/1999	CZ-4B Taiyuan	849×868/ 98.79/ 102.11
HTQH-1 (Qinghua-1)	航天清华一号 qīnghuáyīhào	Earth observation, communications • Bus: SSTL Microsat-70	49/50	Built by Surrey Satellite Technology Limited Corp (SSTL) for Tsinghua/Surrey University	6/28/2000	Plesetsk, Russia	683×706/ 98.13/98.66
HY-1A (Haiyang-1A) Ocean 1-A	海洋 1 号 A hǎiyángyīhào	Marine remote sensing • CAST968B Bus	365	CAST/DFH Aerospace Corporation Components from Satlantic (Canada) and CIMEL (France)	5/15/2002	CZ-4B Taiyuan	793×798 km fixed orbit, 98.80°

HTQH-2/ HTSTL-1/ KT-1PS (Tsinghua-2)	航天清华二号 qīnghuáèrhào	Multispectral Earth imaging, experimental communications payload	50/ 35.8/34	SSTL?/PRC partner	9/15/2002	Kaituozhe-1 Taiyuan	Failed; second stage malfunction before could attain 300 km polar orbit
PS-2		Guidance system, fairing separation and satellite-launcher separation succeeded; fourth stage failed to ignite	40		9/16/2003	Kaituozhe-1 Taiyuan	300 km × 300 km polar orbit
CX-1 (Chuangxin-1) Innovation-1	创新 1 号 chuàngxīnyihào	Digital store and rebroadcast communications •data retransmission	75–99	CAST, Chinese Academy of Sciences, Shanghai (?), Shanghai Academy of Space Technology, Shanghai Telecommunications	10/21/2003	CZ-4B Taiyuan	
DSP-E (Double Star Equator)	探测一号 tàncèyīhào	Plasma science	330	European Space Agency (ESA)/CAST, China Aerospace Science and Technology Corp.	12/30/2003	CZ-2C/CTS Xichang	

(continued)

— Appendix 14.1 PRC Small Satellite Projects (continued)

Abbreviation/English Designation	Chinese name	Equipment/Function	Weight (kg)	Manufacturer	Launch Date (PRC time)	Launcher, Site	Orbit
SY-1 (Shi Yan-1) Experimental Satellite 1 or, Explorer-1	实验卫星一号 shíyànwèixīng yīhào / 探索1号 tànsuǒyīhào	Optical remote sensing land resource survey; stereo mapping: • 10 m resolution observation capacity	204/250*	Harbin Institute of Technology, CAST, China Space Technology Research Institute, Chinese Academy of Sciences, Changchun Light Technology Institute, Xi'an Mapping Research Institute, Astrium?	4/18/2004	CZ-2C	Polar orbit
NX-1 Nano-satellite 1	纳星1号 nàxīngyīhào	Hi-tech experiments: • CMOS camera, inertia survey, data transmission, remote sensing photography, attitude control, track maintenance and axial change	<=25	Tsinghua University, Aerospace Tsinghua Satellite Co. Ltd.	4/18/2004	CZ-2C	
DSP-P (Double Star Polar)	探测二号 tàncèèrhào	Plasma science	270*	EuropeanSpace Agency (ESA)/PRC	7/25/2004	CZ-2C Taiyuan?	
SJ-6A/6-01B (Shi Jian-6A) Practice-6A	实践六号 A shíjiànliùhào A			Prime Contractor: DFH Satellite Co., Ltd., CAST	9/8/2004	CZ-4B Taiyuan	578 km × 593 km, 97.7°

SJ-6B/6-01A (Shi Jian-6B) Practice-6B 实践六号 B shíjiànliùhào B	Probe cosmic environment, radiation, related space experiments, • ELINT technology • CAST968 Bus	350	*Prime Contractor:* DFH Satellite Co., Ltd., CAST *Operator:* China Aerospace Science and Technology Corp.	9/8/2004	CZ-4B Taiyuan	593 km × 602 km, 97.7°
SY-2 (Shi Yan-2) Experimental Satellite 2 实验卫星二号 shíyànwèixīng èrhào	Satellite technology testing, land, resources, and environmental surveying from sun synchronous 700 km orbit • Bus: CAST2000	300	*Operator:* DFH Satellite Company *Prime:* CAST	11/18/2004	CZ-2C Xichang	
SJ-7 (Shi Jian-7) Practice-7	Scientific experiments; nonrecoverable satellite	???		7/5/2005	CZ-2D Jiuquan	547 km × 580 km, 97.6°
BJ-1 Beijing-1 北京一号	Earth observation • 4m, image, 32 m resolution imager in 3 spectral bands • Bus: SSTL Microstat-100 (enhanced) • Resistojet propulsion	166	*Operator:* Beijing Landview Mapping Information Technology Ltd. (BLMIT) *Contractor:* SSTL	10/27/2005	Kosmos-3M Plesetsk, Russia	
SJ-6C/6-02B (Shi Jian-6C) Practice-6C			DFH Satellite Co., Ltd., CAST	1/24/2006	CZ-4B Taiyuan	
SJ-6D/6-02A (Shi Jian-6D) Practice-6D	CAST968 Bus		DFH Satellite Co., Ltd., CAST	1/24/2006	CZ-4B Taiyuan	

(continued)

Appendix 14.1 PRC Small Satellite Projects (*continued*)

Abbreviation/ English Designation	Chinese name	Equipment/Function	Weight (kg)	Manufacturer	Launch Date (PRC time)	Launcher, Site	Orbit
HY-1B (Haiyang-1B) Ocean-1B	海洋 1 号 B hǎiyángyīhào	Ocean color mapping • 10-band ocean color scanner • 4-band CCD imager/ 250 m resolution • infrared water profile radiometer		CAST	4/11/2007	CZ-2C	782×815/ 98.60°/ 100.80 min
HJ-1A (Huanjing-1A) Environment-1A	环境一号 A huánjìngyīhào A	In constellation w/ HJ-B/C; to be 1 of 8 disaster reduction and environmental monitoring constellation satellites • 4 cameras: 2 CCD cameras w/ 30 m resolution & 700 km breadth; IR camera w/ 150 m (near, center infrared) resolution & 720 km breadth; & hyperspectral imager w/ 100m/50km resolution/ breadth & spectrum 5 nm resolution	470	DFH Satellite Co.*	9/6/2008	LM-2C Taiyuan	650 km sun synchronous orbit; 48-hour revisit interval for China & surrounding area

HJ-1B (Huanjing-1B) Environment-1A	环境一号 B huánjìngyīhào B	In constellation w/ HJ-B/C; to be 1 of 8 disaster reduction and environmental monitoring constellation satellites • 4 cameras: 2 CCD cameras w/ 30 m resolution & 700 km breadth; IR camera w/ 150 m (near, center infrared) resolution & 720 km breadth; & hyperspectral imager w/ 100m/50km resolution/ breadth & spectrum 5 nm resolution	470	DFH Satellite Co.*	9/6/2008	LM-2C Taiyuan	650 km sun synchronous orbit; 48-hour revisit interval for China & surrounding area
BX-1 (Banxing-1) Companion Satellite	伴飞小卫星/伴星 bànfēixiǎo wèixīng/bànxīng	2 CCD cameras, orbital maneuvering, liquid ammonia propellant • 0.4m cube	30–40	Shanghai Institute of Technical Physics, under the Chinese Sciences	9/27/2008	Shenzhou 7 spacecraft	
SJ-6E/6-03B (Shi Jian-6E) Practice-6E	实践六号 E shíjiànliùhào	2 yr.+ design life • optical camera • survey space environment, space radiation environment and its effects, parameters of physical space environment, space experiments • CAST 968 or FY-1 bus	~300	SAST group in Shanghai or the DFH Co. in Beijing	10/25/2008	LM-4B Taiyuan	580×604 ×97.7°

(continued)

— Appendix 14.1 PRC Small Satellite Projects (*continued*)

Abbreviation/ English Designation	Chinese name	Equipment/Function	Weight (kg)	Manufacturer	Launch Date (PRC time)	Launcher, Site	Orbit
SJ-6F/6-03A (Shi Jian-6F) Practice-6F	实践六号 F shíjiànliùhào	2 yr.+ design life • optical camera • survey space environment, space radiation environment and its effects, parameters of physical space environment, space experiments • CAST 968 or FY-1 bus	~300	SAST group in Shanghai or the DFH Co. in Beijing	10/25/2008	LM-4B Taiyuan	580×604 ×97.7°
Chuangxin 1-02	创新一号 02 星 chuàngxīnyīhào 02 xīng	Collect and relay hydrological and meteorological data and data for disaster relief		Chinese Academy of Sciences	11/5/2008	LM-2D Jiuquan	
SY-3 (Shi Yan-3) Experimental Satellite 3	试验卫星三号 shìyànwèixīng sānhào	Experiments on new technologies in atmospheric exploration		Harbin Institute of Technology	11/5/2008	LM-2D Jiuquan	
SJ-11-01		Technology demonstration		*Major Contractor:* ADSC *Operator:* CASC	11/12/2009	LM-2C Jiuquan LA-4, SLS-2	Sun-synchronous; 699.9× 690.5× 98.3°

XW-1 (Xiwang-1) Hope-1	China's first public science satellite; provides radio amateurs w/ space communications functions; beacon voice & data transmission, terrestrial panoramic color imaging & observing state of 5-color soil grains in microgravity environment		12/15/2009	
SJ-12	Scientific research; rendezvous technology & satellite inspection; rendezvous w/ SJ-6F 2010.08; possible physical contact	SAST	6/15/2010	CZ-2D Jiuquan
SJ-6G/6-04B (Shi Jian-6G) Practice-6G		DFH Satellite Co., Ltd., CAST	10/6/2010	CZ-4B Taiyuan
SJ-6H/6-04A (Shi Jian-6H) Practice-6H	CAST 968 bus	DFH Satellite Co., Ltd., CAST	10/6/2010	CZ-4B Taiyuan

Note: "Small satellites" are defined here as those weighing less than 500kg. Data in appendix 14.1 are derived from sources cited elsewhere in text. The symbol * indicates that data are uncertain.

15

Disjecta membra, the Kármán Line, and the 38th Parallel

James Schwoch

Disjecta membra: An alteration of Horace's disjecti membra poetae *"limbs of a dismembered poet," used = Scattered remains.*
—Oxford English Dictionary

The lines of national frontiers derive from chance and accidents of history; they have distributed natural resources and men without the slightest economic or global rationality.
—François Perroux

Using a National Geographic *map, we looked just north of Seoul for a convenient dividing line but could not find a natural geographic line. We saw instead the 38th parallel and decided to recommend that.*
—Dean Rusk

Between launch and orbit appear disjecta membra. The scattered remains of rocket stages, boosters, thrusters, canisters, tanks, and other odds and ends of launch-to-orbit debris appear to represent a trail of afterthoughts for most observers of outer space activities. Lacking the visual and sonic spectacle of blastoff, and devoid of the excitement and utility from either machine or human activities in outer space that commence once the desired height above Earth is reached, the debris in between launch and orbit is a lacuna for most people. Space debris is an empty signifier that is also a floating signifier—floating for a while, anyway, until it falls down to Earth, hopefully burned away into nothingness by friction and heat as the debris enters the inner atmospheric layers, with any surviving remnants cast into the oblivion of the deep blue sea.

Once in a great while, space debris gains widespread attention. Some sort of turn of events overtakes that floating and empty signifier of materiality strewn between launch and orbit and charges that material with potential meanings, signs, and interpretations. So there is, so to speak, a kind of fragmentary history of these fragments—a set of scattered narratives of scattered remains, a disjointed disjecta membra discourse. These scattered narratives, on the relatively few occasions they have emerged into mainstream awareness and popular discourse, usually are about objects falling from an orbit down to Earth; elsewhere in this volume, Lisa Parks takes up a few such cases in detail.

Even Sputnik, in 1957, had a famous piece of orbiting debris: as Patrick McCray notes in his book about Operation Moonwatch, *Keep Watching the Skies!*, the network of amateur skygazers who actively monitored satellite activities at the dawn of the space age, "In Indianapolis, scores of people called their local Moonwatch team leader after the rocket body accompanying the first Sputnik satellite appeared overhead."[1] The spectacular collapse-explosion-failure of the first U.S. satellite launch, the Vanguard attempt of December 6, 1957, produced nothing but disjecta membra about four feet above the launchpad while millions of viewers watched on television. On September 5, 1962, Sputnik 4 produced the infamous "Manitowoc fragment": about twenty pounds of Sputnik 4 landed in downtown Manitowoc, Wisconsin, creating local and national headlines. The Manitowoc city council passed a resolution officially naming the Sputnik 4 remnant the "Manitowoc fragment" and hoped to have it on permanent display in a local museum, but the fragment was returned by the United States to the USSR. Manitowoc now celebrates the event with a local Sputnikfest.[2] The de-orbiting and decay of two space stations—Skylab in July 1979 and Mir in 2001—each caused real consternation regarding the location of debris landing until both fell into the Pacific Ocean. Recent satellite shoot-downs and midspace collisions have added more lore to the debris tale. And unfortunately some of this scattered history is truly and humanely tragic, such as Space Shuttle and cosmonaut missions that ended in death and destruction.

What about the space debris that was never intended to reach orbit, that fragmentary trail of materiality between the launch of a rocket and the orbit of a satellite in outer space? For the most part, these scattered remains seem to draw little attention. I remember watching Mercury, Gemini, and Apollo missions on TV as a child, and seeing (thanks to excellent camerawork) the firing of the second (and sometimes the third) rocket stage on the way up beyond the ionosphere, but I gave little if any thought to the fate of those spent boosters, reassured by the conventional wisdom relayed by TV announcers that the boosters would burn up in the atmosphere far out over the Atlantic Ocean. I suppose I gave a little more thought to some of the most unusual space debris out there: used lunar exploratory modules (LEMs) and, by extension, other scattered remains on the moon and some other planets of the solar system. As I grew older, I became more aware of the incredible amount of space debris in Earth orbit—more scattered remains of that fragmented outer space history of a disjointed disjecta membra discourse.

As it turns out, some of those booster stages of my TV youth did not in fact burn up in the sky with remains landing in the Atlantic Ocean. Charred remains of rocket stages, boosters, and other debris from the Mercury astronaut series of the early 1960s routinely made landfall around the rim of the South Atlantic Ocean, from Brazil over to an area of Africa reaching from the Gold Coast to the Cape of Good Hope. These found objects were treated as— well, as something that was a challenge to explain, and even more of a challenge to signify with some sort of importance or relevance. Discussions at the time ranged from returning everything to the United States to placing the remnants in local or national museums to allowing the finders to sell them in local junkyards to—well, just ignoring these fragments, basically avoiding the articulations and discussions of this disjointed disjecta membra discourse that had no comfortable or recognized place as symbolic signs in the overarching global narrative of the space race.[3] These floating signifiers could not be signified as they fell down to Earth and thus basically remain vacant signs in telling the story of the space age.

Space debris remains, in the main, literally relegated to the scrap heap of history. Disjecta membra is therefore, as David Bowie sang, a space oddity: first like Major Tom, floating in a tin can, eventually becoming ashes to ashes strung out in heavens high. When in the course of a mission was space debris truly up in outer space, and when was it no longer in outer space but merely falling through the atmosphere back down to Earth? Where did outer space begin and end? During the Cold War, the question of where the atmosphere ended and outer space began was yet another ambiguous extraterritoriality. There are no international laws, United Nations treaties, or other forms of global legalities that define the border between the atmosphere and outer space. However, despite the vacuum of legal decrees, many flight and outer space enthusiasts eventually settled on the Kármán line as the functional and convenient border between the planetary atmospheres and outer space. The Kármán line was named in honor of its "inventor," Theodore Von Kármán. According to the Fédération Aéronautique Internationale (FAI), "In the early 1950s, Aeronautics and Astronautics were considered the same thing . . . [b]ut Von Kármán had the feeling that there was a difference between the two. If such was the case, a line could be defined to separate them."[4] The Kármán line is 100 km (62 miles) above the surface of the planet. The line is not recognized by any international treaties as a legal dividing line between the atmosphere and outer space, but is nevertheless used by many as a line of functional convenience.

Among the biggest advocates of the Kármán line are the members of the FAI, the best-known global organization devoted to the study and pursuit of air-based sports and explorations, including aeronautical and astronautical human (and unmanned) flight world records and breakthrough achievements. FAI certified, for example, the authenticity of the solo-pilot Charles Lindbergh New York–to–Paris 1927 flight. Ultimately, however, the Kármán line is an arbitrary convenience akin to lines of latitude and longitude, or the measurement of time via the frequency of a cesium oscillator (atomic clock). There is great sensibility in "Kármánizing" at 100 km above mean sea level, as that is the general area where the ability to orbit begins (above the line) and the tendency to decay begins (below the line). However, the Kármán line is not a nation-state or international legal standard codified by treaty or other international agreement. It is an extralegal line of functional convenience that emerged during the Cold War. For satellites and manned missions as well as for scattered fragments, rocket debris, and other disjecta membra space oddities, above the Kármán line it's Major Tom floating in a tin can, and below the Kármán line it's nothing but ashes to ashes and funk to funky.

The Kármán line is not the only line of functional convenience utilized on land, sea, and skies. Beyond the marked physical borders of nation-states, provinces, counties, cities, villages, and private property, many extraterritorial lines of functional convenience beyond political borders emerged in response to global and extraterritorial concepts, such as latitude and longitude; time zones; magnetic fields; metrological isobars; concepts of territorial integrity extended into oceans, seas, and the skies (such as 3-, 12-, or 200-mile offshore border concepts, sea lanes and air traffic corridors, or no-fly zones above nations); or the engineered signal-reception perimeters of terrestrial broadcast stations and of satellite footprints. These are lines of functional convenience because, in the main, they are not constrained or otherwise shaped by the conveniences and inconveniences of the planet's landmass or its relative habitability, but by the conveniences and inconveniences of templating the globe and then using those extraterritorial templates of functional convenience to enact some sort of system of discourse, knowledge, and power over the planetary environs. For example, it is not a physically driven requirement to have time "begin" at Greenwich, but rather the convenience of a long and often contentious process of social, scientific, and political shaping at global scales of discourse, knowledge, and power.

Sometimes the extraterritorial lines of functional convenience and the political lines of territorial demarcation are conflated. A line of extraterritorial functional convenience is thus transformed and does double duty as a political line of territorial demarcation. One of the longest nation-state borders in the world—the border between the United States and Canada—uses for much of that border the functional convenience of the 49th parallel north. The lack of a seemingly natural or geophysical demarcating line, particularly west of the Great Lakes ecosystem, led to the benign functional convenience of the 49th parallel for Canada and the United States (a benign functional convenience for the two federal governments, but not benign or convenient for the long-term Native American inhabitants of the Great Plains of North America). Or another example: the southwest corner of Egypt is the intersection of latitude 22° north and longitude 25° east—sharing much of 25° east as a border with Libya, and much of 22° north as a border with Sudan, part of which is contested by these two nations (the Hala'ib Triangle). Sometimes these transformed double-duty lines are contentious, but eventually disappear with the end of contentiousness, such as the 17th parallel north that formerly divided Vietnam. Of all such situations, likely the most geopolitically contentious of all is the functional convenience of the 38th parallel north, which divides Korea.[5]

Since August 1945, the 38th parallel has been an extraterritorial line of global functional convenience transformed into a continuous geopolitical flashpoint, dividing North and South Korea. A line on the globe—across the Korean peninsula—now further transformed into a demilitarized zone (DMZ) strewn with weaponry, militaries, checkpoints, and tensions. To capriciously cross that 38th parallel, that Korean DMZ—well, it might set off a war. Everyone knows the risks in crossing the Korean DMZ, particularly for North or South Korea—a risk anchored in a line of functional convenience for more than sixty years. As much as anything, the continued impossibility of successfully resolving and removing that line of functional convenience, the 38th parallel DMZ, perpetuates the Korean crisis, prevents unification of the Korean peninsula, and makes Northeast Asia one of the planet's most geopolitically volatile regions. One wonders how history would be different had *National Geographic not* printed the 38th parallel on its map of Korea.

In 2009, the geopolitical risk of North Korea attempting unilaterally to cross a different line of functional convenience—the Kármán line—burst into global consciousness with the North Korean announcement of an attempted satellite launch. More accurately, the geopolitical risk—the threats of war and retaliation—were bizarrely centered not on the consequences of North Korea actually crossing the Kármán line, but on the consequences of North Korea failing to cross the Kármán line. The specific threat was the disjecta membra of the North Korean launch, the unpredictability of North Korean rocket boosters as space debris falling down out of the sky to land . . . where? No one could say with any certainty, precision, or accuracy. However, the disjecta membra of fifty years of global space history was no longer a floating, empty signifier in search of a signified to become a sign. The North Korean satellite attempt of 2009 shifted the disjointed disjecta membra discourse of the space age away from a space oddity and into a narrative home by becoming the sign of a clear and present danger to global security. After more than fifty years, space debris in 2009 became—above and beyond all else—a global security threat that demanded action, counter-response, armaments, and the coordinated, incremental, and graduated escalation of global security discourse.

Whether the North Korean launch was a genuine satellite launch, a disguised missile test, or a combination of both is an interesting question, and, no matter the answer, the launch was dual-use: a testing of satellite capabilities as well as a testing of missile capabilities. However, the question of debris, the scattered remains of the North Korean 2009 launch, became a major issue even before the launch itself took place. The question of space debris, particularly preorbit debris, had never so captured global attention. Thus the 2009 North Korea launch represents a transformative moment in the disjointed disjecta membra of the space era, filling a lacuna with a discourse of global security.

On March 12, 2009, the International Civilian Aviation Organization (ICAO) posted information received from North Korea about two potential "danger zones" for an upcoming North Korean satellite launch between April 4 and April 8. U.N. Secretary-General Ban Ki-moon said North Korea's planned launch of a rocket would "threaten regional peace and security" and urged North Korea to "abide by the rules" in any such activities.[6]

On March 8, North Korea had also warned that any attempts to intercept or interfere with its satellite launch would bring retaliation and "prompt counterstrikes by the most powerful military means. . . . Shooting our satellite for peaceful purposes will precisely mean a war" and would bring "merciless retaliatory blows" from North Korea, thus launching one of the most bellicose and unusual rhetorical battles of global security discourse in recent memory.[7] On March 13, Japan announced it would use its new missile defense system to shoot down the North Korean satellite if it showed any signs of striking Japanese territory: "Under our law, we can intercept any object if it is falling towards Japan, including any attacks on Japan, for our security."[8] Over the rest of the month of March, this warning from Japan was refined, moving more specifically to the targeted destruction, if necessary for security purposes, of North Korean rocket booster debris.[9] Additionally, the rest of the nations of the world joined in the rhetorical conflict and denounced the North Korean launch plans. Although making no claims or plans to shoot down the North Korean launch, the United States and South Korea both sent Aegis-equipped destroyers into the Pacific Ocean to monitor the outcome.[10]

Even though, as March turned into April, more observers were willing to entertain the possibility the launch was indeed a satellite attempt, the

---

FUKUOKA NOTAM

J0732/09 - ALL ACFT INTENDING TO FLY WI FUKUOKA FIR ARE
ADVISED TO PAY SPECIAL ATTENTION TO THE FOLLOWING
INFORMATION. A SATELLITE IS EXPECTED TO BE LAUNCHED FROM
NORTH KOREA.

THE LAUNCHING IS EXPECTED TO OCCUR AS FLW,
1. LAUNCHING DATE AND TIME : BTN 0200UTC AND 0700* UTC DAILY
FM APR 4 TO APR 8
2. POSSIBLE FALLING AREA OF THE FLYING OBJECTS AND/OR THEIR
DEBRIS
AREA(1): AREA BOUNDED BY STRAIGHT LINES
            CONNECTING FOLLOWING POINTS.
            404140N1353445E 402722N1383040E
            401634N1383022E 403052N1353426E
AREA(2): AREA BOUNDED BY STRAIGHT LINES
            CONNECTING FOLLOWING POINTS.
            343542N1644042E 312222N1721836E
            295553N1721347E 330916N1643542E
WHEN FURTHER INFORMATION ABOUT THIS SATELLITE LAUNCHING
OBTAINED, REPLACED NOTAM WILL BE ISSUED. SFC - UNL, 12 MAR
23:40 2009 UNTIL 08 APR 07:00 2009 ESTIMATED. CREATED: 12 MAR
23:41 2009

──────── Figure 15.1 NOTAM (Notice to Airmen) from Fukuoka (Japan) ────────
Air Traffic Control.
Source: North Korean Economy Watch website,
http://www.nkeconwatch.com/2009/03/15/musudan-launch-flight-path/.

distinction and ambiguity between a satellite launch and missile attempt
were causing a bit of discursive confusion for world leaders.[11] This was particularly apparent at the United Nations, where it was not clear that previous
U.N. resolutions triggering sanctions against North Korea for a missile launch
included "peaceful outer space activities."[12] The sanctions clearly prohibited
missile tests, but did not necessarily prohibit North Korea from the explo-

ration and use of outer space—codified as a right of all nations in the 1967 U.N. treaty on peaceful uses of outer space.[13]

Why would North Korea even want a communication satellite? This is the most paranoid, reclusive, sequestered, autarkic, isolated nation in the world when it comes to the global flow of information. Yet . . . there is a tantalizing trail . . . no, not really a trail . . . little more than some bits and pieces of news items, scattered remains . . . disjecta membra, perhaps . . . perhaps not . . . in any event, these scattered remains here and there do possibly suggest a remote possibility for North Korea launching a communication satellite.[14] In December 2008, North Korea launched a new mobile phone service called Koryolink in partnership with Orascom Telecom, a telecommunications multinational whose major holdings are in the Middle East and North Africa. The day after launching Koryolink, North Korea opened a new bank in Pyongyang, and the directorship of the new bank and the leadership of Koryolink were completely intertwined. This suggests a possible indirect link between the new North Korean mobile phone service and Orascom. Earlier in 2008, Orascom and Western Union announced a new pilot program aimed at global finance, specifically at remittances (the portion of wages an expatriate worker in the global labor pool returns to his or her nation of origin).

A precise tally of remittances at a global level (in other words, how much money flows globally because of remittances) is difficult to calculate because some remittances are carried as cash when a worker visits home (or a family member visits the worker abroad), but 2008 estimates for the global flow of remittances range in the area of US$300 billion, with rapid growth in remittances projected for the coming decade. Remittances are now estimated to account for 10 percent of the value of the annual gross national product (GNP) of the Philippines, for example. Currently, online and mobile personal banking services in much of the developing world (Africa and Asia in particular) are underdeveloped to nonexistent. Most remittances are returned to the worker's homeland via Western Union services: the money is literally wired, or telegraphed, home. Orascom and Western Union are now working on pilot projects to take global remittances off of a global circulation platform based on the Western Union telegraph and on to a global circulation platform based on personal banking via mobile telephony. North Korea does in fact send expatriate labor abroad, closely monitors those laborers, and forces those expatriate laborers to return virtually all of their remittances to the North Korean government. So—perhaps—a North Korean communication satellite could be part of a larger North Korean effort to gain a toehold in the global flow of e-finance in the coming decade. Given the relative paucity of global satellite launch services at present, it may also be that North Korea hopes to turn its missile systems into dual-use technologies for Orascom and other multinational telecommunications firms, portending a hypothetical future of North Korean launch-vehicle exports into North Africa, for use as satellite launch systems.

As North Korea made final preparations for launch, the global TV news screens carried more and more stories about the event, until North Korea—at a level of attention not reached before by Pyongyang—dominated the bottom-of-the-screen news updates and became the lead story in the news cycles of CNN, BBC, CCTV, and Al-Jazeera. The U.N. Security Council prepared for emergency sessions. A few hours before President Obama was to deliver what was, to date, the most important speech of his presidency, North Korea launched the rocket-satellite. As predicted, stage I of the booster rocket fell harmlessly into the danger zone identified in the Sea of Japan (East Sea) as stage 2 aimed for the Kármán line. Somewhere above the vast reaches of the Pacific Ocean, stage 2, with the final stage and satellite still attached, went into a downward trajectory, came down to Earth, and entered the abyss. The North Korean satellite launch never crossed the Kármán line. In the end, it was nothing but disjecta membra, at least for the scattered remains of North Korean outer space technology.

Speaking before a world audience in Prague a few hours later, President Obama delivered his speech, calling for global nuclear disarmament. Noting that "in a strange turn of history, the threat of global nuclear war has gone down, but the risk of a nuclear attack has gone up," Obama added that the North Korean missile-satellite test demonstrated "the need for action."[15] The Security Council issued condemnations (as did, for all intents and purposes, every world leader) instructing that North Korea "not conduct any further launch" without ever specifying what kind of launch had just taken place.[16] North Korea claimed a successful satellite orbit, but without any confirmation from other governments, scientists, or amateur satellite watchers.[17] All North Korea did was produce scattered remains across the Pacific and below the Kármán line.

How might one assess the impact of this North Korean outer space disjecta membra? For most, this probably appears to be nothing more than another reckless example of North Korean missile technology, another sign that North Korea is among the greatest risks to global security, another sign of the global frustration in dealing with North Korea among the "family" of nations, and just another spectacular, headline-grabbing story that quickly fell out of the news cycle to make room for everything else in the flotsam and jetsam of global news. Indeed, that is a sensible reading, or interpretation, when one constructs a narrative of this particular space oddity that casts the scattered remains of the North Korean satellite aside as disjecta membra, as empty and floating signifiers that have nothing but the most marginal of roles in this saga.

But what happens to this tale of a North Korean space oddity if the remains of the North Korean satellite launch assume a more central role in that narrative? By way of conclusion, is there a way to tell this story that makes narrative sense of disjecta membra? Perhaps this long passage from the past—written by Stefan Possony and Leslie Rozenzweig in 1955, just before the dawn of the space age—provides an entry:

> Let us conclude with the remark that the continuing progress of flying cannot be halted. As man reaches upwards to the outer atmosphere, new political problems arise. . . . Henceforth, international relations will be geared to the more difficult geometry of the interior of a large spheroid enveloping in its core a smaller and impenetrable spheroid, the earth. But even more confusing: the radius of the outer spheroid—symbolizing the aerospace or the altitude which man has attained at given time—is expanding. The technologically most advanced nation will operate within the highest aeropause, while the spheroids circumscribing the aerial capabilities of the more backward nations will have shorter radii. Hence, in the future, the geometry of power will be described by several enveloping spheroids of different sizes. . . . [T]ruly, a new Weltbild is emerging.[18]

For the first fifty years of the space age, the uneasy marriage of outer space and global security was understood as directly related to the ability of a nation-state to cross the Kármán line. Crossing the Kármán line was an unofficial yet functionally convenient extraterritorial border for outer space that conferred global power and prestige as well as, perhaps, a sense of global responsibility. Or, put another way, the story of outer space and global security was always a story told above the Kármán line. What went on in that story below the Kármán line was—well, it was the scattered remains of the space age story, the disjecta membra of the discourse. The uneasy marriage of outer space and global security began when, like the bridegroom carrying his bride in his arms across the threshold, a nation-state transported an object from the surface of the planet through the atmosphere into outer space and crossed the threshold of the Kármán line. When in 1955 Possony and Rosenzweig hypothesized the "geography of the air," they found global power in outer space accrued above the Kármán line, thus implicitly defining the locale where global security understood via traditional air power handed off to global security understood via outer space power.

Heretofore, if a nation-state could not cross the Kármán line with its rockets and satellites, it was not an outer space event but all just disjecta membra, scattered remains, ashes to ashes—nothing else, narratively speaking. Henceforth, after the North Korea launch of 2009, the scattered remains and disjecta membra of a failed rocket and satellite are in fact a crucial part of the story of outer space and global security. North Korea never crossed the Kármán line and instead brought its space age disjecta membra back down to Earth. For North Korea, the bridegroom never carried the bride across the threshold of outer space, but nevertheless triggered a global security discourse about outer space. The scattered remains and disjecta membra of the North Korea satellite launch of 2009 were narratively anchored, beyond any question or doubt, into the firmament of outer space and global security. Now, finally, the vague zone between this small blue-green planet and the heavens above—that zone that begins at the surface of the planet and ends (more or less as a functional convenience) at the Kármán line is, for better and especially for worse, no longer a lacuna but instead a new locale in the marriage of outer space and global security, courtesy of the North Koreans and the global discourse of counter response.

Space debris both above and below the Kármán line after the North Korean launch is now a matter of global security. Space debris is now a new potential target for missile defense systems and proliferation surveillance networks.[19] With the North Korea launch, the discourse of global security reached back into its Cold War past to conquer yet another planetary extraterritoriality: the lacuna between the planet's surface and the outer reaches of the planet's atmospheric layers. In 2009, this space above the surface of the planet but below the Kármán line became subject to the discourse of outer space security, subsequently to be protected by a variety of means, including missile defense systems. This future scenario of proliferating missile defense systems throughout the planet as a likely long-term outcome of the North Korean launch of 2009 looms over our future—taking ashes to ashes, funk to funky from heaven's high back down to Earth—and hitting an all-time low.

Truly, a new Weltbild is emerging.

NOTES

1 Patrick McCray, *Keep Watching the Skies! The Story of Operation Moonwatch and the Dawn of the Space Age* (Princeton, N.J.: Princeton University Press, 2008), 149.

2 Ibid., 207–208. On the city council resolution, see Arthur Post (city clerk, Manitowoc, Wisconsin) to President John F. Kennedy, with accompanying resolution, September 18, 1962, in Presidential Papers of John F. Kennedy, White House Central Files, Box 653, Folder OS 2 1962–1963, Kennedy Library, Boston; see also the "Sputnikfest!" website, accessed March 27, 2011, http://www.sputnikfest .com/; for a very good website about space debris, see Paul Maley's Space Debris Page, accessed March 27, 2011, http://www.eclipsetours.com/sat/debris.html.

3 On debris from the Mercury launches, see, for example, Secretary of State Dean Rusk to American Embassy, Rome, May 24, 1962, Department of State, Record Group 59, Central Decimal File, 1960–1963, Box 2938, folder 900.8012/8–160, National Archives, College Park, Md. This folder is filled with reports from several American embassies about recovered space debris and the confusion in determining whether the debris was from American launches, from natural objects such as meteorites, or from unknown sources. Rusk added that NASA hoped "to determine why some pieces of space vehicles are not destroyed during the re-entry. . . . One of the incidents involving fragments was the Atlas 109-D booster which launched the Friendship 7 spacecraft carrying John Glenn

into orbit. Four fragments of the booster were found in South Africa. The other involved an Atlas booster, as yet unidentified which impacted in Brazil." Other reports had debris findings in Upper Volta (Burkina Faso), Angola, southwest Africa (Namibia), and some central African locales.

4   S. Sanz Fernández de Córdoba, "Presentation of the Karman separation line, used as the boundary separating Aeronautics and Astronautics," Fédération Aéronautique Internationale website, accessed March 26, 2011, http://www.fai.org/astronautics/100km.asp. The FAI adopted the Kármán line sometime in the 1950s. During the International Geophysical Year (July 1957–December 1958), the scientific and technical challenges presented by using the Kármán line for international legal and jurisdictional purposes were analyzed in Andrew G. Haley, "The Law of Space: Technical and Scientific Considerations," *New York Law Forum* 4 (1958): 262–419. Haley found that the Kármán line was both functionally convenient and, at that time, a technical impossibility for effective international jurisdiction. The United States does *not* use the Kármán line, but prefers to use a different line: 50 miles (80 km) out, perhaps a hangover from early space failures and the first two Mercury manned missions. Those two missions did not reach orbit but were merely (as the Soviets said at the time) brief suborbital probes.

5   General Order No. 1, Military and Naval, quoted in *U.S. Department of State*, Foreign Relations of the United States (FRUS), 1945, Vol. 6, The British Commonwealth, The Far East (Washington, D.C.: Government Printing Office, 1969), 658–659.

6   "Danger Zones Marked for North Korea 'Satellite' Launch," accessed March 27, 2011, http://www.globalpost.com/notebook/south-korea/090313/danger-zones-marked-north-koreas-satellite-launch.

7   "North Korea Warns Intercepting 'Satellite' Will Prompt Counterstrike," AP Newswire, March 8, 2009.

8   "Japan Warns It May Shoot Down North Korean Satellite Launcher," *Guardian*, March 13, 2009; see also James Schwoch, "The North Korea Satellite Launch and Global Financial Restructuring," April 3, 2009, University of Illinois Press Book Blog, accessed March 27, 2011, http://www.press.uillinois.edu/wordpress/?p= 2798.

9   "Japan Readies Missile Interceptors," *New York Times*, March 28, 2009.

10 Aegis-equipped ships have a radar system that enhances the ability to locate, track, and shoot down missiles, and in February 2008 the USS *Lake Erie*, an Aegis-equipped destroyer, shot down the errant satellite US 193; see also "No U.S. Plans to Stop North Korea on ICBM Test" and "Warships Set Sail Ahead of North Korean Rocket Launch," both *New York Times*, March 30, 2009.

11 For an example of this shift in observers' thinking, see "North Korea Threatens to Shoot Down U.S. Surveillance Planes," *New York Times*, April 2, 2009.

12 Ibid.

13 "North Korea Launch a Test for International Law," Associated Press newswire, April 3, 2009.

14 This account of recent North Korea global telecommunications activity possibly suggesting an interest in satellite launch capability—a speculative hypothesis at this writing—is based on the following: Marcus Noland, "Telecommunications in North Korea: Has Orascom Made the Connection?" manuscript, Peterson Institute for International Economics, September 8, 2008, accessed March 27, 2011,

http://www.iie.com/publications/papers/noland1208.pdf; "N. Korea Likely to Provide Internet Service from 2009," *Chosunilbo* (English version), August 7, 2008, accessed March 27, 2011, http://english.chosun.com/w21data/html/news/200808/200808070011.html; Tarek Al-Issawi, "Orascom Telecom of Egypt Opens Bank in North Korea (Update 2)," *Bloomberg.com*, December 16, 2008, accessed March 27, 2011, http://www.bloomberg.com/apps/news?pid=20601087&sid=ayACOoS5c80g&refer=home; "North Korea Allows Cellphone Network," *Los Angeles Times*, March 22, 2009; "2008 Annual Survey of Violations of Trade Union Rights—Korea (Republic of)," accessed March 27, 2011, http://survey08.ituc-csi.org/survey.php?IDContinent=3&IDCountry=KOR&Lang=EN; "North Koreans Toil Abroad under Grim Conditions," *Los Angeles Times*, December 27, 2005; and the following press releases from Orascom Telecom, searchable at http://orascomtelecom.com/media/PressRelease.aspx: "Orascom Telecom Announces Strategic Alliance with Western Union to Pilot Mobile Money Transfer Services," October 22, 2008; "Orascom Telecom Announces Success of the First Call on CHEO Network in Democratic People's Republic of Korea," May 19, 2008; "Orascom Telecom Receives the First Mobile License in the Democratic People's Republic of Korea," January 30, 2008; "Orascom Telecom Inaugurates the First 3G Mobile Network in the Democratic People's Republic of Korea," December 15, 2008.

15 "Obama Seizes on Missile Launch in Seeking Nuclear Cuts," *New York Times*, April 5, 2009.

16 Indicating an ongoing attention to the situation, the Security Council closed its statement with the rather strange comment, "The Security Council will remain actively seized of the matter." See "Statement by the President of the Security Council," April 13, 2009, S/PRST/2009/7, United Nations, accessed March 27, 2011, http://daccess-dds-ny.un.org/doc/UNDOC/GEN/N09/301/03/PDF/N0930103.pdf?OpenElement; see also the U.N. Security Council Resolution 1874 condemning additional nuclear and ballistic missile tests by North Korea, June 12, 2009, S/RES/1874 (2009), accessed March 27, 2011, http://daccess-dds-ny.un.org/doc/UNDOC/GEN/N09/368/49/PDF/N0936849.pdf?OpenElement.

17 I did a close monitoring of the SEE-SAT-L website (http://www.satobs.org/seesat/seesatindex.html) during the time of launch and for two days afterward to see if any amateur spotters reported the North Korean satellite in orbit; no one did. SEE-SAT-L and its members are the descendants of Operation Moonwatch; see McCray, *Keep Watching*. For the notification from the Department of Defense on orbit failure, see "NORAD and USNORTHCOM Monitor North Korean Launch," April 5, 2009: "Stage one of the missile fell into the Sea of Japan/East Sea. The remaining stages along with the payload itself landed in the Pacific Ocean. No object entered orbit and no debris fell on Japan," accessed March 27, 2011, http://www.northcom.mil/news/2009/040509.html.

18 Stefan T. Possony and Leslie Rosenzweig, "The Geography of the Air, 1955," *Annals of the American Academy of Social and Political Science* 299 (1955): 10–11.

19 In addition to activities discussed herein, see also Paul Rincon, "Standing Watch over a Crowded Space," *BBC News*, April 10, 2009, for new activities by the European Space Agency, accessed March 27, 2011, http://news.bbc.co.uk/2/hi/science/nature/7916582.stm. Israel tested its missile defense system about forty-eight hours after the North Korea launch; see Nathan J. Hunt, "Israel Conducts Successful Test of Arrow-2 System," April 7, 2009, accessed March 27, 2011, http://www.interceptorshield.com/2009/04/. For the Missile Defense Advocacy Alliance website, see http://www.missiledefenseadvocacy.org/default.aspx.

CONTRIBUTORS

BEN ASLINGER is an assistant professor of media and culture in the Department of English at Bentley University. His work has appeared in the journal *Popular Communication* and the anthology *Teen Television: Essays on Programming and Fandom*. He has also served as a columnist for the online journal *Flow*. He is currently working on a book manuscript on the licensing of popular music for television series and video games.

MARTIN COLLINS is a curator at the Smithsonian National Air and Space Museum. He is author of *Cold War Laboratory: Rand, the Air Force, and the American State, 1945–1950* and serves as editor of the journal *History and Technology*. His current research interest concerns the relations between satellite communications and globalization.

CHRISTY COLLIS is a senior lecturer in media and communication at Queensland University of Technology (UQ) and has been a researcher in cultural geography and a lecturer in the creative industries faculty there since 2005. In her three theses, her postdoctoral research fellowship at UQ (2000–2003), her research fellowship at the Australian National University (2004), and her visiting fellowships at Scott Polar Research Institute at Cambridge University (2003 and 2004), she studied the cultural and legal ways in which vast and sparsely populated geographies—the Canadian high Arctic, the Australian central deserts, and Antarctica—have been made into possessions. She has also published on the legal geographies of outer space.

ANDREW S. ERICKSON is an assistant professor in the Strategic Research Department at the U.S. Naval War College and a founding member of the department's China Maritime Studies Institute (CMSI). He is an associate in research at Harvard University's Fairbank Center for Chinese Studies, a fellow in the National Committee on U.S.-China Relations' Public Intellectuals Program (2008–2011), and a member of the Council for Security Cooperation in the Asia Pacific (CSCAP). His research on East Asian defense, foreign policy, and technology issues has been published widely in journals such as *Orbis*, *Journal of Strategic Studies*, *Joint Force Quarterly*, and *Proceedings*.

JAMES HAY is a professor in the Institute of Communications Research and the Department of Media & Cinema Studies at the University of Illinois–Champaign-Urbana. He has written extensively about communication media, governmentality, and space. Most recently, he coauthored *Better Living through Reality TV*.

BILL KIRKPATRICK is an assistant professor of media studies in the Communication Department at Denison University in Ohio. He specializes in broadcast history and policy, and is currently working on a book on localism in American thought and media policy.

MICHAEL MURPHY teaches new media production, digital audio, and emerging technologies in the School of Radio and Television Arts at Ryerson University, Toronto, Canada, and has been a visiting professor at the Hochschule der Medien, in Stuttgart, Germany.

BRIAN O'NEILL is head of the School of Media at Dublin Institute of Technology, Ireland. He is a member of the Digital Radio Cultures in Europe research group (www.drace.org) and deputy head of the Audiences Section of the International Association for Media and Communication Research (IAMCR).

TREVOR PAGLEN is a geographer and artist. He earned his Ph.D. at the University of California, Berkeley (2009), and his M.F.A. at the School of the Art Institute of Chicago. Paglen is the author of four books. The first of these, *Torture Taxi: On the Trail of the CIA's Rendition Flights* (coauthored with A. C. Thompson) was the first book to systematically describe the CIA's "extraordinary rendition" program. His second book, *I Could Tell You but Then You Would Have to Be Destroyed by Me*, offered an examination of the visual culture of "black" military programs. His third book was *Blank Spots on the Map: The Dark Secrets of the Pentagon's Secret World*, and in 2010 he published his first photographic monograph, *Invisible: Covert Operations and Classified Landscapes*.

LISA PARKS is a professor in the Department of Film and Media Studies at the University of California, Santa Barbara. She is the author of *Cultures in Orbit: Satellites and the Televisual* and coeditor of *Planet TV: A Global Television Reader* and *Undead TV*. She is currently finishing two new books: *Coverage: Media, Space and Security after 9/11* and *Mixed Signals: Media Infrastructures and Cultural Geographies*.

ALEXANDER RUSSO is an associate professor of media studies at the Catholic University of America. His research interests include the cultural history of media technologies, sound studies, and media geographies. He is the author of *Points on the Dial: Golden Age Radio beyond the Networks*, and has contributed essays to the *Historical Journal of Film, Radio, and Television*, *The Velvet Light Trap*, and *The Radio Reader*.

NAOMI SAKR is a professor of media policy at the Communication and Media Research Institute (CAMRI), University of Westminster (London), and director of the CAMRI Arab Media Centre. She is the author of *Arab Television Today* and *Satellite Realms: Transnational Television, Globalization and the Middle East*, and has edited two collections, *Women and Media in the Middle East: Power through Self-Expression* and *Arab Media and Political Renewal: Community, Legitimacy and Public Life*. Her research interests center on the political economy of Arab media, including relationships between corporate cultural production, media law, and human rights.

JAMES SCHWOCH is the senior associate dean for the School of Communication, a professor at Northwestern University–Qatar, and a professor in the Department of Communication Studies at Northwestern's main campus in Evanston, Illinois. His research and teaching explore global media, media history, diplomacy and security, international relations, and research methodologies. Previous books include *Global TV: New Media and the Cold War, 1946–69*, *Questions of Method in Cultural Studies*, *Writing Media Histories: Nordic Views*, *Media Knowledge*, and *The American Radio Industry*. Twice a Fulbright scholar (Germany, Finland), Schwoch has also received research support from the Ford Foundation, the National Science Foundation, and the National Endowment for the Humanities, and was a resident fellow at the Center for Strategic and International Studies in Washington, D.C.

RICK W. STURDEVANT (B.A. and M.A., University of Northern Iowa; Ph.D., University of California, Santa Barbara) is the deputy director of history at Headquarters Air Force Space Command in Colorado Springs, Colorado. An American Astronautical Society (AAS) Fellow, Sturdevant is editor of the AAS History of Rocketry and Astronautics series, which incorporates proceedings of International Academy of Astronautics history symposia. He also serves on the editorial board of *Quest: The History of Spaceflight Quarterly* and on the advisory staff of *High Frontier: The Journal of Space and Missile Professionals*.

PAUL TORRE is an assistant professor in media industries at the College of Mass Communications and Media Arts at Southern Illinois University, Carbondale. He has entertainment-industry experience in film and television development, production management, and global marketing and distribution. He researches and publishes on entertainment-industry structures, technologies, and management; on the interplay between U.S. and global media markets; and on how new technologies are shaping the media business models of the future.

BARNEY WARF is a professor of geography at the University of Kansas. His research and teaching interests lie within the broad domain of human geography. Much of his research concerns economic geography, emphasizing services and telecommunications. His work straddles contemporary political economy and social theory, on the one hand, and traditional quantitative, empirical approaches on the other. He has studied a range of topics that fall under the umbrella of globalization, including New York as a global city, telecommunications, offshore banking, international networks of financial and producer services, and the geographies of the Internet. He has authored, coauthored, or coedited seven books, two encyclopedias, thirty-two book chapters, and roughly 100 refereed journal articles. Currently, he serves as editor of *Professional Geographer* and coeditor of *Growth and Change* and co–book review editor for *Dialogues in Human Geography*, and edits a series of geography texts for Rowman and Littlefield publishers.

INDEX

CPSIA information can be obtained
at www.ICGtesting.com
Printed in the USA
LVHW090326290819
629297LV00001B/3/P